This is a timely, readable and comprehensive survey of recent work in the area of mobile networks, providing a good background reader for students and practitioners alike. It highlights some of the opportunities and challenges that accompany the deployment of both current wireless LANs as well as next-generation mesh and multihop networks.

Henning Schulzrinne
Professor and Chair, Dept. of Computer Science
Columbia University, New York, USA

This book complements other publications in an area that is of significant interest in research and deployment these days and as such is welcome and timely. By covering both technology and applications it is a useful hand-book for people interested in both aspects of this field. The various topics are represented in nicely connected chapters, each representing previously unpublished work. The chapter authors are authorities.

Imrich Chlamtac
President, CreateNet Research Consortium
Bruno Kessler Honorary Professor
University of Trento, Italy

This book will get the reader started with what is happening on the research front in the area of wireless LANs, multihop ad hoc and sensor networks. And surely, once the reader goes through some of the chapters, she will want to explore deeper the subjects covered and the issues addressed in the book. These include performance analysis, design methodology, and middleware, energy management, security, and applications.

Bijendra Jain
Professor, Dept. of Computer Science and Engg.
Indian Institute of Technology, Delhi, India

With contributions from several experts in the field, this collection of papers provides an easily accessible, comprehensive and up-to-date coverage of the vast and ever growing literature in several important areas of wireless net-working, particularly, wireless LANs, multihop wireless networks, and wire-less ad hoc sensor networks. The educator, the practitioner, and the researcher will all find this book very useful.

Anurag Kumar
Professor and Chairman, Dept. of Electrical Communication Engg.
Indian Institute of Science, Bangalore, India

This book comprehensively addresses the fundamental problems in the emerging area of mobile, wireless and sensor networks. The careful choice of articles, and the lucid and technically sound way in which they are authored, make this book particularly valuable for researchers who wish to work in this exciting area.

Nisheeth Vishnol
IBM Research, India and
Georgia Institute of Technology, Atlanta, USA

With the proliferation of all sorts of wireless technologies empowering a wide spectrum of personal to enterprise applications and services, research and development in the related areas has experienced a very healthy growth since the early 1990s and shows no signs of slowing down. With this as a factual backdrop, this book presents a collection of contributions from leading experts from the industry and academia presenting the latest and forward looking trends in research and development in this area. The book logically organizes the presentation of wireless, ad-hoc, and sensor network activities and then ties them with value-add middleware which exposes the essence of these networks to an intelligent set of application. The thesis of this book, organized in 15 chapters, is presented in an advanced and rigorous yet approachable manner that will surely benefit newcomers and seasoned researchers and practitioners alike. I commend the editors for putting together a well-thought and well-prepared book that further enriches the intellectual capital of this maturing yet persistently young and challenging research area.

Chatschik Bisdiklan, PhD and IEEE Fellow
IBM T. J. Watson Research Center, New York, USA

This book, written by experts in the field, does an excellent job of covering the latest developments in wireless and sensor networks. It is a must-read for both researchers and industry professionals working in this emerging area.

Kumar Sivarajan
CTO, Tejasnetworks
Bangalore, India

The advent of mobile wireless and sensor networks is likely to have a huge impact on society. This collection of contributed papers from eminent experts in the field is indeed timely, and will serve as an excellent reference to students, researchers, and practitioners in the field.

Sanjay Shakkottal
Assistant Professor, Department of Electrical and Computer Engg.
University of Texas at Austin, USA

The editors of *Mobile, Wireless, and Sensor Networks* have done a fabulous job in compiling the latest advances in wireless networking research into a single volume that will be of great use to students, faculty, and researchers alike. Each chapter provides a good overview of the subject area, with a heady mix of algorithms, results, and the underlying theory, woven tightly with simple explanations and supported by a large number of references. It is a must-have text for anyone considering research in this field.

Thyaga Nandagopal
Communication Protocols and Internetworking Research
Bell Laboratories, Lucent Technologies, New Jersey, USA

MOBILE, WIRELESS, AND SENSOR NETWORKS

MOBILE, WIRELESS, AND SENSOR NETWORKS

TECHNOLOGY, APPLICATIONS, AND FUTURE DIRECTIONS

Edited by

Rajeev Shorey
IBM Research
Indian Institute of Technology

A. Ananda
Mun Choon Chan
Wei Tsang Ooi
National University of Singapore

IEEE PRESS

A JOHN WILEY & SONS, INC., PUBLICATION

Published by John Wiley & Sons, Inc., Hoboken, New Jersey.
Published simultaneously in Canada.

For general information on our other products and services please contact our Customer Care Department within the U.S. at 877-762-2974, outside the U.S. at 317-572-3993 or fax 317-572-4002.

Wiley also publishes its books in a variety of electronic formats. Some content that appears in print, however, may not be available in electronic format.

Library of Congress Cataloging-in-Publication Data:

Mobile, wireless, and sensor networks: technology, applications, and future directions/edited by Rajeev Shorey... [et al.].
    p.   cm.
    Includes bibliographical references.
    ISBN-13 978-0-471-71816-1 (cloth: alk. paper)
    ISBN-10 0-471-71816-5 (cloth: alk. paper)
  1. Wireless communication systems.   2. Wireless LANs. 3. Mobile communication systems. 4. Sensor networks. I. Shorey, Rajeev.
    TK103.2.M634 2006
    621.382′1–dc22   2005010792

Printed in the United States of America

10 9 8 7 6 5 4 3 2 1

Dedicated to my mother, wife and daughters, and,
my teachers and friends — Rajeev Shorey

Dedicated to the lineage of gurus — A. L. Ananda

To my wife and daughter — Mun Choon Chan

To my parents — Wei Tsang Ooi

CONTENTS

FOREWORD   ix

PREFACE   xi

CONTRIBUTORS   xv

PART I   RECENT ADVANCES IN WLANs AND MULTIHOP WIRELESS NETWORKS

1. **Measuring Wireless LANs**   5
   Tristan Henderson and David Kotz

2. **Understanding the Use of a Campus Wireless Network**   29
   David Schwab and Rick Bunt

3. **QoS Provisioning in IEEE 802.11 WLAN**   45
   Sunghyun Choi and Jeonggyun Yu

4. **A Perspective on the Design of Power Control for Mobile Ad Hoc Networks**   73
   Alaa Muqattash, Marwan Krunz, and Sung-Ju Lee

5. **Routing Algorithms for Energy-Efficient Reliable Packet Delivery in Multihop Wireless Networks**   105
   Suman Banerjee and Archan Misra

PART II   RECENT ADVANCES AND RESEARCH IN SENSOR NETWORKS

6. **Detection, Energy, and Robustness in Wireless Sensor Networks**   145
   Lige Yu and Anthony Ephremides

7. **Mobile Target Tracking Using Sensor Networks**   173
   Ashima Gupta, Chao Gui, and Prasant Mohapatra

8. **Field Gathering Wireless Sensor Networks**   197
   Enrique J. Duarte-Melo and Mingyan Liu

9. **Coverage and Connectivity Issues in Wireless Sensor Networks**   221
   Amitabha Ghosh and Sajal K. Das

10. **Storage Management in Wireless Sensor Networks**   257
    Sameer Tilak, Nael Abu-Ghazaleh, and Wendi B. Heinzelman

11. **Security in Sensor Networks**   283
    Farooq Anjum and Saswati Sarkar

**PART III   MIDDLEWARE, APPLICATIONS,
AND NEW PARADIGMS**

12. **WinRFID: A Middleware for the Enablement of Radiofrequency
    Identification (RFID)-Based Applications**   313
    *B. S. Prabhu, Xiaoyong Su, Harish Ramamurthy, Chi-Cheng Chu,
    and Rajit Gadh*

13. **Designing Smart Environments: A Paradigm Based on
    Learning and Prediction**   337
    Sajal K. Das and Diane Cook

14. **Enforcing Security in Mobile Networks: Challenges
    and Solutions**   359
    Feng Bao, Robert H. Deng, Ying Qiu, and Jianying Zhou

15. **On-Demand Business: Network Challenges in a Global
    Pervasive Ecosystem**   381
    Craig Fellenstein, Joshy Joseph, Dongwook Lim, and J. Candice D'Orsay

INDEX   409

FOREWORD

Wireless networking could possibly be at the cusp of a takeoff. Scarcely a day passes without an email announcement of another wireless networking conference. This, of course, is a manifestation of the intense research activity in the filed. In such a dynamic environment there is a definite need for carefully organized collections of papers that organized the research output in a coherent way. This is useful not only for new entrants but also for active researchers in the area in keeping abreast of developments not within the immediate sphere of their investigations. The present book is just such an effort, and includes contributions from several well-known researchers in wireless networking. It features carefully written articles covering topics of much current interest in the areas of wireless local area networks (WLANs), multihop networks, sensor networks, and middleware. Through these expert contributions, it serves a useful and timely role in the reduction to entropy, and thus the assimilation of the research output, thereby furthering evolution of the field of wireless networking.

P. R. KUMAR

University of Illinois
Urbana–Champaign
May 13, 2005

■■■■■ PREFACE

Objectives

The market for wireless communications has enjoyed tremendous growth. Wireless technology now reaches or is capable of reaching virtually every location on the face of the earth. Hundreds of millions of people exchange information every day using laptops, personal digital assistants (PDAs), pagers, cellular phones, and other wireless communication devices. Success of outdoor and indoor wireless communication networks has led to numerous applications in sectors ranging from industries and enterprises to homes and universities. No longer bound by the harnesses of wired networks, people are able to access and share information on a global scale nearly anywhere they venture.

Some of the most remarkable growth has occurred in the deployment of wireless LANs (WLANs) where IEEE 802.11 based wireless networks have been used to provide connectivity not only as hotspots but also to substantial portions of a city. At the same time, we are also witnessing a significant trend in the emergence of small and low-cost computation and communication devices. While these tiny devices, called *sensor nodes*, are tightly constrained in terms of energy, storage capacity, and data processing capability, they have the potential to serve as a catalyst for a major change in how we communicate and interact with the environment.

This book aims to address challenging issues in wireless networks, in particular, wireless LANs, multihop wireless networks, sensor networks and their applications. In addition, the book discusses emerging applications and new paradigms, for example, middleware for RFID, smart home design, and "on-demand business" in the context of pervasive computing.

The objective of the book is twofold: to bridge the gap between practice and theory and between different types of related wireless networks.

We believe that the theme and focus of the book is timely. The book focuses on important topics of current interest in wireless, mobile, and sensor networks and covers issues ranging from architecture, protocols, modeling and analysis to new applications, solutions, and emerging paradigms.

To the best of our knowledge, there is no existing book that brings together closely related topics in a single volume and that discusses key technical challenges along with important applications. The chapters of the book are written by researchers and practitioners from academia and industry who are experts in the field. In most of the chapters, the authors begin with a broad survey of the topic and then move on to discuss the technical challenges and solutions.

Intended Audience

This book should serve as an excellent source of information for students, researchers, and practitioners who want to track new research and developments in wireless communication but do not have time or patience to read numerous papers and specifications. The book will be very useful to research students who are looking for open problems in this space. In addition, the book also targets professionals in the field of wireless/mobile communications, designers, and networking managers.

How Did the Idea of the Book Originate?

The idea of this book was conceived in March of 2004, when we conducted a highly successful international workshop in Singapore titled "MOBWISER" (Mobile, Wireless and Sensor Networks: Technology and Future Directions). The readers are encouraged to see the Website `http://mobwiser.comp.nus.edu.sg/` for more details.

The workshop had 13 invited experts from all over the world to present their work in areas ranging from multihop wireless networks to sensor networks and their applications. The workshop was a great success in every respect. It was attended by more than 125 delegates. Encouraged by the success of the workshop and the enthusiasm of the attendees, the editors decided to put together a book in this area that would address core issues in wireless networks, with emphasis on wireless LANs, multihop wireless networks, and sensor networks, along with their applications.

When we approached MOBWISER workshop speakers to contribute a paper to the book, many of them generously agreed. We are extremely grateful to Sunghyun Choi, Sajal Das, Robert Deng, Anthony Ephremides, Craig Fellenstein, Marwan Krunz, Mingyan Liu, Archan Misra, and Prasant Mohapatra for their enthusiasm and interest in this project from a very early stage.

As the book became a reality, we realized the need to include additional topics in the text in order to present a more complete and broader scope. The new topics added were wireless LAN measurements, security in mobile networks, security and storage management in sensor networks, and sensor-network-related middleware and applications. We are extremely thankful to Farooq Anjum, Rick Bunt, Rajit Gadh, Wendi Heinzelman, and David Kotz for their contribution to the book.

Acknowledgments

A good book is always a collective effort of many people. In addition to the authors of the 15 chapters, we would like to especially thank Professors Sajal Das and

Anthony Ephremides, who were instrumental in initiating the idea of this book. We greatly appreciate their constant encouragement and support in this project.

Since the idea of this book originated during the MOBWISER workshop, we would like to thank Mr. Yap Siang Yong and Ms. Sarah Ng at the National University of Singapore for their excellent support in organizing the MOBWISER workshop in Singapore in March 2004. We would like to thank our organizations — the National University of Singapore and IBM India Research Laboratory, New Delhi for their generous support for this project and for allowing us to use the resources in the organizations.

A special thanks to the staff at John Wiley in Hoboken, New Jersey (USA), who have been wonderful. Our thanks go to Val Moliere and Emily Simmons for their encouragement, patience, and excellent support throughout this project. This book would not have been possible without their help, and we are greatly indebted to them. Working with Val and Emily has been a very pleasant and memorable experience, and we thank them for an outstanding job.

Organization of the Book

This book is split into three logically distinct parts. Part I describes more recent advances in wireless LANs and multihop wireless networks. Part II is devoted to recent advances and research in wireless sensor networks. The topics covered in the Part III are RFID middleware, smart home environments, security in mobile networks, and on-demand business.

It is expected that the reader will follow the chapters in each part in sequence so as to gain a better understanding of the subject. The three parts of the book can be read in any order.

We hope the readers will find this unique book interesting and useful in gaining a deeper insight into the fascinating subject of "mobile, wireless, and sensor" networks.

RAJEEV SHOREY

IBM, New Delhi
November 1, 2005

AKKIHEBBAL L. ANANDA,
MUN CHOON CHAN, AND
WEI TSANG OOI

NUS, Singapore
November 1, 2005

■■■■ CONTRIBUTORS

Nael Abu-Ghazaleh, Department of Computer Science, Watson School of Engineering and Applied Sciences, Binghampton University, Binghampton, NY 13902

Farooq Anjum, Applied Research, Telcordia Technologies, One Telcordia Drive, Piscataway, NJ 08854

Suman Banerjee, Department of Computer Sciences, University of Wisconsin, Madison, WI 53706

Feng Bao, Institute for Infocomm Research, 21 Heng Mui Keng Terrace, Singapore 119613

Rick Bunt, University of Saskatchewan, E234 Administration Building, 105 Administration Place, Saskatoon S7N 5A2, Canada

Sunghyun Choi, School of Electrical Engineering, Seoul National University, Kwanak, P. O. Box 34, Seoul 151-600, Korea

Chi-Cheng Chu, University of California at Los Angeles, Wireless Internet for the Mobile Enterprise Consortium (WINMEC), 420 Westwood Place, Los Angeles, CA 90095

Diane Cook, Department of Computer Science and Engineering, University of Texas at Arlington, Arlington, TX 76019

Sajal K. Das, Department of Computer Science and Engineering, University of Texas at Arlington, Arlington, TX 76019

Robert H. Deng, School of Information System, Singapore Management University, 469 Bukti Timah Road, Singapore 259756

J. Candice D'Orsay, IBM Global Services, Network Services (NS) Organization, 6 Hunting Ridge Road, Brookfield, CT 06804-3710

Enrique J. Duarte-Melo, Department of Electrical and Computer Engineering, University of Michigan, Ann Arbor, MI 48109-2122

Anthony Ephremides, Department of Electrical and Computer Engineering, University of Maryland, College Park, MD 20742

Craig Fellenstein, IBM Global Services, Network Services (NS) Organization, 6 Hunting Ridge Road, Brookfield, CT 06804-3710

Rajit Gadh, University of California at Los Angeles, Wireless Internet for the Mobile Enterprise Consortium (WINMEC), 420 Westwood Place, Los Angeles, CA 90095

Amitabha Ghosh, Department of Computer Science and Engineering, University of Texas at Arlington, Arlington, TX 76019

Ashima Gupta, Department of Computer Science, 2063 Kemper Hall, University of California at Davis, Davis, CA 95616

Chao Gui, Department of Computer Science, 2063 Kemper Hall, University of California at Davis, Davis, CA 95616

Wendi B. Heinzelman, Department of Electrical and Computer Engineering, University of Rochester, Hopeman Building, P.O. Box 270126, Rochester, NY 14627

Tristan Henderson, Department of Computer Science, 6211 Sudikoff Laboratory, Dartmouth College, Hanover, NH 03755-3510

Joshy Joseph, IBM Global Services, Network Services (NS) Organization, 6 Hunting Ridge Road, Brookfield, CT 06804-3710

David Kotz, Department of Computer Science, 6211 Sudikoff Laboratory, Dartmouth College, Hanover, NH 03755-3510

Marwan Krunz, Department of Electrical and Computer Engineering, University of Arizona, Tucson, AZ 85721

Sung-Ju Lee, Mobile and Media Systems Lab, Hewlett-Packard Laboratories, Palo Alto, CA 94304

Dongwook Lim, IBM Global Services, Network Services (NS) Organization, 6 Hunting Ridge Road, Brookfield, CT 06804-3710

Mingyan Liu, Department of Electrical and Computer Engineering, University of Michigan, Ann Arbor, MI 48109-2122

Archan Misra, IBM T.J. Watson Research Center, 19 Skyline Drive, Hawthorne, NY 10532

Prasant Mohapatra, Department of Computer Science, 2063 Kemper Hall, University of California at Davis, Davis, CA 95616

Alaa Muqattash, Department of Electrical and Computer Engineering, University of Arizona, Tucson, AZ 85721

B. S. Prabhu, University of California at Los Angeles, Wireless Internet for the Mobile Enterprise Consortium (WINMEC), 420 Westwood Place, Los Angeles, CA 90095

Ying Qiu, Institute for Infocomm Research, 21 Heng Mui Keng Terrace, Singapore 119613

Harish Ramamurthy, University of California at Los Angeles, Wireless Internet for the Mobile Enterprise Consortium (WINMEC), 420 Westwood Place, Los Angeles, CA 90095

Saswati Sarkar, Applied Research, Telcordia Technologies, One Telcordia Drive, Piscataway, NJ 08854

David Schwab, University of Saskatchewan, E234 Administration Building, 105 Administration Place, Saskatoon S7N 5A2, Canada

Xiaoyong Su, University of California at Los Angeles, Wireless Internet for the Mobile Enterprise Consortium (WINMEC), 420 Westwood Place, Los Angeles, CA 90095

Sameer Tilak, Department of Computer Science, Watson School of Engineering and Applied Sciences, Binghampton University, Binghampton, NY 13902

Jeonggyun Yu, School of Electrical Engineering, Seoul National University, Kwanak, P. O. Box 34, Seoul 151-600, Korea

Lige Yu, Department of Electrical and Computer Engineering, University of Maryland, College Park, MD 20742

Jianying Zhou, Institute for Infocomm Research, 21 Heng Mui Keng Terrace, Singapore 119613

■■■■■■ PART I

RECENT ADVANCES IN WLANs AND MULTIHOP WIRELESS NETWORKS

Wireless LANs (WLANs) are becoming increasingly popular and are being widely deployed in academic institutions, corporate campuses, and residences. WLAN hotspots are now a common sight at airports, hotels, and shopping malls in many parts of the world. Despite this growth, the WLAN market is still young and immature. Issues such as user behavior, security, business models, and the types of applications that WLAN can support remain unclear. This leads to a constantly evolving mix of applications and performance demands on WLANs.

Measurement studies of WLANs are important for understanding the characteristics of a WLAN. Information about user session behavior, mobility pattern, network traffic, and the load on the access point can often help engineers identify bottlenecks and improve performance. In Chapter 1, Henderson and Kotz describe the tools, metrics, and techniques for measuring WLANs. The chapter provides extensive survey on methodologies and results of existing measurement studies, and outlines challenges for collecting future wireless traces. Similarly, in Chapter 2, Schwab and Bunt seek to understand the current usage patterns of the WLANs deployed in the University of Saskatchewan campus. They determine where, when, how much, and for what their network is being used, and how that usage is changing over time. The chapter also describes the methodology that has been employed to study usage patterns and some of the results that have been obtained from studies to date.

Measurement studies of WLAN traffic reveal that while Web traffic is still the dominant traffic, streaming media traffic has increased significantly in usage. Unlike bulk data transfer applications, streaming media applications such as Voice over Internet Protocol (VoIP) requires low latency, low delay jitter, and low packet loss rate to ensure reasonable playback quality. The current 802.11 WLAN, however, does not provide any quality-of-service (QoS) guarantee because its distributed coordination function (DCF) mandates the use of contention-based channel access. In Chapter 3, Choi and Yu present solutions to provide QoS provisioning

Mobile, Wireless, and Sensor Networks: Technology, Applications, and Future Directions
Edited by Rajeev Shorey, Akkihebbal L. Ananda, Mun Choon Chan, and Wei Tsang Ooi
Copyright © 2006 John Wiley & Sons, Inc.

in WLANs using a combination of short-term and long-term solutions. The short-term solution adopts a novel scheme called *dual queues* (MDQ) that can be implemented in software as part of the NIC's driver, and therefore is compatible with existing hardware. The emerging IEEE 802.11e standard, which provides QoS through differentiated channel access on the basis of packet priority [enhanced distributed channel access (EDCA)], provides a long-term solution. This chapter also presents comparative evaluations of the various QoS provisioning mechanisms in WLANs.

While the first three chapters deal with single-hop 802.11 deployments, the next two chapters present more recent advances in power-aware multihop wireless networks, focusing on transmission power control and routing algorithms for energy-efficient reliable packet delivery.

Wireless ad hoc networks (or multihop wireless networks) consist of mobile nodes communicating over a shared wireless channel. Contrary to cellular networks, where the nodes communicate with a set of carefully placed basestations, there are no basestations in wireless ad hoc networks; any two nodes are allowed to communicate directly if they are within each other's communication range, and nodes must use multihop routing to deliver their packets to distant destinations. These infrastructureless networks have many potential applications, from personal area networks, to search and rescue operations, to massive networks of millions of sensors.

In Chapter 4, Muqattash et al. investigate transmission power control (TPC) in mobile ad hoc networks (MANETs) and examines various TPC approaches proposed in the literature. The authors argue that TPC has a great potential to address the challenge of simultaneously providing high network throughput and low-energy communications between mobile nodes.

The authors discuss the factors that influence the selection of the transmission power, including the important interaction between the routing (network) and the medium access control (MAC) layers. Protocols that account for such interaction are presented. The authors argue that using the minimum transmission power does not deliver the maximum throughput in MANETs. The impact of mobility on the design of power-controlled MAC protocols is also addressed. Various complementary approaches and optimizations are highlighted and discussed, including the use of rate control, directional antennas, spread-spectrum technology, and power saving modes. The chapter outlines several directions for future research in this area. An important conclusion in the paper is that the design of efficient TPC schemes in MANETs should take into account the interplay between the routing, MAC, and the physical layer.

Wireless-enabled devices are primarily energy constrained. As a result, various energy aware routing protocols have been proposed to lower the communication energy overhead in multihop wireless networks.

In Chapter 5, Banerjee and Misra address energy-efficient communication in multihop wireless networks. The authors show why the effective total transmission energy, which includes the energy spent in potential retransmissions, is the proper metric for reliable, energy-efficient communications.

The energy efficiency of a candidate route is critically dependent on the packet error rate of the underlying links, since they directly affect the energy wasted in retransmissions. Analysis of the interplay between error rates, number of hops, and transmission power levels reveals several key results. The authors show that for reliable energy-efficient communication, the routing algorithm must consider both the distance and quality (e.g., in terms of the link error rate) of each link. Thus, the cost of choosing a particular link should be the overall transmission energy (including possible retransmissions) needed to ensure eventual error-free delivery, and not just basic transmission power. This is particularly important in practical multihop wireless environments, where packet loss rates could be high.

CHAPTER 1

Measuring Wireless LANs

TRISTAN HENDERSON and DAVID KOTZ

Department of Computer Science, Dartmouth College, Hanover, New Hampshire

1.1 INTRODUCTION

Wireless local area networks (WLANs) have appeared in many venues, including academic and corporate campuses, residences, and wireless "hotspots." It becomes increasingly important to understand how these networks are used, as they continue to appear in more numerous and varied environments. Measuring and collecting data from production WLANs in a usage study is one way of fulfilling this need for understanding.

Wireless usage studies and usage data are valuable for many aspects of wireless network research. Understanding how and where clients use the network, what applications clients are using, and how applications are using the network can help with network provisioning and deciding where to expand or augment coverage in an existing WLAN. Models of wireless application workloads can aid the design of future network protocols. Measurements of client mobility in a WLAN can help with the design of location-aware applications, or for developing and improving mobile handoff algorithms.

Collecting data on a WLAN can be difficult, however. There are many technical and nontechnical logistical hurdles involved in collecting high-quality wireless measurements. We have been continuously monitoring a campus WLAN for over 3 years in the course of conducting two of the largest wireless measurement studies to date [9,13], and we have encountered many of these hurdles. In this chapter we describe some of the tools that the research community has used for measuring WLANs, and provide hints for their effective use obtained from our real-world experiences. We also discuss some of the usage studies that have been conducted using these tools, both on our own campus and elsewhere. In particular we concentrate

Mobile, Wireless, and Sensor Networks: Technology, Applications, and Future Directions
Edited by Rajeev Shorey, Akkihebbal L. Ananda, Mun Choon Chan, and Wei Tsang Ooi
Copyright © 2006 John Wiley & Sons, Inc.

on the most common type of wireless LAN, the IEEE 802.11 infrastructure network, as this has seen the highest number of deployments, and thus most usage studies have considered infrastructure networks.

This chapter is laid out as follows. In Section 1.2 we examine some of the tools that are available for measuring a WLAN. Section 1.3 surveys various wireless measurement studies, considering both the tools that were used and the insights that were learned. Section 1.4 concludes the chapter with a checklist of items that a potential wireless usage researcher should consider.

1.2 MEASUREMENT TOOLS

The purpose of a wireless usage study is to collect data about the operations of a WLAN. There are several tools available to the researcher for this purpose. The most commonly used tools include syslog, SNMP, network sniffing, authentication logs, and developing client-side applications. Figure 1.1 shows how some of these tools might be deployed in an example WLAN. In this section, we summarize the pros and cons of using each of these tools, and offer some advice from our own experiences.

1.2.1 Syslog

Syslog is a somewhat loosely specified standard [14] for sending and receiving logging messages. Messages can be stored locally or transmitted across a network to another host.

Many 802.11 access points (APs) can be configured to send syslog messages. By choosing appropriate events to be logged, syslog messages can be used to understand the state of clients on the network. For instance, an AP can send a time-stamped syslog message whenever a client authenticates, deauthenticates, associates, disassociates, or roams to that AP. By collecting these syslog messages from all of the APs in a network, it is possible to determine the state of the clients on the network.

Once an AP has been configured to send syslog messages to a particular host, no further information is required from the receiving host. This makes syslog a simple tool to set up. The receiving host, however, must take care to ensure that messages are being received correctly, as network problems, firmware upgrades, or malfunctioning APs, may lead APs failing to send syslog messages.

There is no standard format for a syslog message, and there is also no standard format for an 802.11 syslog message. The messages that APs send can vary in format, and in the amount of information that is contained. Figures 1.2 and 1.3 show two sets of syslog messages. These messages are both taken from the same Cisco Aironet 350 802.11b AP. Figure 1.2 shows messages from the AP when it was running the VxWorks operating system, whereas Figure 1.3 is a set of messages from the AP after it had been upgraded to the Cisco Internetworking Operating System

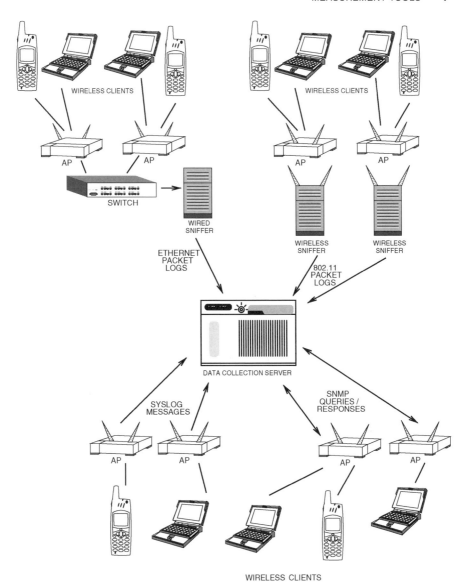

Figure 1.1   Tools for measuring a wireless LAN.

```
Jan  1 04:54:27 example1-ap example1-ap (Info): Station 1234567890ab Reassociated
Jan  1 04:54:27 example2-ap example2-ap (Info): Station 1234567890ab roamed
Jan  1 04:55:22 example3-ap example3-ap (Info): Station 0987654321ef Reassociated
Jan  1 04:55:26 example4-ap example4-ap (Info): Station 0987654321ef Reassociated
Jan  1 04:57:23 example5-ap example5-ap (Info): Deauthenticating abcdef123456, reason "Inactivity"
```

Figure 1.2   Example of Cisco VxWorks AP syslog.

```
Jan  1 04:57:58 example1-ap 382: example1-ap:Jan  1 08:57:57: %DOT11-6-DISASSOC: Interface \
   Dot11Radio0, Deauthenticating Station 1234.5678.90ab Reason: Disassociated because \
   sending station is leaving (or has left) BSS
Jan  1 04:58:01 example2-ap 36723: example2-ap:Jan  1 08:58:00: %DOT11-6-DISASSOC: Interface \
   Dot11Radio0, Deauthenticating Station abcd.ef12.3456 Reason: Previous authentication \
   no longer valid
Jan  1 04:58:01 example3-ap 13031: example3-ap:Jan  1 08:58:00: %DOT11-6-DISASSOC: Interface \
   Dot11Radio0, Deauthenticating Station 0987.6543.12fe Reason: Disassociated because \
   sending station is leaving (or has left) BSS
Jan  1 04:58:08 example2-ap 36724: example2-ap:Jan  1 08:58:07: %DOT11-6-ASSOC: Interface \
   Dot11Radio0, Station  abcd.ef12.3456 Associated KEY_MGMT[NONE]
Jan  1 04:58:10 example4-ap 6882: example4-ap:Jan  1 08:58:09: %DOT11-6-DISASSOC: Interface \
   Dot11Radio0, Deauthenticating Station 0004.2356.5b74 Reason: Previous authentication \
   no longer valid
```

Figure 1.3   Example of Cisco IOS AP syslog.

(IOS). Both sets of messages contain the same basic information: client 802.11 events. They differ, however, in the way that this information is presented; in Figure 1.3 there are multiple timestamps (from the syslog daemon and the AP itself), and the client MAC addresses are formatted differently. Parsing syslog messages can therefore be a tedious process, as the format can change between different AP firmware versions. A long-term measurement study should monitor syslog messages for format changes, and also monitor changes in firmware, either through close communication with network administrators, or by using SNMP (see Section 1.2.2).

A further consideration when parsing AP syslog messages is that not all messages may accurately correspond to 802.11 events. Figure 1.4 shows a set of syslog messages from a "wireless switch." This switch is representative of the newest type of 802.11 infrastructure network, where "dumb" APs are deployed across the area to be covered, and a centralized switch handles authentication, association, and access control. In this setup is the switch that sends syslog messages, not the APs. Rather than sending an individual message for each authenticate, associate, roam, disassociate, and deauthenticate event, the switch sends only two types of message: "station up" and "station down." The types of message available from the APs in the WLAN to be measured may impact the suitability of syslog as a measuring tool, depending on the type of data required for the study.

```
Jan  1 03:11:48 wireless-switch.example.com 2004 [1874327] auth[30927]: <INFO> \
   station up <01:23:45:67:89:0a> bssid 00:11:22:33:44:55, essid Example_ESSID, vlan 12, \
   ingress 4226, u_encr 1, m_encr 1, loc 156.1.1 slotport 4035
Jan  1 03:14:04 wireless-switch.example.com 2004 [1874341] auth[30927]: <INFO> \
   station up <09:87:65:43:21:fe> bssid 00:11:22:44:55:66, essid Example_ESSID, vlan 12, \
   ingress 4258, u_encr 1, m_encr 1, loc 2.2.1 slotport 4035
Jan  1 03:14:07 wireless-switch.example.com 2004 [1874345] auth[30927]: <INFO> \
   station up <09:87:65:43:21:fe> bssid 00:11:22:44:55:66, essid Example_ESSID, vlan 12, \
   ingress 4258, u_encr 1, m_encr 1, loc 2.2.1 slotport 4035
Jan  1 03:14:40 wireless-switch.example.com 2004 [1874359] auth[30927]: <INFO> \
   station up <12:34:56:78:90:ab> bssid 00:11:22:55:66:77, essid Example_ESSID, vlan 12, \
   ingress 4262, u_encr 1, m_encr 1, loc 156.4.1 slotport 4035
Jan  1 03:14:47 wireless-switch.example.com 2004 [1874369] auth[30927]: <INFO> \
   station up <12:34:56:78:90:ab> bssid 00:11:22:66:77:88, essid Example_ESSID, vlan 12, \
   ingress 4296, u_encr 1, m_encr 1, loc 156.4.1 slotport 4035
```

Figure 1.4   Example of wireless switch syslog.

```
1072933205 0123456789ab roamed example1-ap
1072933214 0123456789ab disassociated example1-ap
1072933215 0123456789ab reassociated example1-ap
1072933241 09876543e1ef deauthenticated example2-ap
1072933244 09876543e1ef authenticated example2-ap
1072933244 09876543e1ef reassociated example2-ap
1072933265 0123456789ab roamed example1-ap
1072933269 0123456789ab disassociated example1-ap
1072933270 0123456789ab reassociated example1-ap
1072933307 abcdef123456 reassociated example3-ap
```

Figure 1.5 Parsed syslog messages.

In a mixed AP environment such as ours, with multiple types of AP and thus multiple types of syslog messages, we have found it useful to translate syslog messages into an intermediate format prior to data analysis. Figure 1.5 shows this intermediate format. The time, client MAC address, event, and AP hostname are extracted from the syslog messages. The year is added to the time, as syslog messages do not contain a year, and the time is replaced with a Unix timestamp. Some syslog messages contain only the MAC address of an AP and not the hostname, as in Figure 1.4 (e.g., `bssid 00:11:22:33:44:55`). For these APs, we keep a separate mapping of AP names to AP MAC addresses, and refer to this when translating syslog messages.

Once the syslog messages have been collected and translated into a parsable format, it is possible to create a state machine that can calculate a session for each MAC address observed in the syslog trace. Figure 1.6 shows the session state machine that we have used in our campus wireless traces [9,13]. A session consists of an association, followed by zero or more roam events, and ends with a disassociate or deauthenticate event.

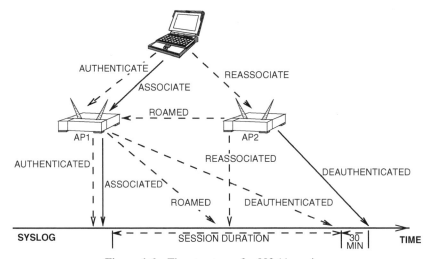

Figure 1.6 The structure of a 802.11 session.

This session structure assumes that a MAC address corresponds to a unique user. This may not be the case in some network environments, for instance, where 802.11 network interface cards (NICs) are shared among several users, or where users tend to alter their MAC addresses. If this is likely to be the case, and the purpose of the study is to track individual's usage, combining syslog data with other data such as authentication logs (Section 1.2.3) may be required.

Our final hint for dealing with syslog messages is to be conscious of holes in the data. As most syslog daemons use a UDP transport, some messages may be lost or misordered in the network. Additional messages may be lost as a result of changes in network configuration or malfunctions. These holes can lead to errors in the estimation of a session length. For instance, if a disassociate message is lost, a simple parser may assume that a client has never disassociated from their last observed AP, and so overestimate the session length. In our studies, we have attempted to alleviate this problem by looking for sessions that are still active at the end of our trace. We assume that these sessions are missing a disassociate message, and we manually terminate the sessions 30 min after the last syslog message recorded for this MAC address. We chose a 30-min window since this is the usual period that an AP uses to time out inactive clients. Advantages and disadvantages of syslog are as follows:

Pros — somewhat passive (no additional traffic sent *to* APs); one-second granularity

Cons — no common data format; UDP transport means that messages can be lost; may need to manually configure every AP to send syslog messages

1.2.2 SNMP

As its name implies, the Simple Network Management Protocol (SNMP [15]) is a means for managing network devices, or more generally, network objects. A network administrator runs a tool known as a manager, which communicates with SNMP agents. Agents run on network devices, and provide an interface between the device and the manager. A network device can contain several managed objects, such as statistics or configuration items, arranged in a database known as a Management Information Base (MIB).

For the purposes of measuring a wireless LAN, SNMP provides a mechanism for extracting more detailed information out of an access point than syslog provides. The level of data depends on the extent of the particular AP's SNMP support. The IEEE 802.11 standard includes a MIB [11], but this is sparse, and concentrates on client-side variables. In keeping with the intent of RFC 1812 [2], which requires "the ability to do anything on the router through SNMP that can be done through a console," many AP vendors have written their own vendor-specific MIBs. These MIBs contain many variables that are useful for measuring a wireless LAN. These may be client-specific variables, such as the MAC address, signal strength, or power saving mode of each client associated with the AP. Or they may be AP-specific variables, such as the number of clients currently associated with the AP, or the number of clients that have recently roamed away from the AP.

Even if an AP lacks a vendor-specific wireless MIB, there remain many useful data that can be obtained from general MIBs. Most APs support the standard network interface MIBs [8]. By querying these MIBs, it is possible to determine some interface-specific variables, for instance, the number of inbound and outbound bytes and packets that have passed the AP's wired interface. This can be used as an indicator of the amount of wireless traffic, although it may not include traffic between two wireless hosts on the same AP, whose traffic may not traverse the wired interface.

As with syslog, SNMP data collection can be impacted by different WLAN setups. If a centralized wireless switch is deployed, it may be necessary to query this switch in addition to, or instead of, individual APs. Some networks may prevent SNMP for security reasons, or allow SNMP queries only from particular subnets.

Once the variables to be queried have been determined, a script is required to query these variables on a periodic basis. If querying a large number of APs, a tool that can perform simultaneous asynchronous queries, without having to wait for previous queries to complete, is highly recommended. In our studies, we have had success using the open-source net-snmp suite of SNMP tools [18] and the related Perl modules.

By collecting the MAC addresses of the associated clients at each AP over time, SNMP can also be used for the identification of client sessions. The accuracy of these sessions, however, will depend on the chosen poll interval, that is, the period between queries. If the poll interval is too high, then the SNMP queries may fail to observe those clients who associate and disassociate with an AP between two polls. On the other hand, if polls are too frequent, the resulting additional traffic to and from the APs may impact the performance of the network by overloading the APs or links. Previous studies (see Section 1.3) have used poll intervals ranging from 1 to 15 min. In our studies, where SNMP was used to query over 500 APs, we found that a 5-min poll interval was required to prevent overloading the network with SNMP traffic. Advantages and disadvantages of SNMP are as follows:

Pros — detailed information easily retrievable from many APs; data can include link, network, and transport layers

Cons — coarse temporal granularity (a poll interval below 5 min may saturate a LAN); vendor-specific MIBs means additional effort required to measure different types of AP

1.2.3 Authentication Logs

Wireless LANs are popular because of the ease with which a client can connect. This presents new security vulnerabilities, however, and as such, many deployed WLANs require some form of authentication before a client is permitted to access the network. Analysis of the logs from an authentication server is another mechanism for determining user behavior; user sessions can be calculated by recording login and logout times. Since an individual user will always use the same login name, irrespective of the host being used to access the network, these sessions may be more accurate for studies where individual usage patterns are of interest. On the other hand, these sessions may not necessarily correspond to actual WLAN behavior; they may lack details of the APs that a user visits, or the timestamps may differ from actual 802.11 authentication and deauthentication times. Nonetheless, in a network that uses authentication, authentication logs are a source of data that are easy to collect, as they will typically be stored in a single central authentication server. Advantages and disadvantages of authentication are as follows:

Pros — accurate session-level information for each individual user; easy to collect from a single source

Cons — not all networks use authentication; authentication sessions may not necessarily correspond to wireless sessions

1.2.4 Network Sniffing

Network or packet "sniffing" refers to the act of capturing network traffic. By placing a network interface into promiscuous mode, the interface will ignore its assigned address and accept all frames. It is then possible to observe any packets that pass this interface. A program such as tcpdump [25] can capture these packets to disk, and a protocol analyzer such as ethereal [6] can dissect these packets to determine such useful data as the source and destination, the protocol, and in many cases the application being used.

By placing a network sniffer near a router or switch that connects a WLAN's APs to the wired network, it is possible to record the traffic that is traversing the wireless portion of the network. If MAC addresses are being used to represent individual users, then care must be taken to place the sniffer before the first router, so that the original wireless client MAC addresses are preserved. Some switches offer a "port mirroring" mode, which can bounce the traffic seen on some ports to another port. This can be useful for sniffing, as a sniffer could be connected to a mirrored port and thus monitor any number of ports on that switch. This requires a sniffer with two Ethernet interfaces: one interface connected to the mirrored port, and another to the wired LAN for remote access. Tcpdump can then be run on the interface that is connected to the mirrored port. If port mirroring is not used, and a sniffer with only one interface is used, then it is necessary to remove any traffic to and from the sniffer (e.g., remote logins) from the packet traces. Furthermore, we have found that a sniffer intended to monitor a wireless subnet may sometimes end up seeing traffic from wired hosts on that subnet because the switches have been misconfigured or are malfunctioning. It is useful to correlate sniffer data with data from other sources, and we use a list of the MAC addresses observed through syslog to remove any nonwireless data.

Since sniffers need to be located before the first router to capture client MAC addresses, they may need to be physically located near the APs being sniffed. For our studies, we have deployed 18 sniffers among 11 buildings around our campus.

These sniffers are located in locked switchrooms, and physical access requires contacting a network sysadmin. Management of these sniffers is therefore more challenging than for a syslog collecting machine or a SNMP poller, both of which have no restrictions on physical location. To minimize the need for physical access, our sniffers are connected to an uninterruptible power supply (UPS) and configured to automatically reboot after a power failure. Our central data collection server periodically runs a script to check that all the sniffers are reachable via the network, and that they are correctly collecting packet traces. As well as making sure that the sniffers are alive and running, they need to be kept secure. While there exist fully automated mechanisms for keeping the software on a machine up-to-date and patched (e.g., "Windows Update" or "RedHat Up2Date"), we have found that automatically applying updates may interfere with the sniffing process. Instead, our scripts signal the presence of updated software, which are then tested on a sniffer in our laboratory before being manually applied to the deployed sniffers.

One important consideration with network sniffing is that the amounts of data involved are much larger than with syslog and SNMP. Monitoring an 11-Mbps (megabits per second) 802.11b WLAN can quickly create hundreds of gigabytes of packet traces, and even more storage space is required to sniff a higher-throughput 802.11a or 802.11g WLAN. It is vital to ensure that sufficient disk space is available for a trace, and it is useful to perform test sniffing to estimate space requirements before the actual start of the measurement study. Even then, some studies have seen machines run out of disk space because of unexpectedly high levels of traffic [27]. Our sniffers collect packet traces 24 hours a day, and then compress and transfer these to a central data collection server in the middle of the night, when network activity is low. We use a feature in tcpdump to ensure that when a trace file reaches a particular size, the file is closed and a new file is created, to prevent large files from exceeding filesystem limits. In addition, we periodically run scripts to monitor free disk space on both the sniffers and the central collection server.

A further consideration is privacy. The packets that are captured through sniffing may contain sensitive data, especially if the LAN being monitored does not use encryption. Most academic institutions will require a study to be approved by their Institutional Review Board for human-subjects research. Some privacy concerns may be alleviated by only capturing packet headers, which may be sufficient for a study that is only concerned with header-level data (packet sizes, interarrival times, and so forth). Advantages and disadvantages of sniffing are as follows:

Pros — detailed packet capture information, including Ethernet headers and data; microsecond temporal granularity

Cons — easiest to capture traffic only on the wired side of the AP, which misses some wireless traffic; ease of sniffing depends on network topology; lots of disk space is required; potential privacy concerns; if a sniffer is monitoring several APs, it can be difficult to determine which AP delivered a particular packet in a trace

1.2.5   Wireless Sniffing

SNMP, syslog, and network sniffing are useful tools for measuring the wired side of the wireless LAN, that is, the wireless traffic that APs bridge on to the wired network. In most wireless LANs, this might be preferred, since the wireless side of the network is likely to be more bandwidth-constrained, and so any active measurement should take place in the less utilized wired network. The disadvantage of only looking at the wired side of the WLAN is that not all wireless data are observable on the wired network. Wireless hosts who are communicating with each other, while both associated with the same AP will not send their traffic via the wired network. IEEE 802.11 management frames and beacons, retransmissions, and collisions are not sent on the wired network, as they are specific to the wireless side. Users that fail to associate with an AP, for instance, rogue wireless clients attempting to gain access to a closed WLAN through MAC address spoofing, or clients that have been misconfigured, will also not be seen on the wired network.

To measure all of this additional traffic and observe the 802.11 PHY/MAC layer, it is necessary to "sniff" the wireless side of the network, that is, to scan the RF spectrum. Fortunately this can be accomplished using relatively simple hardware. Certain 802.11 NICs are capable of being placed into "monitor" mode. With a card in this mode, a packet sniffer will capture 802.11 headers and management frames as well as data packets. These stored frames can be analyzed in a fashion similar to those for wired sniffing. Not all NICs support this mode; popular chipsets with monitor support include the Intersil Prism, Orinoco, and Atheros.

Another measurement option is to use dedicated wireless monitoring hardware, such as a "wireless intrusion protection system" [19]. These typically involve small low-powered wireless devices, designed to be placed in monitor mode and monitor the RF spectrum for specific behavior, such as rogue clients. These devices are similar to APs, and with one of the many APs that run Linux, such as the Linksys WRT54G, it is possible to flash a new firmware on to the AP to turn it into a wireless sniffer [22]. Using these systems can be cheaper than using PCs as sniffers. They lack dedicated storage, however, and a measurement study that intends to store 802.11 frames would require frames to be transmitted from these devices to a central server. To transfer frames from these devices, the 802.11 frames need to be encapsulated into an Ethernet packet for transmission across the wired network. There are several different formats for this encapsulation, depending on the tool being used [1,12,24]; to facilitate data analysis, it is useful to ensure that all the measuring devices use the same format.

Wireless sniffing has several challenges that are not present in wired sniffing. Yeo et al. [26] define three instances where a wireless sniffer might not capture all the traffic on the network. *Generic loss* is where frames are lost because of lack of signal strength, for instance, if a sniffer is too far away from the AP or the client being sniffed. *Type loss* is where frames are not captured as a result of device driver failure, or the inability of a particular card to be placed in monitor mode. The third type of loss, *AP loss*, occurs when firmware incompatibilites cause a particular 802.11 NIC to be incapable of capturing all the packets from a particular type of AP. Some of these losses can be minimized by using multiple sniffers, or sniffers with different 802.11 chipsets. Experimentation in the area to be measured with various antennas and sniffer positioning may also help.

In addition to inadvertently missed frames from type, generic, and AP losses, a wireless sniffer may also miss frames if it is on the incorrect channel. Most wireless NICs can monitor only one channel at a time. With three nonoverlapping channels in the 2.4 GHz band, and 12 non-overlapping channels in the 5 GHz band, monitoring just one channel may potentially miss a large amount of traffic. Mishra et al. [17] find that it is possible to sniff three adjacent channels simultaneously, although 12% of the frames are lost. To resolve this problem, one could choose to either (1) monitor only the channels on which the WLAN's APs are operating, thereby missing any misconfigured client traffic; (2) cycle the sniffer's NICs through all the available channels, which may miss traffic on the channels not currently being monitored; or (3) install one sniffer for each 802.11 channel, at a greater expense.

Whereas wired sniffing can use a relatively small number of sniffers to measure several APs by placing a wired sniffer near an appropriately located router, a wireless sniffer needs to be physically collocated with the APs that it is monitoring, as it needs to be able to "hear" the same frames as the AP. This means that the number of sniffers is proportional to the number of APs, and so a wireless measurement study of a large WLAN could prove expensive. Advantages and disadvantages of wireless sniffing are as follows:

Pros — can capture *all* wireless traffic, including management frames, as opposed to just the traffic that traverses the wired side of an AP

Cons — capturing every packet can be difficult, and is highly dependent on antennas, 802.11 card firmware, and the positioning of sniffers; not all cards support monitoring; no common data format; privacy issues

1.2.6 Client-Side Tools

The previously discussed tools are all designed to monitor from the network perspective. Another measurement method is to directly measure what a wireless client is doing, by installing software on the client. This offers many advantages. A client-side tool can accurately determine exactly what the client is seeing. While syslog provides the AP at which a client is associated, a client-side tool could list all the additional APs that a client can see, which can be useful for mobility tracing. A client-side tool can list all the applications that a wireless device is using, rather than just those applications that are generating network traffic.

Writing a client-side tool can be challenging, however, if it is to run on a variety of client devices, with different operating systems and different device drivers. In addition, a tool will need to be installed on end devices. Some users may find this intrusive, and choose to disable the tool, and there may be privacy implications to consider. Advantages and disadvantages of client-side tools are as follows:

Pros — the best way to accurately capture exactly what the client is seeing

Cons — can be difficult to write a tool that supports multiple platforms, device types, and device drivers; difficult to deploy and maintain tool on a large number of devices; privacy issues

1.2.7 Other Considerations

As well as the software and hardware required for a wireless LAN measurement study, there are some data that require manual nonautomated collection. Much of this requires collection before a study should commence.

The first item that needs to be manually obtained is a list of the APs to be measured. The AP MAC addresses will be needed to make sense of syslog data. If the APs have been assigned IP addresses or hostnames, then these should also be collected. If the APs have dynamically assigned IP addresses, then access to a DHCP server may be required to collect SNMP data, as the AP's IP address needs to be known to query it via SNMP. If syslog or SNMP are to be used, then all the APs to be measured will need to be configured for syslog and/or SNMP, and the SNMP community string will need to be ascertained.

Collecting mobility traces requires knowledge of the physical location of the APs. This can be done with the aid of GPS units, but in most WLAN installations, the APs are indoors, where GPS is of little use. Maps of buildings are generally the best way of plotting the location of APs.

Long-term measurement studies must also keep track of changes in the network. In the course of monitoring our campus WLAN for over 3 years, we have found that APs will be moved to improve coverage, additional APs are introduced over time, or new security measures are introduced that interfere with data collection. In some scenarios it may be possible to automatically determine these changes, for instance, through a wireless sniffer detecting frames from new APs. In most cases, however, close interaction with network administrators will be needed to track these changes. Moreover, if syslog and SNMP are being used, new APs will need to be configured appropriately as they are added to the network.

1.3 MEASUREMENT STUDIES

Having described some of the techniques that can be used to measure a wireless LAN, we now discuss some of the measurement studies that have been conducted, and the methods that these studies have employed.

1.3.1 Campus WLANs

Most measurement studies have taken place in a university campus setting. This is not surprising, as for an academic researcher, it is typically easier to get permission to measure one's own network.

One of the first wireless LANs to be measured was at Stanford University. Tang and Baker measured 74 users on the Stanford Computer Science departmental WLAN for 12 weeks in 2000 [23]. They used network sniffers, authentication logs, and SNMP with a 2-min poll period. With only 12 APs in the wireless subnet, a 2-min poll period was feasible, as each poll generated only approximately 50 kB (kilobytes) of traffic. Moreover, the SNMP polls were small, as they only queried one specific variable: the list of MAC addresses associated with a particular AP.

Tang and Baker's study looked at user behavior, mobility, and traffic. They found that usage peaks in the middle of the day. Users were not highly mobile, and on average only 3.2 users visited more than one AP in a day. The sniffer analysis indicated that the most popular applications were WWW browsing and ssh or telnet sessions. The latter is unsurprising given the computer scientist population. Half of the users used interactive chat applications such as ICQ and IRC.

Hutchins and Zegura [10] traced a subset of the Georgia Tech campus WLAN, comprising 109 APs in 18 buildings, for a 2-month period in 2001. The methods used included network sniffers, SNMP, and Kerberos authentication logs. Their SNMP polls had a relatively large interval of 15 min. The authentication logs provided a basis for calculating user sessions, from the time that a user logged in, until the network's firewall timed out an idle user.

This study was again concerned with user behavior, mobility, and traffic patterns. Strong diurnal usage patterns were found, and there was a peak in usage around 4 P.M. which they suggest was due to the end of the workday. The number of users each day grew almost linearly over time, falling only during university holidays. From the sniffer traces, they examined flow counts and flow lengths, rather than the absolute amounts of traffic. Short flows (less than 5 min) dominated, although some long flows of almost 9 hours were observed. The longest flows were ssh or telnet, but the largest number of flows were HTTP. Over the course of the study, 228 out of 444 users were seen in more than one building. They calculated mobility on an aggregate basis, rather than a per user basis, and users who "ping-ponged" between nearby APs may skew these data.

Chinchilla et al. [5] conducted a WLAN measurement study focusing on WWW users at the University of North Carolina. They used syslog and network sniffers to trace 222 APs over an 11-week period. Rather than collect every single wireless packet, this study chose to collect only HTTP requests. The authors used tcpdump to look for TCP traffic on any port, and recorded any packet where the payload began with the ASCII string GET.

This study was interested in the locality of WWW behavior and mobility; 13% of the unique URLs being requested accounted for 70% of the HTTP requests, and 8% of requests were for WWW objects that a nearby client had requested within the last hour. This suggests that caching at APs might have some benefit, and they estimate that a cache at each AP would have been useful for 55% of the requests over the entire trace. Student residences were found to have the most wireless associations, and most clients were nonmobile, which may be due to students leaving their laptops connected to the WLAN in their dorms. A Markov chain was used to develop an algorithm to predict the next AP that a user will visit; this was capable of predicting the correct AP 87% of the time over the trace.

Schwab and Bunt measured the WLAN at the University of Saskatchewan in 2003 [20]. They used a network sniffer and Cisco LEAP authentication logs to trace 18 APs over a one-week period. Unlike most other measurement studies, this study did not use the tcpdump sniffer, but an alternative program called EtherPeek [7].

The Saskatchewan study examined user behavior, mobility, and traffic. This is a nonresidential WLAN, and so again the diurnal patterns mirrored the workday. Web traffic accounted for ≈30% of the traffic, but there was little ssh or telnet usage, which may be due to most WLAN users being law students, as opposed to computer scientists. Users were nonmobile, and the APs in the law school saw significantly more use than did other APs. This led the authors to conclude that APs should be deployed with a view to providing network access in a specific location, rather than providing ubiquitous mobile access.

McNett and Voelker [16] used a client-side tool to measure mobility on the University of California San Diego WLAN. A tool was installed on 272 PDAs, which were equipped with a 802.11b CompactFlash adapter. The tool periodically recorded the client's signal strength for each visible AP, the AP at which the client was associated, the device type, and whether the PDA was using AC or battery power. As the PDAs lacked large storage capabilities, the PDAs would contact a central server to upload collected data.

This PDA study looked at user session behavior and mobility. There were regular diurnal patterns, and less usage at the weekends. Usage was bursty, which may be due to the difficulty of using a PDA for long periods of time. Interestingly, there was a steady decline in the number of users over the trace period. This may be due to the user population (students) becoming bored with the devices. They defined two types of session: (1) the *AP session*, the amount of time that a given PDA spends associated with an AP, and (2) the *user session*, the contiguous time period in which a PDA is switched on and connected to the WLAN. AP sessions were significantly shorter than user sessions, indicating that roaming was taking place while the PDA is in use. Despite the difficulty of using a PDA, there were some long sessions, with 20% of user sessions over 41 min long. Over the course of the trace, 50% of the users visited more than 21 APs. As in the Saskatchewan study, AP load was uneven, and 50% of the APs only saw 5 users or less, and 10% of the APs saw 84 users or more. The mobility traces were used to develop a *campus waypoint* mobility model, which incorporated knowledge of specific geographic locations on campus. Comparing the trace-based mobility model to traditional synthetic mobility models indicated three significant differences. In the trace-based model (1) only a small number of users (11%) were actually mobile at any given time, compared to most nodes in a synthetic model; (2) users were walking at lower speeds (1 m/s) than synthetic models (0–20 m/s); and (3) users appeared and disappeared from the network, which is not considered in most synthetic models.

At Dartmouth College, we have conducted the largest studies of an academic WLAN. We have collected syslog messages from most of the APs on campus since their installation in 2001. We have also used SNMP and tcpdump wired sniffers for two extensive studies covering 476 APs for 11 weeks in 2001/02 [13] and 566 APs for 17 weeks in 2003/04 [9].

In our 2001/02 study we examined user behavior and traffic patterns. The Dartmouth campus differs from those in other studies as it covers a wide range of locations: academic areas, sporting grounds including a ski slope, residential dormitories and houses, communal eating and social areas, and parts of the town in which the college is based, including some shopping areas, a hotel, and restaurants. In terms of the amount of traffic, the residential areas dominated all other areas. The diurnal usage patterns observed elsewhere were also present at Dartmouth, although the residential nature of the campus meant that usage did not stop at the end of the workday, with many students using the WLAN late at night. User sessions were short, with a median of 16.6 minutes, and 71% of sessions were shorter than one hour. As in other studies, WWW traffic was the most popular application, although some clients also used backup programs over the WLAN, which contributed to a large proportion of the overall traffic.

In our 2003/04 study, we chose to examine changes in user behavior on the WLAN. After 3 years of deployment, the WLAN could be considered a mature network, and an integral part of college life. The college had also begun to replace the analog telephone system with a Voice over IP (VoIP) telephone system, and some students were issued with VoIP clients, which could be used over the WLAN. We found that the types of application used on the WLAN changed dramatically between 2001/02 and 2003/04; while HTTP was still the most popular application in terms of the amount of traffic, peer-to-peer file sharing and streaming media saw significant increases in usage. Wireless VoIP did not appear to be a popular application, with most VoIP calls being made on the wired network. As a result of the increase in file-sharing, local (on-campus) traffic exceeded off-campus traffic, a reversal of the 2001/02 situation. Residences still continued to generate the most traffic, and usage remained diurnal, between our two studies.

Our 2003/04 study also examined mobility. The syslog data indicated that many users "ping-ponged" between APs in range, and so when examining the mobility of a session, we considered the *session diameter*, that is, the maximum distance between any two APs visited in a session. Sessions with a diameter below 50 m were considered to be nonmobile, as they were assumed to consist of ping-ponging clients. From the tcpdump logs, we used a tool for analyzing TCP flows to estimate the operating system being used by a device (by looking for differences in window sizes, ACK values and so on). We used this information to classify the device by type: Mac or Windows laptop, VoIP phone, PDA, and so forth. This information was used to characterize mobility among different device types. Devices such as VoIP phones, which are always switched on, were found to visit significantly higher numbers of APs and have longer session durations than laptops, which are typically powered down before a user moves between locations. Overall, users were found to be nonmobile, with 50% of users spending 98% of their time in a *home location*, that is, a group of one or more APs within a 50 m^2 area with which a user is most often associated. In separate work, we have also used the association and disassociation times in our 3 years of syslog traces to create a mobility history for each user, which were then used to develop and evaluate mobility prediction models [21]. For user histories containing less than 1000 movements, most predictors performed badly. For histories longer than this, however, the best predictors had accuracies of around 65–72% for the median user; that is, they were able to correctly predict the next AP with which a user would associate 65–72% of the time. Interestingly, simple Markov-based predictors performed just as well as more complex compression-based predictors. In particular, an order 2 Markov predictor, with a "fallback" to a shorter order 1 predictor when encountering a new context not seen in the user's history, performed the best overall.

1.3.2   Nonacademic WLANs

One of the few WLAN measurement studies to take place outside an academic campus setting was conducted at a corporate research facility by Balazinska and Castro in 2002 [4]. They used one method, SNMP with a polling interval of 5 min, to query 117 APs over 4 weeks.

This corporate study concentrated on AP loads and user mobility. As seen in other studies, some APs were little used, with 10% of the APs seeing less than 10 simultaneous users. The most highly utilized APs in terms of the number of simultaneous users were in communal locations such as cafeterias and auditoriums. In terms of traffic levels, however, the most highly utilized APs were in laboratories and conference rooms.

Users were found to be predominantly nonmobile, with 50% of the users visiting less than three APs in a given day. Two metrics were introduced to characterize mobility: *prevalence*, the amount of time that a user spends at a given AP over the course of a user's trace; and *persistence* which measures the amount of time that a user stays associated with a given AP before moving to the next AP. Using the prevalence data, users were categorized by varying degrees of mobility, from "stationary" to "highly mobile." Stationary users had a high maximum prevalence, as they spent most of their time associated to a single AP, while highly mobile users had low maximum and median prevalences, spending their time at different APs. The persistence metric complements prevalence by accounting for the amount of time spent at each AP, and unsurprisingly, persistence was lower at guest locations.

Also outside the academic setting, Balachandran et al. [3] used SNMP and sniffers to analyze 195 wireless users at the 2001 ACM SIGCOMM conference. They chose a polling interval of one minute. Such a short polling interval was possible because of the small number of APs involved in the study—there were only four APs used at the conference.

This study examined user behavior, traffic patterns, and AP loads. Given the conference setting, usage closely followed the conference schedule, and the number of users rose when sessions were taking place, and fell during meals and breaks. Arrival times were modeled using a Markov modulated Poisson process, where arrivals vary randomly during an ON period (the conference sessions). Session durations were Pareto distributed, with most sessions under 5 minutes in length, and many of the longest sessions idle and transferring little data. The most popular application was again WWW browsing, and since SIGCOMM participants are predominantly computer scientists, ssh was the second most popular application. Unlike most studies, the majority of users (over 80%) were seen at more than one AP in a day, although this may be conference-specific, where an attendee does not have a designated seat, and so they would associate with a different AP depending on where they are sitting in a given conference session. AP loads were found to vary not with the number of users, but rather with the applications that individual users are using.

1.3.3   Wireless-Side Measurement Studies

As we have described in Section 1.2, wireless sniffing is complicated, and as such, there have been few large measurement studies of the wireless side of a WLAN.

In two studies, Yeo et al. [26,27] looked at the difficulties of conducting wireless-side measurement. To estimate the amount of loss incurred in wireless measurement, three wireless sniffers were compared to a wired sniffer and SNMP polls with an interval of one minute. A packet generator was used to send UDP packets, marked with sequence numbers, between hosts, all on the same channel. The three sniffers were found to have different viewpoints of the wireless medium. All the sniffers were more successful at capturing traffic from the AP, rather than from the clients, as APs tend to have larger and more powerful antennas, and clients may move around and end up out of sight of a sniffer. On average, the sniffers saw 99.4% of the packets from the AP, but only 80.1% of the packets from clients. By merging the traces from the three wireless sniffers, this capture rate was improved, to 99.34% of the traffic that the wired sniffer observed. One recommendation from this study is that one sniffer should be placed near to the AP being monitored, with any other sniffers placed as near as possible to the predicted location of clients.

In a subsequent experiment, Yeo et al. considered seven APs in the University of Maryland's Computer Science department. Three wireless sniffers were used, equipped with Orinoco 802.11b NICs placed into monitor mode, locked to one channel (6), and configured to capture 802.11 frames using the Prism2 file format. This enabled the monitoring of the three APs that were using channel 6. The study took place over 2 weeks, although there was one hole because the sniffers ran out of disk space.

This study concentrated on the PHY/MAC layer, as this can be examined only using wireless sniffers. The maximum throughput seen on a single AP was only 1.5 Mbps, due to contention on the channel that was shared between the three APs. The level of transmission errors, that is, the number of retransmitted frames divided by the total number of frames, varied by day, but there were more transmission errors in the data being sent to an AP, rather than from an AP. Examining the types of frames, they found that dataframes made up 50.7% of the frames sniffed, and beacon frames made up 46.5%. Association and reassociation response frames tended to be sent at the highest data rate, 11 Mbps, whereas the corresponding request frames were sent at 1 Mbps. The 802.11 standard does not specify a behavior for response frames, and by sending responses at a high data rate, many response frames did not reach the client and needed to be retransmitted. Other management frames, including probe response and power-save polls, were also often retransmitted. For data frames, multiple data rates were common, and the average data rate was 5.1 Mbps.

Mishra et al. [17] used wireless sniffing to look solely at the 802.11 MAC layer handoff process. Eight machines were used as wireless sniffers, with a total of 14 802.11b NICs installed across the machines. Each NIC was set in monitor mode and locked to one individual channel, which allowed the monitoring of all 11 2.4 GHz channels. One user with a laptop then walked around the University of Maryland Computer Science department, which had three WLANs comprising some 60 APs using Cisco, Lucent, and Prism2 chipsets. As this study concentrated on handoffs, the only frames that the sniffer recorded were probe requests and probe responses, reassociations, and authentication frames. To examine variations in handoffs between device drivers, three different 802.11b NICs were used in the client laptop: Lucent Orinoco, Cisco 340, and a ZoomAir Prism2.5.

Across all the devices, probe delay (probe request and probe response frames) was found to account for over 90% of the overall handoff latency. There was a large variation in the handoff latency between devices, with a Lucent client and Cisco AP taking an average of 53.3 ms, and a Cisco client and Cisco AP taking 420.8 ms. With the same device and AP configuration, there was a large variation in handoff latency, and the higher the latency, the higher the standard deviation. Some of the differences in latency between devices could be explained by the different behaviors between devices. The Lucent and Prism NICs would send a reassociate request prior to authenticating with a new AP, and a second reassociate request after authentication. There were also large differences between each device's probe wait time (the amount of time that a scanning client waits before moving on to scanning the next channel). The Cisco client sent 11 probes on each channel, and spent 17 ms on channels with traffic, and 38 ms on channels with no traffic; the Lucent sent three probes on channels 1, 6, and 11, and spent almost the same amount of time on channels irrespective of traffic; the ZoomAir sent only three probes on channels 1, 6, and 11, and spent an additional 10 ms after the three probes on selecting the AP with which to associate. Using the empirical data from the sniffer logs, the authors suggest that device manufacturers could choose to lower these probe wait times.

1.3.4 Discussion

Table 1.1 lists the methods used in the studies that we have discussed above.

In summary, Table 1.1 shows that there have been several studies of academic campus WLANs, and fewer studies of nonacademic WLANs. Common methods include syslog, SNMP, and sniffing. The results of these studies show that the most common applications used on a WLAN are not necessarily mobile applications, with HTTP accounting for most traffic, and telnet and ssh used in computer science environments. Short flows and sessions are common, and this should be kept in mind when choosing poll intervals for a measurement study. Users tend

TABLE 1.1 Wireless Studies and Methods Used

Study	Location	Duration	APs	Syslog	SNMP [Poll Interval (min)]	Methods Used Sniffers	Authentication Logs	Client Tools	Wireless Sniffing
Balachandran et al. [3]	Conference	52 hours	4		1	✓			
Balazinska and Castro [4]	Corporate	4 weeks	117		5	✓			
Chinchilla et al. [5]	Academic	11 weeks	222	✓		✓			
Henderson et al. [9]	Academic	17 weeks	566	✓	5	✓			
Hutchins and Zegura [10]	Academic	2 months	109		15	✓	✓		
Kotz and Essien [13]	Academic	11 weeks	476	✓	5	✓			
McNett and Voelker [16]	Academic	11 weeks	>400					✓	
Mishra et al. [17]	Lab	30 min	60			✓			✓
Schwab and Bunt [20]	Academic	1 week	18			✓	✓		
Tang and Baker [23]	Academic	12 weeks	12		2	✓	✓		
Yeo et al. [26,27]	Lab	2 weeks	3		1	✓			✓

to be nonmobile, although the introduction of new always-on devices is leading to increased mobility. APs tend to be unevenly used across a WLAN, with certain locations accounting for high levels of traffic. Trace-based mobility models and predictors have been developed, and it will be interesting to see how these perform with traces of newer, more mobile, clients.

Wireless sniffing is still a new area, and one that presents many challenges. The studies that have used wireless sniffing are much smaller than those that have used wired sniffing, syslog and SNMP, and have concentrated on specific channels in specific locations. Although small, these studies have yielded insights into 802.11 MAC behavior, and highlighted the differences between chipsets and devices. For instance, 46.5% of the frames observed in one study were 802.11 beacons, which indicates the large amount of data that can be missed in a wired sniffing study. Larger-scale wireless sniffing, with a variety of chipsets and device types, could prove useful for future wireless protocol development.

1.4   CONCLUSIONS

Wireless LANs are becoming increasing popular, and it is useful to be able to measure various characteristics of these WLANs. In this chapter we have discussed the tools available for measurement, and the studies that have already been conducted using these tools. To conclude, we present a checklist that we hope will be useful for those intending to carry out a wireless measurement study.

1.4.1   Wireless Measurement Checklist

- Determine which tools are most appropriate for the purposes of the study. Syslog is useful for mobility, while SNMP is an easy method for extracting traffic statistics. Client tools and wireless sniffing provide the most detail, but incur the greatest costs in terms of setup time and equipment. It is also useful to use multiple tools and correlate the data, such as using MAC addresses observed in syslog messages to verify that sniffer logs are accurately capturing wireless client traffic.

- Gain approval from the appropriate Institutional Review Board for human-subjects research. Wireless data collection can involve potentially sensitive information, such as the location of wireless users or the data that they are transferring.

- Decide how much of the WLAN will be monitored. Different tools may be able to monitor different parts; for instance, it may be easy to use SNMP to monitor every AP, but sniff only a subset of the WLAN.

- Draw up a list of all the APs to be measured. If required, determine the physical location of these APs, using a building plan and/or GPS.

- Ensure that all the APs that are to be measured are configured correctly, for example, that they are configured to send syslog messages, or to allow SNMP queries, and that the network security policies (if any) allow this syslog and SNMP data to be transmitted to the host that is storing the data. Do not rely on a sysadmin to do this, but confirm it for yourself.

- Test the data collection and analysis software in a "dry run" before the actual measurement study begins. Checking that the analysis software works will help to determine whether sufficient data are being collected and if the appropriate tools are being used.

- Closely monitor the data collection. Keep track of changes in output, such as syslog messages changing as a result of AP firmware changes. Measuring devices may malfunction or run out of disk space, which also requires careful monitoring.

- Keep in touch with the WLAN's sysadmins. It is important to know when new APs are installed, or when existing APs are moved or decommissioned.

- Minimize disruption on the network being measured. Most of the tools described here are *active* measurement tools, in that they generate additional network traffic. It is vital not to impact the network being monitored. For instance, on one particular type of AP, we have found that frequent SNMP queries could cause the AP to stop forwarding packets.

- Expect the unexpected! Measurement of a live network, with large numbers of real wireless network users, may encounter many surprising events. We have had our measurement studies impacted by viruses, worms, misconfigured wireless clients, firewalls, changes in network subnetting and VLANs and more. With a comprehensive monitoring system as discussed above, however, we have been able to detect most of these problems and reconfigure the measurement infrastructure where required.

Readers who are interested in conducting a wireless measurement study, or who would like access to data from some of the studies discussed here, are directed to our Websites at `http://www.cs.dartmouth.edu/~campus` and `http://crawdad.cs.dartmouth.edu/`.

REFERENCES

1. Apware project, `http://nms.csail.mit.edu/projects/apware/software/`.
2. F. Baker, *Requirements for IP Version 4 Routers*, IETF RFC 1812, June 1995.
3. A. Balachandran, G. M. Voelker, P. Bahl, and P. Venkat Rangan, Characterizing user behavior and network performance in a public wireless LAN, *Proc. Int. Conf. Measurements and Modeling of Computer Systems* (*SIGMETRICS*), Marina Del Rey, CA, June 2002, ACM Press, pp. 195–205.

4. M. Balazinska and P. Castro, Characterizing mobility and network usage in a corporate wireless local-area network, *Proc 2003 Int Conf Mobile Systems, Applications, and Services (MobiSys)*, San Francisco, May 2003, USENIX Assoc., pp. 303–316.

5. F. Chinchilla, M. Lindsey, and M. Papadopouli, Analysis of wireless information locality and association patterns in a campus, *Proc. 23rd Annual Joint Conf. IEEE Computer and Communications Societies (InfoCom)*, Hong Kong, March 2004, IEEE.

6. Ethereal protocol analyzer, `http://www.ethereal.com`.

7. EtherPeek protocol analyzer, `http://www.wildpackets.com`.

8. J. Flick and J. Johnson, *Definitions of Managed Objects for the Ethernet-like Interface Types*, IETF RFC 2665, Aug. 1999.

9. T. Henderson, D. Kotz, and I. Abyzov, The changing usage of a mature campus-wide wireless network, *Proc. 10th Annual ACM Int. Conf. Mobile Computing and Networking (MobiCom)*, Philadelphia, Sept. 2004, ACM Press.

10. R. Hutchins and E. W. Zegura, Measurements from a campus wireless network, *Proc. IEEE Int. Conf. Communications (ICC)*, New York, April 2002, IEEE Computer Society Press, Vol. 5, pp. 3161–3167.

11. IEEE 802.11 MIB, `http://standards.ieee.org/getieee802/download/MIB-D6.2.txt`.

12. Kismet wireless sniffing software, `http://www.kismetwireless.net`.

13. D. Kotz and K. Essien, Analysis of a campus-wide wireless network, *Wireless Networks* **11**: 115–133 (2005).

14. C. Lonvick, *The BSD Syslog Protocol*, IETF RFC 3164, Aug. 2001.

15. K. McCloghrie, D. Perkins, and J. Schoenwaelder, *Structure of Management Information Version 2 (SMIv2)*, IETF RFC 2578, April 1999.

16. M. McNett and G. M. Voelker, *Access and Mobility of Wireless PDA Users*, Technical Report CS2004-0780, Dept. Computer Science and Engineering, Univ. California, San Diego, Feb. 2004.

17. A. Mishra, M. Shin, and W. A. Arbaugh, An empirical analysis of the IEEE 802.11 MAC layer handoff process, *ACM SigComm Comput. Commun. Rev.* **33**(2):93–102 (April 2003).

18. Net-snmp SNMP tools, `http://net-snmp.sourceforge.net`.

19. Network Chemistry RFProtect wireless intrusion protection system, `http://www.networkchemistry.com`.

20. D. Schwab and R. Bunt, Characterising the use of a campus wireless network, *Proc. 23rd Annual Joint Conf. IEEE Computer and Communications Societies (InfoCom)*, Hong Kong, March 2004, IEEE.

21. L. Song, D. Kotz, R. Jain, and X. He, Evaluating location predictors with extensive Wi-Fi mobility data, *Proc. 23rd Annual Joint Conf. IEEE Computer and Communications Societies (InfoCom)*, Hong Kong, March 2004, IEEE.

22. Sveasoft alternative Linksys WRT54G firmware, `http://docs.sveasoft.com/`.

23. D. Tang and M. Baker, Analysis of a local-area wireless network, *Proc. 6th Annual ACM Int. Conf. Mobile Computing and Networking (MobiCom)*, Boston, Aug. 2000, ACM Press, pp. 1–10.

24. Tazmen Sniffer Protocol, `http://www.networkchemistry.com/support/appnotes/an001_tzsp.html`.

25. Tcpdump packet capture software, `http://www.tcpdump.org`.

26. J. Yeo, S. Banerjee, and A. Agrawala, *Measuring Traffic on the Wireless Medium: Experience and Pitfalls*, Technical Report CS-TR 4421, Dept. Computer Science, Univ. Maryland, Dec. 2002.

27. J. Yeo, M. Youssef, and A. Agrawala, *Characterizing the IEEE 802.11 Traffic: The Wireless Side*, Technical Report CS-TR 4570, Dept. Computer Science, Univ. Maryland, March 2004.

CHAPTER 2

Understanding the Use of a Campus Wireless Network

DAVID SCHWAB and RICK BUNT

University of Saskatchewan, Saskatoon, Canada

2.1 INTRODUCTION

The University of Saskatchewan campus covers a large physical area, with more than 40 buildings distributed over 147 hectares of land on the banks of the South Saskatchewan River. Our geography has a significant impact on our approach to delivery of information technology. The campus wireless network is one of several new projects that we have introduced since 2001 to enhance the computing environment for our 18,000 students. Our approach is to provide mobile users with access to our wireline network through high-speed wireless access points located in very public areas.

Our initial deployment began in the 2001/02 academic year with a pilot project, consisting of a small number of access points (18) placed strategically in a number of locations. This trial deployment demonstrated that wireless technology would be an effective way to give students greater access to network resources and the Internet. The demand for wireless networking has grown steadily since then. After the trial rollout was completed, wireless access points were fully integrated into the campus network and are now regularly used by a growing number of wireless users. We continue to expand the network to meet that demand, and now have close to 80 access points. Further wireless installations are being planned, both for new buildings and as part of ongoing expansion.

In order for us to plan for any expansion, it is important that we understand current usage patterns — that we understand where, when, how much, and for what our wireless network is being used. It is also important to understand how usage patterns are changing and what future usage can be projected from current trends. This

Mobile, Wireless, and Sensor Networks: Technology, Applications, and Future Directions
Edited by Rajeev Shorey, Akkihebbal L. Ananda, Mun Choon Chan, and Wei Tsang Ooi
Copyright © 2006 John Wiley & Sons, Inc.

chapter describes the methodology we are employing to collect data on usage, and what we are learning. Authentication logs were collected in cooperation with our Information Technology Services Division (ITS) over the 2003/04 academic year. In our analysis, these usage data are supplemented with short-term wireless packet traces gathered at specific campus locations.

The chapter is organized as follows. Section 2.2 reviews related work in wireless network measurement. Section 2.3 describes the wireless network at the University of Saskatchewan, including its initial deployment, results from early user measurement research conducted on it and the production network configuration in use during the current study. In Section 2.4 we describe the methodology followed when gathering and analyzing the data, and compare it to the methodology employed during our earlier research. Section 2.5 contains the results of our analysis, and offers comparisons between current and past results. We conclude in Section 2.6 with a summary of our key findings.

2.2   RELATED WORK

The design of this study was motivated by work done by Balachandran et al. [1]. Their analysis and characterization of the traffic generated by attendees at a popular ACM conference in the summer of 2001 provided many useful insights. They employed two mechanisms to gather wireless traffic traces during the conference. One trace was gathered by periodically polling each of four access points positioned in the conference hall with SNMP requests. This trace revealed usage statistics at the access point level, including the number of users currently connected and the number of transmission errors. The second trace was gathered at a router that connected the access points to the campus network. This tracing was done using tcpdump [2] to gather anonymized TCP packet headers. The analysis of those headers revealed access-point-independent statistics, such as the total amount of traffic on the wireless network and the application mix of that traffic.

Although the conference trace was gathered successfully and analyzed thoroughly, the findings from its analysis have limited applicability to a full campus setting. The conference had a set schedule, which caused readily apparent traffic patterns as all attendees moved from event to event. Furthermore, the access points were all placed in the same conference hall area, which resulted in almost identical usage patterns being observed at each access point.

The analysis of the Dartmouth College wireless network by Kotz and Essien [3] is more relevant to campus-wide networks. Dartmouth's wireless network is made up of 476 access points providing coverage in 161 buildings for almost 2000 users. The Dartmouth study used a combination of three forms of trace gathering: event-triggered log messages, SNMP polling, and packet header recording. Because of the decentralized structure of the Dartmouth network, however, packet headers could be gathered from only a small number of locations, and because the SNMP and log messages were sent by each access point individually via UDP packets, some of the data were lost or misordered. Also, some of the access points experienced power failures or misconfiguration problems that resulted in gaps in the trace.

Both these studies were based on earlier research done at the Stanford University Computer Science Department. Tang and Baker [4] used tcpdump and SNMP polling to gather statistics on 74 wireless users over a 12-week period. While their study did establish the methodology used by subsequent wireless network traces, the scope of their work was limited to a single department in a single building and does not fully reflect the activities of the broad spectrum of campus wireless users.

More recently, Papadoupouli et al. [5] studied wireless usage on the University of North Carolina campus at Chapel Hill. Their investigation focused on user mobility patterns, specifically the predictability of roaming behavior and the correlation between association patterns and Web access. Papadoupouli et al. believe that wireless users would benefit from localized, peer-to-peer, and predictive caching systems, especially with regard to location-specific information and services. Although the Web is not a location-based service, their study suggests that a significant percentage of all Web requests — a larger percentage of requests from highly mobile users — could be considered location-dependent.

2.3 NETWORK ENVIRONMENT

As of Fall 2004, our campus wireless network consists of close to 80 access points that provide service to over 700 clients, including students, faculty, and staff. New buildings, such as our new Kinesiology building, are being constructed with wireless access points from day one. Older buildings are rapidly being added to the wireless network as new access points go online every month.

From the beginning, we have used a mix of Cisco AP350 and AP1200 access points. Both models support 802.11b [or Wireless Fidelity (WiFi)] connections, and the AP1200 is upgradable to support 802.11g and/or 802.11a connections. Using the proprietary Cisco Lightweight Extensible Authentication Protocol (LEAP), each connection is authenticated by verifying the username and password specified with a Cisco Secure Access Control Server (ACS). This allows users to connect to the wireless network using the same username and password as they use to log in to laboratory machines and Internet services. The ACS records every authentication and deauthentication that occurs on the wireless network [6].

Clients can connect to the wireless network using any wireless network adapter with drivers that support LEAP — such as the Cisco Aironet 350 or Apple Airport. To encourage early adoption of wireless technology, Aironet 350 wireless adapters were made available at a subsidized price to students, faculty, and staff through our campus computer store during the first year.

Our initial wireless network was deployed as a pilot project during the summer of 2001 on a virtual subnet of our extensive wireline network. The use of a subnet enabled us to distinguish wireless traffic from nonwireless traffic and helped ensure that unauthorized wireless users would not have access to campus services.

In early 2003, we conducted an initial study of the fledgling wireless network and we reported the results of this study [7]. Traffic on the entire wireless pilot project subnet was mirrored at the central campus router from January 22 to 29, 2003. The mirrored traffic was recorded using the network analysis package EtherPeek [8]. Anonymized ACS log data from the period was also made available for this preliminary study. Although the week-long trace could not be seen as representative of average wireless user behavior, our work established a useful methodology and our analysis provided a statistical snapshot of the status of the early campus wireless network. These preliminary observations are compared to our current results in Section 2.5. During the summer of 2003 the campus shifted from a switched network to a routed network, and wireless access points were integrated into the common campus subnet. This necessitated some adjustment to our methodology.

We are currently pursuing several new applications for 802.11 wireless networking technology on our campus. In the spring of 2004, ITS began installing long-range point-to-point wireless links between the campus and remote research facilities located some distance outside the city. Trials are underway to allow low-end devices (such as handheld computers using the Palm OS and wireless inventory tracking devices) that do not support LEAP authentication to connect to the campus wireless network. These devices will be authenticated by comparing their machine (MAC) addresses to a list of allowed devices. We are also deploying Voice over IP (VoIP) phone service in some newly constructed buildings, and this may also soon be usable over the campus wireless network as an alternative to cellular service.

2.4 METHODOLOGY

2.4.1 Authentication Logging

The Cisco Secure ACS keeps track of every wireless user currently connected to the network, and this information is logged for security monitoring purposes. The ACS log includes a record of each authentication and deauthentication that occurs on the wireless network. The information recorded includes the date and time, username, client card address, session identification, and access point address associated with each event. In addition, each deauthentication record includes the number of packets transmitted and received and the amount of data those packets contained. ITS has been saving anonymized copies of these log files for use in our research since late August 2003.

2.4.2 Trace Collection

While the ACS logs reveal overall usage patterns over the entire campus, they do not give the specific information needed to characterize the applications and traffic patterns associated with wireless users. To gain this more detailed information requires a trace of wireless traffic.

In our earlier study of the wireless network [7], the network topology enabled us to mirror and trace the wireless traffic over the entire campus. Once we converted to a routed network, however, such mirroring was no longer possible. To capture only the wireless traffic now, it is necessary to mirror the traffic at each individual access point.

ITS agreed to use a trace gathering system that we developed for this project to capture packet headers from a number of wireless access points on campus. Our trace gathering system was developed using a customized NetBSD kernel [9] and standard trace gathering utilities (described below). It is configured to begin a new trace gathering session automatically on each startup, operate continuously for long periods without direct monitoring, and safely terminate the trace gathering process when shut down. This minimized the time and resources that ITS needed to allocate to this project during the trace gathering period. Since our custom trace gathering system was specifically tailored for wireless user measurement research, the results recorded were far more detailed than those gathered using commercial network analysis tools [8] in our earlier study.

Our trace gathering system was deployed by ITS staff in three high-traffic campus locations between March 5 and May 3, 2004. The data we collected form the basis of Section 2.5.

2.4.3 Anonymization

ACS logs were sanitized by ITS using a custom-built anonymization tool. These anonymized log files contain the same information as the actual ACS log, with two exceptions: (1) the usernames contained in the anonymized logs were replaced with unique identifiers generated by the SHA1 one-way hashing algorithm and (2) events in the log that were not related to activity at wireless access points were removed. During our earlier study, the ACS logs were anonymized by simply stripping them of all private data fields, a process that greatly reduced the usefulness of the anonymized logs in our research. By hashing private identifiers, we can maintain user privacy without sacrificing any of the log's value to our research.

The trace gathering system we developed was also designed with user anonymity in mind. Tcpdpriv [10] anonymizes tcpdump-formatted traces by stripping them of all packet payload information, leaving only the header fields for later analysis. Instead of gathering packet traces using tcpdump and then anonymizing them afterward, we opted to gather our traces using tcpdpriv directly. Although this restricts the depth of analysis that we can perform on the trace data, it ensures that no private information contained in the wireless packets is ever recorded.

2.4.4 Analysis

We analyzed the ACS log and packet header data using a combination of preexisting and custom-written data analysis tools.

Storing the ACS log entries in a relational database allows for far more flexible and efficient analysis than was possible using the analysis scripts from our earlier study. We used a custom-written data input tool to parse each of the nearly 400,000 log entries into individual fields. These fields were then inserted into database

TABLE 2.1   ACS Log Summary

Attribute	Value
Total events	399,103
Authentications	199,327
Deauthentications	199,776
Unique usernames	710
Unique machine addresses	651
Access points	78

tables designed to store and analyze the log information efficiently. The database can then be used to select those table entries that match specific criteria quickly and return them to our analysis tools. The database can also perform more advanced queries, which join, group, and summarize the data according to the values of particular fields. Entries in the log table can also be joined to other tables, such as a building–access point relation, to analyze the log data according to other criteria.

Analysis of the packet header trace data was performed using the CoralReef analysis package [11]. This software was used in previous studies [1] to analyze tcpdump-formatted traces.

2.5   RESULTS

2.5.1   ACS Log Results

The ACS log data for this study consisted of almost 400,000 events collected over a 9-month period in the 2003/04 academic year (see Table 2.1). These events came from 710 users connecting at 78 different access points, installed in over 20 build-ings. This represents substantial growth from the time of our earlier study when only 134 users were logged connecting at 18 access points. Authentication and deauthentication events occur with equal frequency throughout the log.

Figure 2.1 shows usage by access point, expressed as both the number of users and the number of machine addresses seen. The discrepancy between the number of users and the number of machine addresses (Table 2.1) is due largely to the availability of a number of wireless cards that are loaned to students working in our main Library (13 cards were used by five or more distinct usernames over the course of the year). Three Library access points had the highest ratio of usernames to machine addresses. Of the 710 logged users, 155 connected to the wireless network through more than one com-puter. The access point with the highest average of machine addresses per username was located at a help desk, where student laptops are configured by ITS staff who often use a single username to test numerous machines.

Figure 2.2 shows the number of users per access point and the total number of users for each month in the study.[1] The usage levels were low during the summer,

---

[1]ACS data covers dates between August 20, 2003 and April 17, 2004. August and April averages are based on available data.

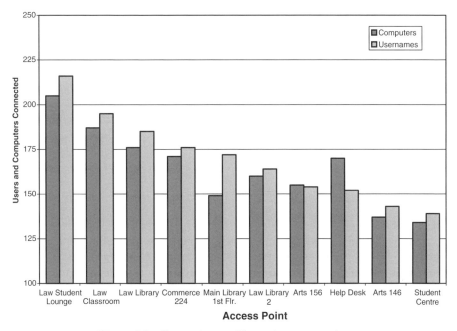

Figure 2.1 Connections at 10 popular access points.

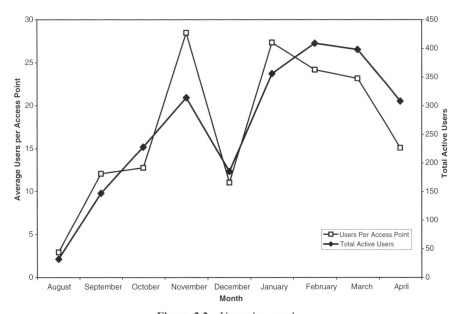

Figure 2.2 Usage by month.

TABLE 2.2   ACS Log Comparison

Statistic	January 22–29, 2004	January 22–29, 2003
Active users	265	134
Active access points	48	18
Mean APs per user	3.12	2.99
Median APs per user	3	3
Mean users per AP	17.25	22.28
Median users per AP	5	14

and climbed to a peak number of users per access point in November (which was also the month with the highest total number of authentication events). Usage dropped significantly (by a factor of ~ 2) in December because of exams and holiday closures.

The total number of users continued to climb in the second term, but the average number of users per access point fell. We attribute this to both a decrease in roaming usage and an increase in the number of available access points as the wireless network expanded. Both usage measures fell late in April as exams and summer approached. Comparing the two terms, it is clear that the number of active wireless users grew significantly. Of the 710 total wireless users in the study, 447 were active from August to December 2003, and 609 were active in early 2004.

By selecting only those authentication events that occurred between January 22 and 29, 2004, we get a picture of how wireless network usage changed over the one-year period following our earlier study in January 2003. Table 2.2 shows some basic statistics from this particular week-long January period. The number of active users has doubled and the average number of access points visited per user has risen slightly, while the median number of access points used remains constant.

The number of users connecting at each access point has changed more significantly. The mean number of users has dropped — but since the number of access points has more than doubled, a 22.5% drop in users per access point actually indicates an increase in overall network usage. The severe drop in the median number of users is due to the change in the distribution of users across the active access points, as shown in Figure 2.3. While in the earlier study, nearly 40% of the access points experienced above average usage, current usage is far more skewed, with only 23% of access points being accessed by a greater than average number of wireless users. In particular, the four most popular access points each experienced extremely high usage levels (100 users or greater) in January 2004.

2.5.2   Roaming Patterns

The roaming patterns of our users is something in which we are particularly interested — we want to determine the extent to which our users take advantage of the roaming opportunities wireless access affords them. Figure 2.4 shows the distribution of access points and buildings visited, for both the current data and the data

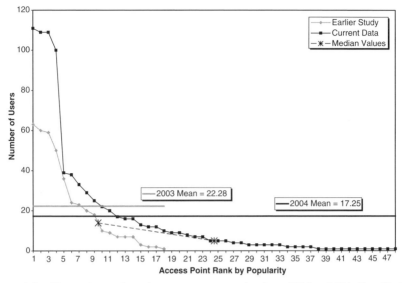

Figure 2.3 Change in number of users per access point from 2003 to 2004 (Jan. 22–29).

from our earlier study. The highest point of both distributions occurs at one building or one access point in the current data. This indicates that many users are connecting to the wireless network either as a wired connection replacement or as a local area network replacement, since they do not connect from other locations on campus.

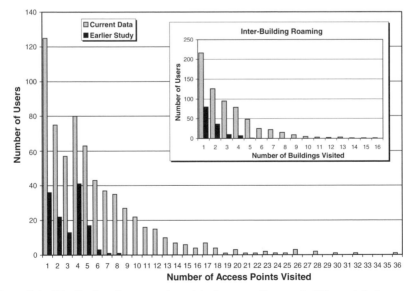

Figure 2.4 Distribution of access points visited per user (*inset* — buildings visited per user).

The second highest point, which occurs at four access points visited, indicates that those users who do roam to multiple access points tend to connect at only a same small number of locations, even over long periods of time. In our earlier study, disproportionately heavy usage in our College of Law (an early adopter of wireless technology) resulted in a mode of four access points visited. Although usage in Law remains high, usage elsewhere on campus has risen significantly. Over 13% of our users visited more than 10 access points (or more than five buildings) over the course of the 2003/04 term. The most actively roaming users on campus (likely ITS service staff) connected to almost half of the 78 access points currently installed.

It is difficult to tell from Figure 2.4 whether the difference between the current data and the earlier data is due merely to the growth in both network size and user population. By normalizing the cumulative distributions of the data over both the number of active access points visited and the total number of users in each dataset, we can factor out the change in network size and popularity when comparing the two studies.

In Figure 2.5 we can see that the overall roaming behavior *has* changed significantly. The fraction of users who do not roam at all (zero on the horizontal axis) has decreased by more than 5% between the two studies. The distributions cross above 30% of active users, as the increase in network size tends to shift lightly roaming users to the left. The most actively roaming users at the top of the two distributions show remarkably similar coverage of the active wireless network. In both studies, the top 3% of roaming users visit more than 22% of the wireless network and the furthest roaming users reach more than 40% of the access points. As the wireless

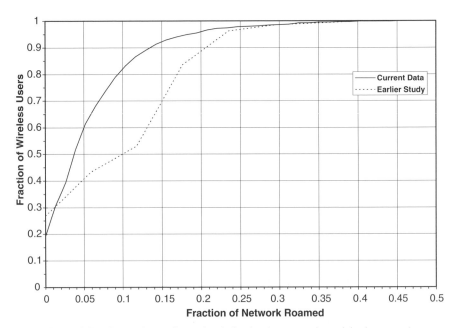

Figure 2.5   Comparison of roaming behavior (access points visited per user).

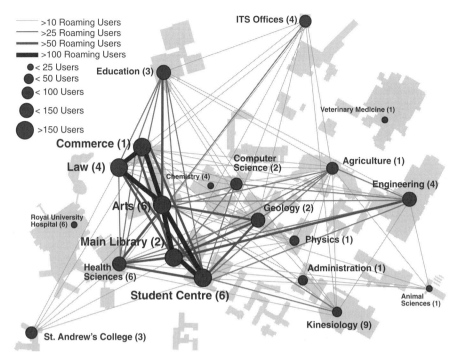

Figure 2.6 Wireless roaming map of the University of Saskatchewan campus.

network continues to expand, we would expect to see a further decrease in the fraction of users who do not roam, but little change in the overall roaming behavior of average and highly roaming users.

Visualizing the ACS log data on a map of the campus (Figure 2.6) gives a clearer picture of where our users are roaming. Each wireless-equipped building is marked with a circle, the radius of which is proportional to the number of unique users seen at that location. The thickness of a line connecting a pair of buildings indicates the number of users who visited both locations. The building names are printed adjacent to their circles, with the number of access points in the building in parentheses.

The five most popular locations (the Arts building, the Commerce building, the Law building, the main Library, and the Student Centre) were each visited by over 150 distinct users. These five most popular locations are connected by the heaviest roaming pattern, forming two triangles that meet at the Arts building. This is consistent with the layout of our campus — the Arts building acts as a hub connecting several other buildings. A similar pattern can be seen on a smaller scale in the roaming map from our earlier study [7]. As described in the comparison of ACS log results (see Figure 2.3), the distribution of the number of users per access point has become skewed since the earlier study, and a small number of highly popular access points (such as those in Law, Commerce, and Arts) are now visited by a disproportionately large fraction of users. The roaming map shows that this skewed distribution corresponds to the underlying layout of the campus, with the most popular access points located in five adjacent, interconnected buildings.

The second-degree roaming links more densely interconnect the five most popular locations and add connections to buildings with 100–150 users (Geology, Engineering, Health Sciences, and Education). Newly installed access points in locations such as Computer Science, Kinesiology, and Agriculture were visited by 25–50 roaming users. More remote buildings, such as the ITS offices, St. Andrew's College, and Animal Sciences, saw even fewer roaming users. Newly installed access points in Royal University Hospital, Chemistry, and Veterinary Medicine were used by only a small number of users, fewer than 10 of whom roamed to other buildings. The relative disuse of the newest access points is consistent with the skewed distribution of users per access point observed in the ACS log comparison.

The roaming map suggests that a building's popularity with wireless users depends on user familiarity and location (newer and less accessible access points saw less usage) rather than the amount of wireless coverage available in a building. The single access point in Commerce, for example, was visited by several times more users than Kinesiology's nine.

2.5.3 Trace Data

As described in the methodology section, packet header traces were gathered at a number of campus locations in order to gain more detailed information on wireless usage. We used the CoralReef toolset [11] to determine which protocols and applications are most commonly used.

Figure 2.7 shows the percentages of packets and bytes transmitted using each of the five IP protocols we saw in our trace data. Non-IP traffic accounted for 14% of the packets, but only 4% of the bytes, which indicates that almost all actual user data were transferred as IP packets. The Internet Control Message Protocol (ICMP), commonly used by network utilities and routers, accounted for 1.22% of the packets recorded. A small amount of local multicast management (IGMP) traffic was also observed.

The primary IP protocols, UDP and TCP, were used in 85% of the packets, which carried 95% of the bytes on the wireless network. The significant use of UDP may be attributed to various forms of online entertainment, such as network games, peer-to-peer networks, and streaming media.

Examining UDP and TCP traffic in further detail (Fig. 2.8), we find that Web access (HTTP and HTTPS) is by far the most common use of wireless networking — responsible for 71% of the bytes. Applications that use unallocated ports (above 1024) generated over 17% of bytes and 27% of packets. Audio and video sent via the Real Time Streaming Protocol (RTSP) was the second most common application, followed by network management (SNMP) information. All IP addresses on the campus are assigned dynamically using DHCP. Standalone (non-Web) email applications sent and received just over 1% of the packets and bytes traced, most of which were sent through our IMAP-based student email

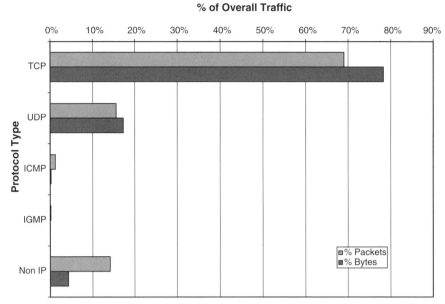

Figure 2.7 Trace traffic by IP protocol.

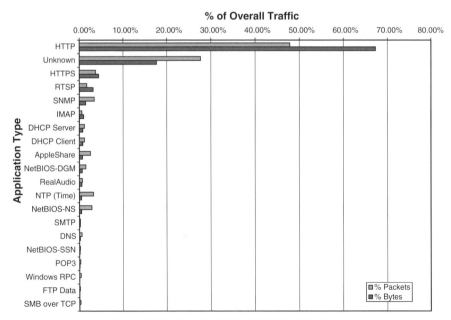

Figure 2.8 Top 20 UDP and TCP applications.

server. Roughly 7% of the packets seen on the wireless network were from various forms of network file access (e.g., NetBIOS, AppleShare, SMB, NFS) and file transfer (e.g., FTP).

2.6 CONCLUSIONS

As we deploy our campus-wide wireless network incrementally, it is important that we understand the needs of our users. Through ongoing analysis we seek to determine where, when, how much, and for what our network is being used, and how that usage is changing over time. In this chapter we have described the methodology we are employing to study our usage patterns and some of the results that we have obtained from our studies to date. Unlike other studies of wireless networks, our data are collected in a centralized manner made possible by the LEAP authentication system and the network environment that we have in place at our university. We augment anonymized ACS log data with localized packet header traces to enable a more complete analysis of user behavior.

Both our wireless network and our wireless usage continue to grow. In the time elapsed between our first study and the present (roughly a year), the number of access points and the number of users have both more than doubled. We now provide at least partial coverage in about half of our campus buildings. The increase we are seeing in user demand certainly warrants continued expansion of this service. We also see a clear need to continue to study usage patterns to guide this expansion.

As we study usage patterns, we are particularly interested in the roaming behavior of our users, now and in the future. Our results to date suggest that the expansion of the wireless network over the past year (at the time of this writing) has changed the roaming patterns of many users, and skewed the distribution of users per access point. Although most of our users still access a limited number of access points in a limited number of buildings, we are seeing an increase in roaming behavior with a larger fraction of our users roaming between buildings and the most active roamers visiting an ever-increasing number of locations. We take this as clear evidence of a growing demand for mobility support. The low use of some of our newest access points emphasizes that popularity is a function of familiarity and underlines our need to be more proactive in publicizing where coverage is available.

In terms of applications, web access via HTTP and HTTPS is by far the most common, accounting for more than 70% of our wireless traffic. Our next most popular application, audio and video via RTSP, is far behind. Non-Web email through our IMAP-based student server contributes very little of our wireless traffic.

The data capture methodology that we have developed is successfully providing us the means to gain valuable insight into the usage of our campus wireless network. Our results are guiding our planning by telling us two important things: (1) which buildings need more wireless coverage and (2) which access points need more promotion. This information will be very useful to us as we continue to evolve our service. Our current research is focusing on a more detailed examination of the roaming patterns of our users, and we hope to have more to say about this soon.

REFERENCES

1. A. Balachandran, G. Voelker, P. Bahl, and V. Rangan, Characterizing user behaviour and network performance in a public wireless LAN, *Proc. ACM SIGMETRICS'02*, Los Angeles, June 2002, pp. 195–205.

2. TCPDUMP, `http://www.tcpdump.org`.

3. D. Kotz and K. Essien, Characterizing usage of a campus-wide wireless network, *Proc. ACM MobiCom'02*, Atlanta, GA, Sept. 2002, pp. 107–118.

4. D. Tang and M. Baker, Analysis of a local-area wireless network, *Proc. ACM Mobi-Com'00*, Boston, Aug. 2000, pp. 1–10.

5. F. Chinchilla, M. Lindsey, and M. Papadoupouli, Analysis of wireless information locality and association patterns in a campus, *Proc. IEEE Infocom 2004*, Hong Kong, March 2004.

6. S. Convery and D. Miller, *SAFE: Wireless LAN Security in Depth — Version 2*, White Paper, Cisco Systems Inc., San Jose, CA, March 4, 2003.

7. D. Schwab and R. Bunt, Characterising the use of a campus wireless network, *Proc. IEEE Infocom 2004*, Hong Kong, March 2004.

8. EtherPeek, `http://www.wildpackets.com/`.

9. NetBSD, `http://www.netbsd.org`.

10. Tcpdpriv, `http://ita.ee.lbl.gov/html/contrib/tcpdpriv.html`.

11. D. Moore et al., The CoralReef software suite as a tool for system and network administrators, *Proc. 15th Systems Administration Conference* (LISA 2001), San Diego, Dec. 2–7, 2001, pp. 133–144.

CHAPTER 3

QoS Provisioning in IEEE 802.11 WLAN

SUNGHYUN CHOI and JEONGGYUN YU

School of Electrical Engineering, Seoul National University, Korea

3.1 INTRODUCTION

Since the early 1990s, IEEE 802.11 WLAN has gained a prevailing position in the market for the (indoor) broadband wireless access networking. The IEEE 802.11 standard defines the medium access control (MAC) layer and the physical (PHY) layer specifications [1]. The mandatory part of the 802.11 MAC is called the *distributed coordination function* (DCF), which is based on carrier sense multiple access with collision avoidance (CSMA/CA). Today, most of the 802.11 devices implement only the DCF. Because of the contention-based channel access nature of the DCF, it supports only the best-effort service without guaranteeing any quality of service (QoS). More recently, the needs for real-time (RT) services such as Voice over IP (VoIP) and audio/video (AV) streaming over the WLANs have been increasing drastically. However, the current 802.11 devices are not capable of supporting the RT services properly, which are delay-sensitive while tolerable of some losses.

The emerging IEEE 802.11e MAC, which is an amendment of the existing 802.11 MAC, will provide QoS [3,5,6,9,15,16,24–27]. The new MAC protocol of the 802.11e is called the *hybrid coordination function* (HCF). The HCF contains a contention-based channel access mechanism, called *enhanced distributed channel access* (EDCA), which is an enhanced version of the legacy DCF, for a prioritized QoS support. With the EDCA, a single MAC contains multiple queues with different priorities that access channel independently in parallel. Frames in each queue are transmitted using different channel access parameters. In this chapter, we focus on the schemes based on contention-based channel access.

In our previous work [8,12] we proposed the "dual queue" scheme with the legacy 802.11 MAC, which is a software upgrade-based approach to provide a

Mobile, Wireless, and Sensor Networks: Technology, Applications, and Future Directions
Edited by Rajeev Shorey, Akkihebbal L. Ananda, Mun Choon Chan, and Wei Tsang Ooi
Copyright © 2006 John Wiley & Sons, Inc.

QoS for real-time applications such as VoIP. This is proposed as a short- and mid-term solution to provide QoS in the 802.11 WLAN since it does not require any WLAN device hardware (HW) upgrade. Note that for many of today's 802.11 MAC implementations, the 802.11e requires a HW upgrade. However, replacing existing 802.11 HW devices to provide QoS should be quite costly, and hence may not be desirable to many WLAN owners, especially hotspot service providers with a huge number of deployed APs [12].

In this chapter, after we briefly overview the research trends for QoS support in the 802.11 WLAN, we introduce legacy DCF, dual-queue scheme as an interim solution, and the emerging IEEE 802.11e as a final solution. Finally, we compare three schemes, namely, legacy DCF, dual-queue, and 802.11e EDCA schemes via simulations.

3.2   RELATED WORK

There has been a remarkable amount of work to provide the QoS in 802.11 WLAN. In this section, we overview such related work briefly.

Several service differentiation mechanisms have been proposed by modifying the DCF. Most of them achieve their goals by assigning different MAC channel access parameters or backoff algorithms to differentiate traffic classes [19–21]. The parameters include those for contention window sizes (i.e., CWmin and CWmax) and interframe space (IFS). In addition to the parameters, Aad and Castelluccia [19] consider the maximum framelength in addition to the above mentioned parameters. However, these approaches are not compliant with the 802.11 standard.

The upcoming IEEE 802.11e provides differentiated channel access to different types of traffic classes via the usage of CWmin, CWmax, and AIFS in the EDCA mode [3]. The service differentiation capability of EDCA has been evaluated in several articles [5,6,9,24–27]. These articles demonstrate, on the basis of various simulations, that the EDCA can well provide differentiated channel access for different priority traffic. Moreover, it is desired to perform a fine-tuning of EDCA parameters based on the underlying network condition to provide QoS better and optimize the network performance [5,6,25]. For EDCA parameter tuning, we need to know how the parameters affect on the performance when they change.

The influences of channel access parameters on the service differentiation as well as throughput and delay performances have been studied analytically [15,28–34] Ge and Hou [28] extend Cali's model [22,23], based on p-persistent approximation of the 802.11 DCF, to derive the throughput in terms of channel access probabilities of different traffic classes. Probability p in Ge and Hou's paper [28] can be roughly translated to CWmin in the standard. Accordingly, the authors study the effect of CWmin. On the other hand, Chou et al. [29] and Bianchi and Tinnirello [33] consider only the effect of AIFS on the service differentiation in terms of throughput and delay performances. Xiao [30] studies the effect of the CWmin, CWmax, and backoff window increasing factor on service differentiation. Other authors [15,31,32] investigate the system performance dependence on

CWmin and AIFS for throughput maximization and service differentiation. Moreover, Xiao [15] and Zhao et al. [32] suggest that although both CWmin and AIFS can be used to provide service differentiation, they have different effects on the differentiation. Finally, Xiao [34] investigates the effect of the CWmin, backoff window increasing factor, and retry limit on service differentiation.

However, the differentiation does not mean the QoS guarantee. Moreover, when the traffic load in each class changes dynamically, the assignment of fixed channel access parameters is not efficient in the context of both QoS guarantee and the optimal channel utilization as mentioned above. Therefore, the channel access parameters should be dynamically adapted according to the network conditions for the two objectives described above. Dynamic adaptation algorithms of the channel access parameters have been proposed in the literature [16,35–38].

Stations dynamically adjust the EDCA parameters according to the behaviors of one frame or multiple frames [16]. Specifically, it proposes a *fast-backoff scheme*, which uses larger window increasing factors as the backoff stage increases, and the *dynamic adjustment scheme* which increases or decreases the values of CWmin and AIFS by some factors based on consecutive unsuccessful or successful transmissions. In the paper by Romdhani et al. [35], based on the EDCA framework, after each successful transmission, the CW values of different classes are not reset to CWmin as in EDCA. Instead, the CW values are updated based on the estimated channel collision rate, which takes into account the time varying traffic conditions. Malli et al. [36] propose the adaptation of backoff timers based on the channel load: that is, they extend the EDCA by increasing the contention window size additively when the channel is busy and by decreasing the backoff counter value exponentially when the channel is idle over a threshold time. A self-adaptive contention-window adjustment algorithm — called *multiplicative increase, multiplicative/linear decrease* (MIMLD) algorithm — similar to TCP congestion window adjustment procedure, is introduced by Pang et al. [37]. It increases or decreases the value of the backoff counter by different mechanisms (i.e., MIMLD) according to the current value of CW and threshold.

The four papers cited above have considered mainly distributed mechanisms for the adaptation of channel access parameters. On the other hand, Zhang and Zeadally [38] introduce a centralized approach for the parameter adjustment. Under this approach, all the access parameters are adapted by the access point (AP), and announced to the stations via beacon frames. The AP exploits a link-layer quality indicator (LQI), namely, delays and drops of real-time traffic, in order to dynamically adjust channel access parameters by utilizing the following two algorithms. One algorithm is used to adjust the relative differences between the channel access parameters of the different classes for QoS guarantee and the other algorithm, to synchronously adapt the channel access parameters of all the classes to achieve high channel utilization.

Admission control is a QoS provisioning strategy to limit the number of new traffic streams into the networks in order to limit network congestion. Without a good admission control mechanism and a good protection mechanism, the existing real-time traffic cannot be protected and QoS requirements cannot be met.

Contention-based admission control schemes for the EDCA are discussed elsewhere in the literature [16,21,38].

Xiao and Choi [16] introduce a distributed admission control, which is a revised version based on the IEEE 802.11e draft 4.3, in which channel utilization measurements are conducted during each beacon interval and available budgets are calculated. When one class' budget becomes zero, no additional new traffic stream belonging to this class is allowed into the network, and existing nodes will not be allowed to increase the rate of the traffic streams that they are already using. Veres et al. [21] propose a distributed admission control approach for IEEE 802.11 WLANs by utilizing a virtual MAC (VMAC) and a virtual source (VS) algorithm to locally estimate the achievable service quality. The admission control algorithm compares the results of the VS and VMAC with the service requirements, and then admits or rejects a new session accordingly. However, it actually considers only the effect of existing flows on the incoming flow, not the effects of the incoming flow on existing flows, which may introduce inaccuracy in making admission decisions.

A centralized contention-based admission control has been introduced [38,39]. According to Zhang and Zeadally [38], when a new real-time flow requests admission, the admission controller at the AP determines to admit or reject in order to guarantee the QoS of each real-time flow and the minimum bandwidth of non-real-time flows, based on LQI data and the requested throughput of the flow. If it turns out that a real-time flow was incorrectly admitted, the AP can choose to drop the latest admitted flow by explicitly informing the corresponding station through an admission response message. Kuo et al. [39] propose an admission control algorithm based on an analytical model. This model evaluates the expected bandwidth and the expected packet delay for each traffic class in order to provide a criterion of admission decision. The stations transmit load conditions to their associated AP via some MAC management frames specified in the 802.11e draft. On the basis of this information, the AP estimates the performance of resource usage and decides whether a new traffic can be admitted into the BSS.

3.3 LEGACY DCF

The IEEE 802.11 legacy MAC [1] defines two coordination functions, namely, the mandatory DCF based on CSMA/CA and the optional point coordination function (PCF) based on the poll-and-response mechanism. Most of today's 802.11 devices operate in the DCF mode only. We briefly overview how the DCF works here as the dual-queue scheme proposed by Yu et al. [8] runs on top of the DCF-based MAC and the 802.11e EDCA is also based on it.

The 802.11 DCF works with a single first-in/first-out (FIFO) transmission queue. The DCF CSMA/CA works as follows. When a packet arrives at the head of transmission queue, if the channel is busy, the MAC waits until the medium becomes idle, then defers for an extra time interval, called the *DCF interframe space* (DIFS). If the channel stays idle during the DIFS deference, the MAC then starts

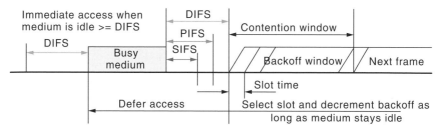

Figure 3.1   IEEE 802.11 DCF channel access scheme.

the backoff process by selecting a random backoff counter. For each idle slot time interval, the backoff counter is decremented. When the counter reaches zero, the packet is transmitted. The timing of DCF channel access is illustrated in Figure 3.1.

Each station maintains a contention window (CW), which is used to select the random backoff counter. The backoff counter is determined as a random integer drawn from a uniform distribution over the interval [0,CW]. If the channel becomes busy during a backoff process, the backoff is suspended. When the channel becomes idle again, and remains idle for an extra DIFS time interval, the backoff process resumes with the suspended backoff counter value. For each successful reception of a packet, the receiving station immediately acknowledges by sending an acknowledgment (ACK) packet. The ACK packet is transmitted after a short IFS (SIFS), which is shorter than the DIFS. If an ACK packet is not received after the data transmission, the packet is retransmitted after another random backoff. The CW size is initially assigned CWmin, and increases to $2 \cdot (CW + 1) - 1$ when a transmission fails.

All the MAC parameters, including SIFS, DIFS, slot time, CWmin, and CWmax, are dependent on the underlying physical layer (PHY). Table 3.1 shows these values for the 802.11b PHY [2], which is the most popular PHY today. The 802.11b PHY supports four transmission rates, namely, 1, 2, 5.5, and 11 Mbps. We assume the 802.11b PHY in this chapter due mainly to its wide deployment base even if the proposed dual-queue scheme and the 802.11e EDCA should work with any PHY.

3.4   DUAL-QUEUE SCHEME FOR QoS PROVISIONING

We proposed a simple dual-queue scheme [8,12] as a short- and mid-term solution to provide a QoS over 802.11 WLAN. The major advantage of this scheme is that it can be implemented in the existing 802.11 hardware. The dual-queue approach is to implement two queues, called *real-time* (RT) and *non-real-time* (NRT) queues,

TABLE 3.1   MAC Parameters for 802.11b PHY

Parameters	SIFS (µs)	DIFS (µs)	Slot (µs)	CW$_{min}$	CW$_{max}$
802.11b PHY	10	50	20	31	1023

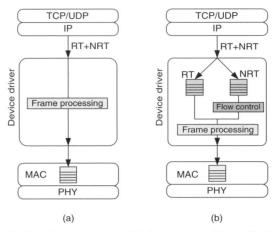

Figure 3.2   Device driver structures: (a) single queue; (b) modified dual queue.

inside the AP as shown in Figure 3.2.[1] In particular, these queues are implemented above the 802.11 MAC controller, specifically, in the device driver of the 802.11 network interface card (NIC), such that a packet scheduling can be performed in the driver level. Packets from the higher layer or from the wireline port (in case of the AP) are classified into either RT or NRT type. The port number as well as UDP packet type is used to classify a RT packet. Note that it is typical to use a set of preconfigured port numbers for specific applications, such as VoIP. Packets in the queues are served by a simple strict priority queuing so that the NRT queue is never served as long as the RT queue is not empty. It turns out that this simple scheduling policy results in surprisingly good performance. We have also implemented the dual-queue scheme in the HostAP driver [13] of Intersil Prism2.5 chipset [12].

The MAC controller itself has a FIFO queue (referred to as "MAC HW queue"). The performance of the dual-queue scheme is compromised by the queuing delay within the FIFO queue when the FIFO queue is large [8]. Unfortunately, the size of the MAC HW queue cannot be configured in many chipsets. To handle this, we have implemented a NRT packet number controller (marked as "flow control" in Fig. 3.2), which restricts the number of outstanding NRT packets in the MAC HW queue. We refer to this modified scheme as *modified dual queue* (MDQ). For the simulation of the modified dual queue in this chapter, we assume that the number of NRT packets in the MAC HW queue is limited to two, owing to the flow control unit. This number is the smallest, which can be practically implemented.

3.5   EMERGING IEEE 802.11e FOR QoS

The emerging IEEE 802.11e defines a single coordination function, called the *hybrid coordination function* (HCF). The HCF combines functions from the DCF

---

[1]Note that we can easily extend this scheme by implementing more queues depending on the desired number of traffic types to support.

and PCF with some enhanced QoS-specific mechanisms and QoS data frames in order to allow a uniform set of frame exchange sequences to be used for QoS data transfers. Note that the 802.11e MAC is backward-compatible with the legacy MAC, and hence it is a superset of the legacy MAC. The HCF is composed of two channel access mechanisms: (1) a contention-based channel access referred to as *enhanced distributed channel access* (EDCA) and (2) a controlled channel access referred to as *HCF controlled channel access* (HCCA). The EDCA is an enhancement of the DCF, while the HCCA is an enhancement of the PCF. We here limit our scope to the EDCA.

One distinctive feature of the 802.11e is the concept of transmission opportunity (TXOP), which is an interval of time when a particular station (STA) has the right to initiate transmissions. During a TXOP, there can be a set of multiple frame exchange sequences, separated by SIFS, initiated by a single STA. A TXOP can be obtained by a successful EDCA contention, and it is called EDCA TXOP in this case.[2] The new concept with TXOP is limiting the time interval during which a STA can transmit its frames. The limit of the TXOP duration is determined by the AP, and is announced to STAs via the beacons in case of EDCA TXOP. The multiple consecutive frame transmissions during a TXOP can enhance the communication efficiency. Moreover, by allocating TXOP other than zero for EDCA, we can reduce the performance anomaly of WLAN [9], which is the performance degradation of all STAs when some STAs use a lower rate than the others due to the bad link [17].

Readers who are interested in the performance of the 802.11e WLAN are referred to papers by Mangold, Choi, and others [5–7]. Even though most of the existing 802.11e papers are based on some old versions of the draft, and hence the exact numbers may not be true, the general tendencies are still valid. The problems of the legacy 802.11 MAC and how the emerging 802.11e fixes those problems are discussed by Choi [6,7]. We briefly explain how the 802.11e EDCA works below.

3.5.1   Enhanced Distributed Channel Access (EDCA)

The 802.11 legacy MAC does not support the concept of differentiating packets with different priorities. Basically, the DCF is supposed to provide a channel access with equal probabilities to all stations contending for the channel access in a distributed manner. However, equal access probabilities are not desirable among stations with different priority packets. The QoS-aware MAC should be able to treat frames with different priority or QoS requirements differently.

The EDCA is designed to provide differentiated, distributed channel accesses for packets with eight different priorities (from 0 to 7) by enhancing the DCF [3]. At the time of writing, we expect the 802.11e standard to be ratified and the first-generation 802.11e products to be introduced in the market by late 2004 or early 2005.

---

[2]A TXOP can be also obtained by receiving a polling frame (called `QoS CF-Poll`) from the AP, but we do not consider this in this chapter.

TABLE 3.2 User Priority to Access Category Mappings

User Priority	Access Category (AC)	Designation (Informative)
1	AC_BK	Background
2	AC_BK	Background
0	AC_BE	Best effort
3	AC_BE	Best effort
4	AC_VI	Video
5	AC_VI	Video
6	AC_VO	Voice
7	AC_VO	Voice

Each packet from the higher layer arrives at the MAC along with a specific priority value. Then, each QoS data packet carries its priority value in the MAC packet header. An 802.11e QoS STA (QSTA) must implement four channel access functions, where each channel access function is an enhanced variant of the DCF. Each frame arriving at the MAC with a user priority is mapped into an access category (AC) as shown in Table 3.2, where a channel access function is used for each AC. Note the relative priority of 0 is placed between 2 and 3. This relative prioritization is rooted from IEEE 802.1d bridge specification [4].

Basically, a channel access function uses AIFS[AC], CWmin[AC], and CWmax[AC] instead of DIFS, CWmin, and CWmax of the DCF, respectively, for the contention process to transmit a packet belong to access category AC. AIFS[AC] is determined by

$$AIFS[AC] = SIFS + AIFSN[AC] \cdot SlotTime$$

where AIFSN[AC] is an integer greater than 1 for non-AP QSTAs and an integer greater than 0 for QAPs. The backoff counter is selected from [0,CW[AC]]. Figure 3.3 shows the timing diagram of the EDCA channel access. One major difference between the DCF and EDCA in terms of the backoff countdown rule is that the first countdown occurs at the end of the AIFS[AC] interval. Moreover, at the end of each idle slot interval, either a backoff countdown or a frame transmission

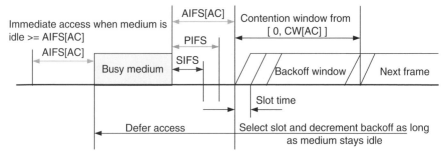

Figure 3.3   IEEE 802.11e EDCA channel access scheme.

occurs, but not both. Note that according to the legacy DCF, the first countdown occurs at the end of the first slot after the DIFS interval, and if the counter becomes zero during a backoff process, it transmits a frame at that moment.

The values of `AIFSN[AC]`, `CWmin[AC]`, and `CWmax[AC]`, which are referred to as the *EDCA parameter set*, are advertised by the AP via beacons and probe response frames. The AP can adapt these parameters dynamically depending on the network condition. Basically, the smaller `AIFSN[AC]` and `CWmin[AC]`, the shorter the channel access delay for the corresponding priority, and hence the more capacity share for a given traffic condition. However, the collision probability increases when operating with smaller `CWmin[AC]`. These parameters can be used in order to differentiate the channel access among different priority traffic.

It should be also noted that the QAP can use the EDCA parameter values different from the announced ones for the same AC. The 802.11 DCF originally was designed to provide a fair channel access to every station including the AP. However, since typically there is more downlink (i.e., AP-to-stations) traffic, AP's downlink access has been known to be the bottleneck of the entire network performance. Accordingly, the EDCA, which allows the differentiation between uplink and downlink channel accesses, can be very useful for controlling network performance.

Figure 3.4 shows the 802.11e MAC with the four independent enhanced distributed channel access functions (EDCAFs), where each access function behaves as a single enhanced DCF contending entity. Each access function employs its own `AIFS[AC]` and maintains its own backoff counter (BC). When there are more than one access functions finishing the backoff at the same time, the highest priority frame among the colliding frames is chosen and transmitted, and the others perform a backoff with increased CW values.

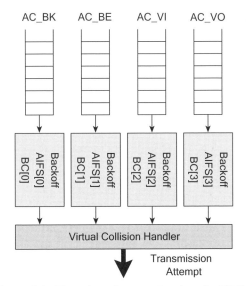

Figure 3.4 Four channel access functions for EDCA.

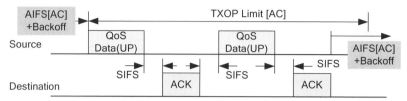

Figure 3.5   EDCA TXOP operation timing structure.

As presented above, during an EDCA TXOP, there can be multiple frame exchange sequences, separated by SIFS, initiated by a single station (or, exactly speaking, by an EDCAF of a station). A TXOP can be obtained by a successful EDCA contention of the corresponding EDCAF. The duration of a TXOP is determined by another EDCA parameter, called *TXOP limit*. This value is determined for each AC, and hence is represented as TXOPLimit[AC]. Figure 3.5 shows the transmission of two QoS data frames of user priority UP during an EDCA TXOP, where the whole transmission time for two data and ACK frames is less than the EDCA TXOP limit. The multiple consecutive frame transmissions during a TXOP can enhance the communication efficiency by reducing unnecessary back-off procedures. Actually, after completing the second frame exchange, the source station could also transmit the next frame partially by transmitting a fragment of the frame. In the example, the source station decides not to transmit the frame partially. Note that the fragmentation is not bound to the fragmentation threshold in the case of the 802.11e. Technically speaking, whether to utilize a TXOP by transmitting multiple frames back-to-back or not is totally up to the transmitting station.

3.5.2   Contention-Based Admission Control

Admission control plays a significant role in providing the desired QoS in IEEE 802.11 network. Admission control is a scheme for regulating the amount of data contending for the medium, that is, the amount of data input into the BSS in order to protect the existing traffic streams (TSs). The EDCA tends to experience severe performance degradation when network is overloaded because of its contention-based channel access nature. In this condition, the contention window becomes large, and more and more time is spent in backoff and collision resolution rather than sending data. As a result, the EDCA cannot guarantee any QoS without a proper admission control even if it provides a differentiated channel access based on different channel access parameters.

The QAP uses the admission control mandatory (ACM) subfield advertised in the EDCA parameter set element of beacon frames in order to indicate whether admission control is required for each of the ACs. Admission control is negotiated by the use of a traffic specification (TSPEC). A QSTA specifies its TS requirements (nominal MSDU size, mean data rate, minimum PHY rate, inactivity interval, and surplus bandwidth allowance) and requests the QAP to set up a TS by sending an ADDTS (add traffic stream) management action frame. The QAP calculatation of the existing load is based on the current set of established TSPECs. According to the current condition, the QAP may accept or deny the new ADDTS request. If the ADDTS request is denied, the particular priority EDCAF inside the QSTA is not permitted to use the corresponding priority access parameters but instead must use the parameters of a lower-priority AC, which does not require admission control. Admission control is recommended not to be used for the access categories AC_BE and AC_BK.

The admission control algorithm, in general, is implementation-dependent and vendor-specific. Moreover, it depends on available channel capacity, link conditions, and retransmission limits of a given TS. All of these criteria affect the admissibility of a given stream.

3.6 VoIP AND 802.11b CAPACITY FOR ADMISSION CONTROL

Voice over IP (VoIP) can be considered a representative real-time application. In this section, we briefly discuss VoIP codec (coding/decoding) in consideration and the capacity of IEEE 802.11b for the VoIP admission control.

3.6.1 Voice over IP (VoIP)

Many types of voice codec are used in IP telephony, including G.711, G.723.1, G.726, G.728, and G.729 [14]. These codecs have different bit rates and complexities. In our work, we consider G.711, the simplest voice codec. G.711 is a standard generating a 64-kbps (kilobits per second) stream, based on an 8-bit pulse-coded modulation (PCM), with the sampling rate of 8000 samples per second. Even though it achieves the worst compression among peer voice codecs, it is often used in practice because of its simplicity. For example, we have observed using a network traffic capturing tool [10] that G.711 is used in Microsoft MSN Messenger for the VoIP application. The number of samples per a VoIP packet is another important factor. The codec defines the size of a sample, but the total number of samples conveyed in a packet affects how many packets are generated per second. There is basically a tradeoff since the larger a packet size (or more samples carried per packet), the longer the packetization delay, but the lower the packetization overheads as analyzed below.

In our work, we assume that a VoIP packet is generated every 20 ms, that is, with 160-byte [= 8 kilobytes per second (KBps)$*$ 20 ms] voice data. We also assume that RTP over UDP is used for the VoIP transfer. When an IP datagram is transferred over the 802.11 WLAN, the datagram is typically encapsulated by an IEEE 802.2 Sub-Network Access Protocol (SNAP) header. Note that all these assumptions are very typical in the real world. Accordingly, the VoIP packet size at the 802.11 MAC Service Access Point (SAP)[3] becomes

160-byte DATA + 12-byte RTP header + 8-byte UDP header + 20-byte IP header + 8-byte SNAP header = 208 bytes per VoIP packet

---

[3]The MAC SAP is the interface between the MAC and the higher layer, namely, IEEE 802.2 logical link control (LLC).

3.6.2 802.11b Capacity for VoIP Admission Control

Apparently, the number of allowable VoIP sessions over WLAN should be limited to maintain an acceptable QoS. The maximum number of VoIP sessions over the WLAN can be approximately calculated as follows. We first calculate the time to transmit a VoIP packet over the 802.11b PHY at 11 Mbps (megabits per second) without any transmission failure assuming that (1) ACK packet is transmitted at 2 Mbps and (2) the long PHY preamble is used. These two assumptions are very valid in the real WLANs. Note that for a successful MAC packet transfer, the following five events happen in order [1]: (1) DIFS deference, (2) backoff, (3) packet transmission, (4) SIFS deference, and (5) ACK transmission. Then, the VoIP packet transfer time is determined to be about 981 μs by adding the following three values as well as one SIFS (= 10 μs) and one DIFS (= 50 μs):

1. VoIP MAC packet transmission time:

    $= 192$-μs PLCP preamble/header + (24-byte MAC header + 4-byte CRC-32 + 208-byte payload) / 11 Mbits/s = 363 μs

2. ACK transmission time at 2 Mbps:

    $= 192$-μs PLCP preamble/header + 14-byte ACK packet / 2 Mbits/s = 248 μs

3. Average backoff duration:

    $= 31$ (CWmin) * 20 μs (One Slot Time) / 2 = 310 μs

A VoIP session consists of two senders, which transmit a packet every 20 ms, since it is interactive. Then, we find that about 20 voice packets (= 20 ms/981 ms) can be transmitted during a 20-ms interval. Accordingly, we estimate that about 10 VoIP sessions can be admitted into IEEE 802.11b WLAN. We discuss this issue further with reference to the simulated results later.

3.7 COMPARATIVE PERFORMANCE EVALUATION

In this section, we comparatively evaluate the performance of the original DCF, the modified dual queue (MDQ) scheme, and the 802.11e EDCA over an infrastructure WLAN environment using an ns-2 simulator [10]. We use the following traffic models for our simulations — two different types of traffic are considered for our simulations, namely, voice and data. The voice traffic is modeled by a two-way constant–bit–rate (CBR) session according to G.711 codec [14]. The data traffic application is modeled by a unidirectional FTP/TCP flow with 1460-byte packet size and 12-packet (or 17,520-byte) receive window size.[4] This application corresponds to the download of a large file via FTP. We use the 802.11b PHY for our simulations. Note that the maximum segment size (MSS) of the TCP across the

---

[4]This window size corresponds to the TCP implementation of MS Windows XP.

TABLE 3.3   Default EDCA Parameter Set

Access Category (AC)	CWmin	CWmax	AIFSN
AC_BK	aCWmin	aCWmax	7
AC_BE	aCWmin	aCWmax	3
AC_VI	(aCWmin+1)/2-1	aCWmin	2
AC_VO	(aCWmin+1)/4-1	(aCWmin+1)/2-1	2

popular Ethernet is 1460 bytes. The 11 Mbps transmission rate (out of 1, 2, 5.5, and 11 Mbps of the 802.11b PHY) is used in the simulations.

Table 3.3 shows the EDCA parameter set used for each traffic type along with the corresponding priorities and ACs. This is the default EDCA parameter set in the IEEE 802.11e/D9.0 draft supplement [3]. We use the default values of Table 3.3 in our simulations unless specified otherwise.

We use the queue size of 500 packets at the AP, which is large enough to ensure that there is no buffer overflow in our simulation environments [8]. The network topology for our simulations is shown in Figure 3.6. Each station involving with a VoIP session generates and receives only voice traffic. The other stations receive only TCP packets, and each of them treats only one TCP flow; thus, the number of TCP flows corresponds to TCP stations. This topology can be often found in the real WLANs with mixed VoIP and Internet traffic.

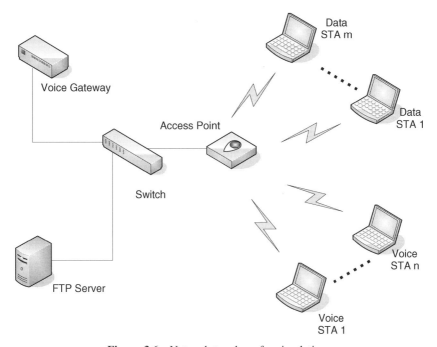

Figure 3.6   Network topology for simulations.

3.7.1   VoIP Capacity for Admission Control

First, we have simulated the pure VoIP situations with legacy DCF (single FIFO queue) in order to evaluate the admission control policy. Figure 3.7 shows both delay[5] and packet drop rate[6] as the number of VoIP sessions increases. The packet drop occurs at the transmitting stations since we have used a limited-size queue (of 50 and 100 packets). We observe from the simulation result that up to 11 VoIP sessions can be admitted into the system since the downlink drop rate is over 0.1 with 12 VoIP sessions, and this high packet drop rate is not acceptable practically. In the previous section we estimated that 10 VoIP sessions can be admitted. Our under-estimation is due to the fact that the average backoff duration is reduced as the number of stations increases up to the point where the packet collision effect becomes dominant. However, we find that our simple calculation-based estimation was quite close. We find that up to 11 VoIP sessions, there is no much difference between 50 and 100 queue sizes since there should not be many queued VoIP packets anyway. Accordingly, we use 50 packets for the RT queue size for the rest of the chapter.

Longer delay and higher drop rate are observed for the downlink transmissions with over 11 VoIP sessions. This discrepancy is due to the fact that the downlink is disadvantaged compared to the uplink (i.e., station-to-AP) since the downlink (or AP transmission) is shared by multiple VoIP sessions. Under the DCF access rule, the AP basically gets the channel access chances as often as other competing stations do for their uplink transfers. It should be also noted that the admission control should be performed more carefully considering the link condition between the AP and each individual station. Note that our simulation results are based on the 11 Mbps data transmission rate. The disadvantage is that the channel condition fluctuates over time, due to station mobility, time-varying interference, and other variables. One possible admission control policy could be admitting up to smaller number of VoIP sessions. For example, when all the stations transmit packets at 2 instead of 11 Mbps, the number of admissible VoIP sessions becomes five from the analysis in the previous section. If all the admitted (up to five) VoIP stations can transmit/receive packets at 11 Mbps owing to good channel conditions, there will be plenty residual bandwidth, which can be utilized by other types of traffic, namely, NRT TCP traffic considered in the following text.

3.7.2   Comparison of Single Queue and MDQ

In this scenario, we simulate with a single VoIP session and various numbers of TCP flows in order to evaluate the effects of different numbers of TCP sessions on VoIP performance. We consider three different queue sizes (i.e., 50, 100, and 500 packets) at the AP. There are always the same numbers of upstream and down-stream TCP flows; that is, the value of one on the *x*-axis represents the case when

---

[5]The delay is the average end-to-end delay.
[6]The *drop rate* is the ratio of the dropped voice packets to all the voice packets sent during the entire simulation.

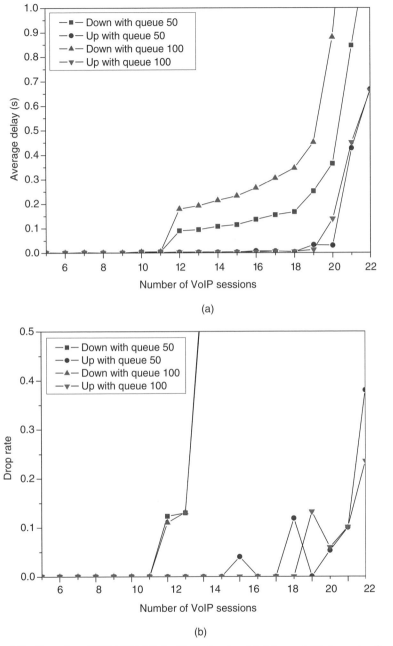

Figure 3.7   Capacity of IEEE 802.11b for VoIP, showing (a) delay and (b) drop rate of voice traffic.

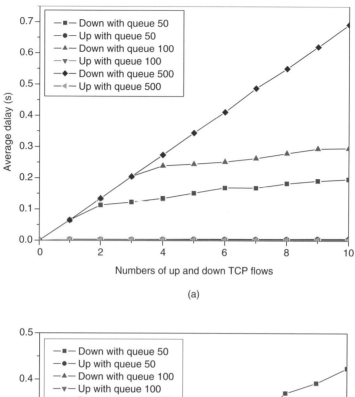

Figure 3.8   Single-queue performances: (a) delay; (b) drop rate.

there is one upstream and one downstream TCP flows. Here, we assume that the MAC HW queue size is equal to one packet.

Figure 3.8 presents both delay and packet drop rate performances with the single queue scheme. We can see that the delay of voice packets with the single queue increases linearly in proportion to the number of TCP flows when the queue size is large enough (i.e., queue size of 500 in our simulations), as shown in Figure 3.8a. This is because the average number of queued packets at the AP linearly increases as the number of TPC flows increases. Note that with TCP, there can be a number of outstanding TCP packets (including both data and ACK packets) inside the network, specifically, between a station and the FTP server in our simulation environment in Figure 3.6. The number of outstanding TCP packets is determined by the minimum of the received window size and the congestion window size. We observe from the simulations that the bottleneck link is the downlink of the WLAN, namely, the AP's downlink transmissions, and hence virtually all the outstanding packets are queued in the AP queue. This is the reason why the VoIP packet delay increases linearly as the number of TCP flows increases with the queue size of 500. However, the situation is a bit different in case of queue sizes of 50 and 100; that is, the delay increases very slowly or is almost saturated while beginning a specific number of TCP flows. This slow delay increase occurs because the packet drops out of the buffer overflow as confirmed from Figure 3.8b.

Figure 3.9 shows the VoIP delay performance as the TCP flow number increases with the proposed MDQ scheme for three different NRT queue sizes. It should be noted that we do not show the packet drop rate performance since no packet drop has been observed in this case. We first observe a significant reduction of the delay with the MDQ; the worst-case delay now is about 11 ms. We can imagine that the delay of downlink voice traffic is due mainly to the queuing delay with the single queue but mainly to the wireless channel access delay in the MAC HW queue with the MDQ. Figure 3.9 shows that delays of both uplink and downlink voice packets with 50- and 100-packet NRT queues increase as the number of TCP flows increases. However, in cases of the queue size of 500 packets, there is almost no change in delay irrespective of the number of TCP flows. This is somewhat counterintuitive since the TCP is known to be aggressive, and hence there should be more uplink contentions as the number of TCP flows increases, thus degrading the voice delay performance.

However, if we delve into the TCP behavior more carefully, the observed delay performance looks very reasonable. As discussed above, when the queue size is large enough, most of TCP packets (either ACK or data) are accumulated at the AP because of the bottleneck WLAN downlink. Therefore, for example, the source stations of upstream TCP flows can transmit a TCP data packet only when it receives a TCP ACK packet from the AP, provided a timeout does not occur. This basically results in only one or two stations with TCP flows actively contending for the channel irrespective of the number of total TCP flows. This is the reason why the delay performance is rather stable across all the TCP flow numbers when the NRT queue size is 500. Note that with 10 upstream and downstream TCP flows, there will be up to 240 ($= 10 * 2 * 12$) TCP packets enqueued at the NRT queue since the receive window size is 12, and 500 is more than enough in this case. On the other hand, when the queue size is small such that some TCP packets are dropped because of the buffer overflow of the NRT queue, there will be retransmit timeout events with some TCP flows, and this will result in more stations with upstream TCP flows actively contending for the channel in order to retransmit

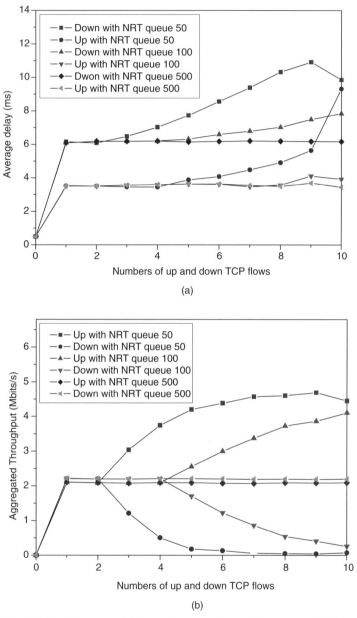

Figure 3.9 MDQ performances: (a) delay of voice packets; (b) aggregated TCP throughput.

TCP data packets. This is the reason why the delay performance gets worse with the NRT queue sizes of 50 and 100 as the TCP flow number increases. This kind of TCP behavior still exists with the single-queue situation, but it is not observed since the delay performance is dominated by the queuing delay discussed above.

Figure 3.9b shows the aggregated throughputs of upstream and downstream TCP flows, which are measured at the AP, with the MDQ scheme. We basically observe the unfairness between upstream and downstream TCP flows with queue sizes 50 and 100, while the unfairness is not observed with queue size 500. This is again because some TCP packets are dropped in case of 50 and 100 queue sizes. For example, with queue size 100, the unfairness is observed beginning the number of TCP flows equal to 5, in which the maximum outstanding TCP packets becomes 120 ($= 5 * 2 * 12$) larger than the queue size. Because the TCP ACK is cumulative, which makes up the loss of previously dropped TCP ACK packet, upstream TCP flows are less affected by the packet drops at the AP, thus achieving a higher throughput than the downstream TCP flows. This observation is consistent with that of pilosof et al. [18], and implies that the queue size for the AP should be large enough in order to avoid the unfairness between uplink and downlink. This is good for us since our MDQ scheme performs better in terms of the delay with large NRT queues. We do not show the TCP throughput performance with the single queue, but the same behavior is basically observed, where the throughput values are a bit larger than in the MDQ case since the VoIP and TCP packets are treated in the same manner with the single queue.

3.7.3 Comparison of MDQ and EDCA

We simulate with a single VoIP session and various numbers of downstream TCP flows in order to compare the VoIP performance of single-queue (i.e., the original DCF), MDQ, and EDCA. Moreover, we assume that the MAC HW queue size for the MDQ scheme is equal to two packets as discussed in Section 3.4. Figure 3.10

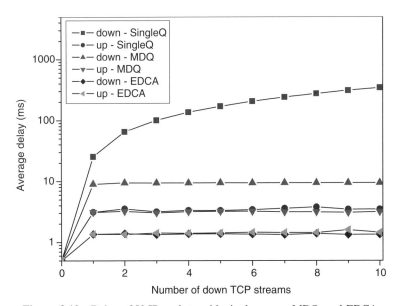

Figure 3.10 Delay of VoIP packets with single queue, MDQ, and EDCA.

presents the delay performance of these three schemes. As observed by Yu et al. [8], the downlink delay of the single queue increases linearly as the number of TCP flows increases, and hence cannot be used for VoIP in the mixed traffic environments. On the other hand, both MDQ and EDCA provide reasonable delay performance virtually independent of the TCP flow number. This is because both schemes provide higher priority to the VoIP packets over the TCP packets. To understand the detailed behavior of the MDQ scheme, the reader is be referred to the paper by Yu et al. [8].

Figure 3.10 shows that the voice delay of EDCA is superior to that of the MDQ, although both of them show good delay performance. The reason can be understood as follows:

1. The EDCA uses smaller values of the channel access parameters than does the MDQ, based on the legacy DCF, namely, $CWmin[AC\_VO] = 7$ and $CWmax[AC\_VO] = 15$ for EDCA AC_VO [3], and $CWmin = 31$ and $CWmax = 1023$ for the legacy DCF [2], respectively, in the case of the 802.11b PHY. Smaller channel access parameters imply a faster channel access.

2. In the case of the MDQ, the MAC HW queue of two packets introduces an extra delay for the downlink VoIP packets since a VoIP packet at the head of the RT queue in the MDQ scheme should wait until all the preceding packets in the MAC HW queue are transmitted. This causes the VoIP downlink delay of the MDQ to exceed the uplink delay.

It should also be noted that the uplink delay performance of both single queue and MDQ are the same since there is no difference between two schemes in case of uplink in our simulation scenarios. Specifically, in our simulations, a station transmits only a single type of traffic, either VoIP or TCP ACK.

Figure 3.11 presents the delay performance comparison for the MDQ and EDCA as the number of VoIP sessions increases. We simulate with 10 downstream TCP flows and various numbers of VoIP sessions. The EDCA provides a better VoIP delay performance than does the MDQ scheme with multiple VoIP sessions, while both of them still provide acceptable delay performances. As discussed above, the main reasons should be the smaller EDCA access parameter values and the MAC HW queue of the MDQ scheme. However, we observe that the delay of EDCA, especially the downlink delay, increases as the VoIP session number increases. This must be a negative effect of small EDCA access parameters, namely, that these small values result in some collisions among VoIP packets from different STAs.

Figure 3.11b shows the aggregated throughput performance of downstream TCP flows, which are measured at the AP, with both MDQ and EDCA. We observe that the EDCA provides a better throughput performance than does the MDQ. Note that it takes a shorter time for the EDCA to transmit a VoIP packet, due to a lower channel access delay. As a result, the EDCA allows more time resource for TCP packet transmissions. Moreover, TCP under EDCA can get more transmission opportunities than possible under the MDQ because it contends in parallel with VoIP under

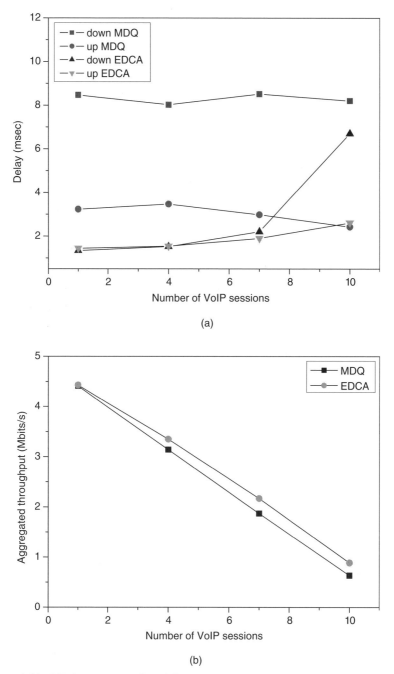

Figure 3.11 MDQ versus EDCA default access: (a) VoIP delay performances; (b) aggregated TCP throughputs.

EDCA. On the other hand, with MDQ scheme, TCP packets are not served when a VoIP packet exists in the RT queue by strict priority queuing. This is why the TCP throughput with the EDCA is a bit larger than that with the MDQ.

3.7.4 EDCA Default Access versus PIFS Access

Although we have observed (Fig. 3.10 and Fig. 3.11) that the EDCA provides a good VoIP delay performance, the performance of EDCA can be further enhanced via the access parameter control. As described in Section 3.5.1, AIFSN[AC] is an integer greater than one for non-AP QSTAs and an integer greater than zero for QAPs. Moreover, the values of CWmin[AC] and CWmax[AC] can be set to zero [3]. Therefore, we can use access parameter values of AIFSN[AC_VO] = 1, CWmin[AC_VO] = 0, and CWmax[AC_VO] = 0, which are the smallest access parameter values for the QAP, for the downlink EDCA AC_VO of the QAP. This allows the QAP to transmit its pending VoIP packet after a PIFS channel idle time. The other parameter values are subject to the guidelines listed in Table 3.3, namely, the default values. We refer to the QAP's channel access after PIFS for VoIP packets as "PIFS access" in the rest of this chapter. On the other hand, the usage of the default access parameters for both non-AP QSTAs and QAP, as we have simulated thus far, is referred to as "default access."

Figure 3.12 compares the VoIP delay performance of the default access and the PIFS access as the number of VoIP sessions increases. We simulate with 10 downstream TCP flows and various number of VoIP sessions. As expected, we observe that the downlink delay performance of VoIP is enhanced significantly through the PIFS access in Figure 3.12. In all cases except for the downlink VoIP delay with the PIFS access, the delay of VoIP packets increases as the number of VoIP sessions increases, because downlink VoIP packets with the PIFS access hardly experience channel access delay because the AP can transmit its pending VoIP packets after a PIFS idle time without any contention from STAs. On the other hand, with the default access as well as the uplink case with the PIFS access, the value of CWmin[AC_VO] is equal to 7, which is too small to avoid collisions in our simulations. This results in large channel access delays. Moreover, as discussed above, TCP under the EDCA is more aggressive than the MDQ with strict priority queuing because it contends in parallel with VoIP packets. At the same time, the number of contending TCP STAs to transmit TCP ACK packets is more than the MDQ case because of aggressive downlink AC_BE for TCP data packets. Accordingly, downlink VoIP packets experience larger queuing delays with the default access.

Figure 3.13 shows the effect of TCP aggressiveness to VoIP delay performance under the EDCA. We simulate with 10 downstream TCP flows, 11 VoIP sessions, and various CWmin[AC_BE] values in order to evaluate the effect of CWmin[AC_BE] on VoIP delay performance. The default values for the access parameters of AC_VO are used for both QAP and non-AP QSTAs. We observe that the larger the value of CWmin[AC_BE], the more conservative TCP (i.e., AC_BE) is. Accordingly, the average VoIP packet delay decreases dramatically in Figure 3.13.

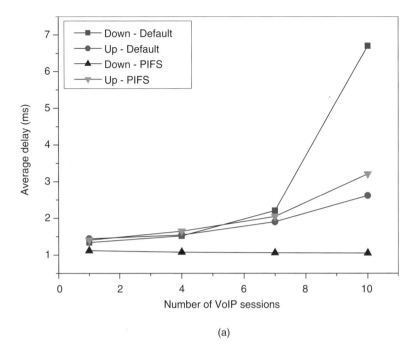

(a)

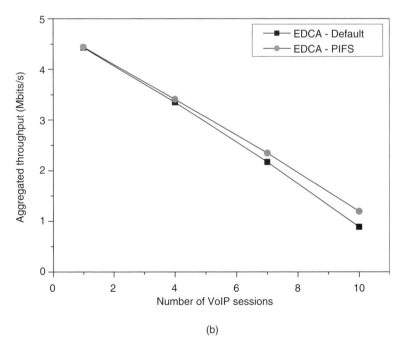

(b)

Figure 3.12   EDCA default access versus PIFS access: (a) VoIP delay performance;
(b) aggregated TCP throughputs.

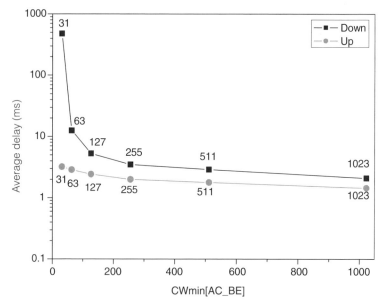

Figure 3.13   The effect of TCP aggressiveness.

Note that with `CWmin[AC_BE]` $= 31$, the downlink VoIP packet delay is about 500 ms, which implies an acceptable VoIP performance. This means that 11 VoIP sessions should have not been admitted with this channel access parameter value for `AC_BE`. We can easily imagine here that the capacity for VoIP, that is, the maximum number of VoIP to be admitted into the network with the satisfactory delay performance, is dependent on the channel access parameters of other traffic types, such as `AC_BE` here. Here, we conclude that the EDCA access parameter adaptations as well as the admission control depending on the network condition are needed in order to provide QoS for VoIP.

3.7.5   Jitter Performance Comparison

The jitter of VoIP is another important performance measure along with the delay. In order to evaluate the jitter performance of four different access schemes, namely, DCF, MDQ, EDCA default access, and EDCA PIFS access, we simulate with 1 or 10 VoIP sessions along with five downstream TCP flows.

Figure 3.14 shows the jitter performance for both 1 and 10 VoIP session cases. We can imagine that there are two major factors, which increase the VoIP jitter in the 802.11 WLAN, namely, contention/collision with other stations and random delay inside the queue. First, when one VoIP session exists (Fig. 3.14a), EDCA schemes (i.e., EDCA default and PIFS accesses) demonstrate better jitter performances than do DCF schemes (i.e., DCF and MDQ) owing to their smaller channel access parameter values. The reason is explained as follows. With 1 VoIP session, TCP flows can use a large fraction of the total bandwidth, and hence more TCP

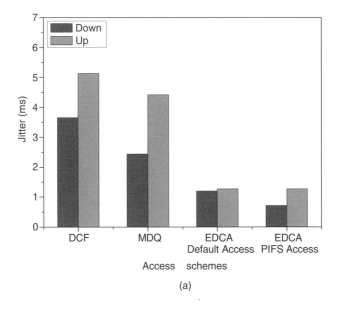

(a)

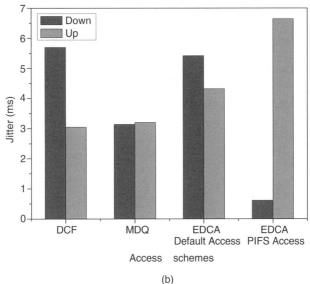

(b)

Figure 3.14 Jitter performances of four access schemes: (a) 1 VoIP session; (b) 10 VoIP sessions.

stations contend for channel. In this situation, AC_VO of EDCA schemes, which use small channel access parameter values, can reduce the contention with TCP stations and AC_BE in the QAP. Accordingly, the jitter becomes smaller. On the other hand, with 10 VoIP sessions as shown in Figure 3.14b, we find the result quite different from 1 VoIP session case. In this situation, CWmin[AC_VO] value of EDCA schemes is not large enough for collision avoidance. Accordingly, many collisions can occur, thus increasing the jitter considerably. However, the jitter of downlink VoIP packets in EDCA PIFS access still remains small because downlink AC_VO can perfectly avoid the contention with TCP stations and AC_BE in AP. The jitter of downlink VoIP packets in all schemes except for EDCA PIFS access increases is because VoIP packet generation times of each VoIP sessions are randomized in our simulations, and hence a VoIP packet arriving at the AP queue experiences random queuing delay.

From the jitter performance evaluation thus far, we can conclude that when there are a smaller number of VoIP sessions, the jitter performance of the EDCA is better than that of the DCF/MDQ while they perform about the same when there are many VoIP sessions.

3.8 SUMMARY

In this chapter, we have introduced various QoS provisioning schemes for IEEE 802.11 WLAN based on contention-based channel access, and then compared them via simulations. Considering the VoIP delay/jitter and TCP throughput, the simulation results show that the EDCA surely provides the best performance because of the flexible channel access parameter control depending on the underlying network condition, such as, the traffic load. However, the MDQ also provides a good performance, which is acceptable in most situations, and is comparable with that of the EDCA. Accordingly, the MDQ scheme can be practically a good solution in order to provide QoS for VoIP services when the 802.11e is not available or where the hardware upgrade for the 802.11e is not desirable. Moreover, the simulation results suggest that we need to develop the algorithm for the optimal channel access parameter adaptation of the EDCA as well as the admission control algorithm for acceptable QoS provisioning depending on the network condition and applications in service.

REFERENCES

1. IEEE Std. 802.11-1999, *Part 11: Wireless LAN Medium Access Control (MAC) and Physical Layer (PHY) Specifications*, ISO/IEC 8802-11:1999(E), IEEE Std. 802.11, 1999 ed., 1999.

2. IEEE 802.11b-1999, *Supplement to Part 11: Wireless LAN Medium Access Control (MAC) and Physical Layer (PHY) Specifications: Higher-Speed Physical Layer Extension in the 2.4 GHz Band*, 1999.

3. IEEE 802.11e/D9.0, *Draft Supplement to Part 11: Wireless Medium Access Control (MAC) and Physical Layer (PHY) Specifications: Medium Access Control (MAC) Enhancements for Quality of Service (QoS)*, Jan. 2004.

4. IEEE Std. 802.1d-1998, *Part 3: Media Access Control (MAC) Bridges*, ANSI/IEEE Std. 802.1D, 1998 edition, 1998.

5. S. Mangold, S. Choi, G. R. Hiertz, O. Klein, and B. Walke, Analysis of IEEE 802.11e for QoS support in wireless LANs, *IEEE Wireless Commun.* **10**(6) (Dec. 2003).

6. S. Choi, J. del Prado, S. Shankar, and S. Mangold, IEEE 802.11e contention-based channel access (EDCF) performance evaluation, *Proc. IEEE ICC'03*, Anchorage, AK, May 2003.

7. S. Choi, Emerging IEEE 802.11e WLAN for quality-of-service (QoS) provisioning, *SK Telecom Telecommun. Rev.* **12**(6):894–906 (Dec. 2002).

8. J. Yu, S. Choi, and J. Lee, Enhancement of VoIP over IEEE 802.11 WLAN via dual queue strategy, *Proc. IEEE ICC'04*, Paris, June 2004.

9. J. del Prado and S. Shankar, Impact of frame size, number of stations and mobility on the throughput performance of IEEE 802.11e, *Proc. IEEE WCNC'04*, Atlanta, GA, March 2004.

10. Airopeek, `http://www.airopeek.com`, `online link`.

11. The network simulator — ns-2, `http://www.isi.edu/nsnam/ns/`, online link.

12. Y. Choi, J. Paek, S. Choi, G. Lee, J. Lee, and H. Jung, Enhancement of a WLAN-based Internet service in Korea, *Proc. ACM Int. Workshop on Wireless Mobile Applications and Services on WLAN Hotspots* (*WMASH'03*), San Diego, USA, Sept. 19, 2003.

13. J. Malinen, Host AP driver for Intersil Prism2/2.5/3, `http://hostap.epitest.fi/`, online link.

14. D. Collins, *Carrier Grade Voice over IP*, 2nd ed., McGraw-Hill, Sept. 2002.

15. Y. Xiao, Enhanced DCF of IEEE 802.11e to support QoS, *Proc. IEEE WCNC'03*, March 2003.

16. Y. Xiao, H. Li, and S. Choi, Protection and guarantee for voice and video traffic in IEEE 802.11e wireless LANs, *Proc. IEEE InfoCom'04*, Hong Kong, March 2004.

17. M. Heusse, F. Rousseau, G. Berger-Sabbatel, and A. Duda, Performance anomaly of 802.11b, *Proc. IEEE InfoCom'03*, San Francisco, March 2003.

18. S. Pilosof et al., Understanding TCP fairness over wireless LAN, *Proc. IEEE Info-Com'03,* March 2003, Vol. 2, pp. 863–872.

19. I. Aad and C. Castelluccia, Differentiation mechanisms for IEEE 802.11, *Proc. IEEE InfoCom'01*, Anchorage, AK, April 2001.

20. D.-J. Deng and R.-S. Chang, A priority scheme for IEEE 802.11 DCF access method, *IEICE Trans. Commun.* **E82-B**(1):96–102 (Jan. 1999).

21. A. Veres, A. T. Campbell, M. Barry, and L. Sun, Supporting service differentiation in wireless packet networks using distributed control, *IEEE J. Select. Areas Commun.* **19**(10) (Oct. 2001).

22. F. Cali, M. Conti, and E. Gregori, IEEE 802.11 wireless LAN: Capacity analysis and protocol enhancement, *Proc. IEEE InfoCom'98*, March 1998.

23. F. Cali, M. Conti, and E. Gregori, Dynamic tuning of the IEEE 802.11 protocol to achieve a theoretical throughput limit, *IEEE/ACM Trans. Networking* **8**(6): 785–799 (Dec. 2000).

24. S. Mangold, S. Choi, P. May, O. Klein, G. Hiertz, and L. Stibor, IEEE 802.11e wireless LAN for quality of service, *Proc. European Wireless'02*, Florence, Italy, Feb. 2002.

25. P. Garg, R. Doshi, R. Greene, M. Baker, M. Malek, and X. Cheng, Using IEEE 802.11e MAC for QoS over wireless, *Proc. IPCCC'03*, Phoenix, AZ, April 2003.

26. D. Gu and J. Zhang, QoS enhancement in IEEE 802.11 wireless area networks, *IEEE Commun.* **41**(6):120–124 (June 2003).

27. H. L. Truong and G. Vannuccini, Performance evaluation of the QoS enhanced IEEE 802.11e MAC layer, *Proc. IEEE VTC'03 — Spring*, Jeju, Korea, April 2003.

28. Y. Ge and J. Hou, An analytical model for service differentiation in IEEE 802.11, *Proc. IEEE ICC'03*, Anchorage, AK, May 2003.

29. C. T. Chou, K. G. Shin, and S. Shankar, Inter-frame space (IFS) based service differentiation for IEEE 802.11 wireless LANs, *Proc. IEEE VTC'03 — Fall*, Orlando, FL, Oct. 2003.

30. Y. Xiao, Backoff-based priority schemes for IEEE 802.11, *Proc. IEEE ICC'03*, Anchorage, AK, May 2003.

31. J. Zhao, Z. Guo, Q. Zhang, and W. Zhu, Throughput and QoS optimization in IEEE 802.11 WLAN, *Proc. 3G Wireless'02 + WAS'02*, San Francisco, CA, May 2002.

32. J. Zhao, Z. Guo, Q. Zhang, and W. Zhu, Performance study of MAC for service differentiation in IEEE 802.11, *Proc. IEEE GlobeCom'02*, Taiwan, Taipei, Nov. 2002.

33. G. Bianchi and I. Tinnirello, Analysis of priority mechanisms based on differentiated inter frame spacing in CSMA-CA, *Proc. IEEE VTC'03 — Fall*, Orlando, FL, Oct. 2003.

34. Y. Xiao, Performance analysis of IEEE 802.11e EDCF under saturation condition, *Proc. IEEE ICC'04*, Paris, June 2004.

35. L. Romdhani, Q. Ni, and T. Turletti, Adaptive EDCF: Enhanced service differentiation for IEEE 802.11 wireless ad hoc networks, *Proc. WCNC'03*, Louisiana, March 2003.

36. M. Malli, Q. Ni, T. Turletti, and C. Barakat, Adaptive fair channel allocation for QoS enhancement in IEEE 802.11 wireless LANs, *Proc. IEEE ICC'04*, Paris, June 2004.

37. Q. Pang, S. C. Liew, J. Y. B. Lee, and S.-H. G. Chan, A TCP-like adaptive contention window scheme for WLAN, *Proc. IEEE ICC'04*, Paris, June 2004.

38. L. Zhang and S. Zeadally, HARMONICA: Enhanced QoS support with admission control for IEEE 802.11 contention-based access, *Proc. Real-Time and Embedded Technology and Applications Symp.*, May 2004.

39. Y. Kuo, C. H. Lu, E. H. Wu, and G. Chen, An admission control strategy for differentiated services in IEEE 802.11, *Proc. IEEE GlobeCom'03*, San Francisco, Dec. 2003.

CHAPTER 4

A Perspective on the Design of Power Control for Mobile Ad Hoc Networks

ALAA MUQATTASH and MARWAN KRUNZ

Department of Electrical and Computer Engineering, University of Arizona, Tucson

SUNG-JU LEE

Mobile and Media Systems Lab, Hewlett-Packard Laboratories, Palo Alto, California

4.1 INTRODUCTION

The design of mobile ad hoc networks (MANETs) has attracted a lot of attention. The interest in MANETs is driven mainly by their ability to provide instant wireless networking solutions in situations where cellular infrastructures do not exist and are expensive or unfeasible to deploy (disaster relief efforts, battlefields, etc.). Furthermore, because of their distributed nature, MANETs are more robust than their cellular counterparts against single-point failures, and have the flexibility to reroute packets around congested nodes. While wide-scale deployment of MANETs is yet to come, extensive research efforts are currently underway to enhance the operation and management of such networks [14,21,55].

Two of the most important challenges in designing MANETs are the needs to provide high throughput and low-energy wireless access to mobile nodes. Power management solutions addressing one or both of these challenges can loosely be classified into three categories:

- *Transmission Power Control* (*TPC*). TPC adapts the transmission power (TP) to the propagation and interference characteristics experienced by the link. Theoretical studies [24] and simulation results [46] have demonstrated that TPC can provide significant benefits in capacity and energy consumption. TPC can also be used as a means of admission control and quality-of-service (QoS) provisioning [5].

Mobile, Wireless, and Sensor Networks: Technology, Applications, and Future Directions
Edited by Rajeev Shorey, Akkihebbal L. Ananda, Mun Choon Chan, and Wei Tsang Ooi
Copyright © 2006 John Wiley & Sons, Inc.

- *Power-Aware Routing (PAR)*. Additional energy saving can be obtained by routing packets over energy-efficient paths. While the TPC protocol aims at making each link as energy-efficient as possible, a PAR protocol decides which of these links to use for the end-to-end path. Design of PAR schemes can be based on various power-related link metrics, such as, the transmission energy consumed per packet [16], mobile node's battery level [22], or a combination of these metrics [63].

- *Power-Saving Modes (PSMs)*. The power consumption of the node's wireless interface can be greatly reduced by putting the interface into sleep.[1] Once in the sleep state, a node is not able to transmit, receive, or even sense the channel. Thus, it is important to decide when to enter the sleep state and for how long to stay in this mode. This has been the topic of extensive research [11,12,32,62,69,76,78,79].

Although these solutions may, at first, seem orthogonal, they are actually interdependent, which makes the task of integrating them in one framework quite challenging. For example, PAR protocols have no basis for favoring one path over another when TPC is not employed. Since packets cannot be routed through sleeping nodes, PAR decisions are affected by the state of the node, which itself is decided according to the PSM protocol. Furthermore, a node that has just awakened up has out-of-date information about the channel state, and hence, may not be able to decide on the required TP.

The main goal of this chapter is to review and analyze the major approaches for TPC that have been proposed in the literature. In addition, we will briefly review several PAR and PSM schemes and explain their interdependences. As demonstrated later, it follows naturally that "cross-layering" is a key design principle for efficient operation of MANETs.

We start by pointing out several deficiencies in the IEEE 802.11 scheme. The tradeoffs in selecting the transmission range are discussed in Section 4.1.2. A class of energy-oriented power control schemes is introduced in Section 4.2. The adverse impact of this class of protocols on network throughput is explained. TPC schemes that are designed with the goal of increasing network throughput (by increasing spatial reuse) are presented in Section 4.3. These schemes include a class of algorithms that use TPC primarily to control the topological properties of the network (connectivity, node degree, etc.), and another class of interference-aware TPC schemes that broadcast interference information to bound the power levels of subsequent transmissions. This later class is found to achieve both goals of energy conservation and throughput enhancement (compared with the IEEE 802.11 scheme). Other protocols that are based on clustering or that combine scheduling and TPC are presented in Section 4.3.5. PSM protocols are surveyed in Section 4.5. The chapter is concludes in Section 4.7 with several open research issues.

---

[1]For example, the Cisco Aironet 350 Series Client Adapter [2] consumes 2.25 W and 1.35 W in the transmit and receive modes, respectively, but consumes only 0.075 W in the sleep mode.

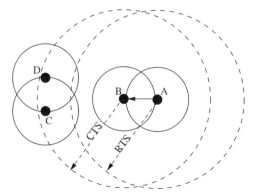

Figure 4.1 Inefficiency of the standard RTS-CTS approach. Nodes *A* and *B* are allowed to communicate, but nodes *C* and *D* are not. Dashed circles indicate the maximum transmission ranges for nodes *A* and *B*, while solid circles indicate the minimum transmission ranges needed for coherent reception at the respective receivers.

4.1.1 Deficiencies in the IEEE 802.11 Approach

The ad hoc mode of the IEEE 802.11 standard is by far the most dominant MAC protocol for ad hoc networks. This protocol generally follows the CSMA/CA (carrier sense multiple access with collision avoidance) paradigm, with extensions to allow for the exchange of RTS/CTS (request-to-send/clear-to-send) handshake packets between the transmitter and the receiver. These control packets are needed to reserve a *transmission floor* for the subsequent data packets. Nodes transmit their control and data packets at a *fixed* (maximum) *power level*, preventing all other potentially interfering nodes from starting their own transmissions. Any node that hears the RTS or the CTS message defers its transmission until the ongoing transmission is over.

The problem of this approach is that it can be "too conservative" in many scenarios. Take, for example, the situation in Figure 4.1, where node *A* uses its maximum TP to send its packets to node *B* [for simplicity, we assume omnidirectional antennas, so a node's reserved floor is represented by a circle in two-dimensional 2D space]. Nodes *C* and *D* hear *B*'s CTS message and therefore refrain from transmitting during *A* → *B*. However, it is easy to see that both transmissions *A* → *B* and *C* → *D* can, in principle, take place at the same time if nodes are able to select their transmission powers appropriately, hence, increasing the network throughout and reducing the per packet energy consumption.

The roots of this problem lie in the fact that the IEEE 802.11 scheme is based on two nonoptimal (in terms of throughput and energy) design decisions: (1) an *overstated* definition of a collision — according to the IEEE 802.11 scheme, if node *i* is currently receiving a packet from node *j*, then *all* other nodes in *i*'s transmission range[2] must defer their own transmissions to avoid colliding with *i*'s ongoing

---

[2]The transmission range of node *i* is defined as the maximum range over which a packet can be successfully received when there is no interference from other nodes.

reception; and (2) the IEEE 802.11 scheme uses a *fixed* common TP approach, which leads to reduced channel utilization and increased energy consumption.

To explain the inefficiencies of the first design principle, consider a network with a fixed TP. Let P_j be the TP used by node j. Let G_{ji} be the channel gain from node j to node i. Then the signal-to-interference-and-noise ratio (SINR) at node i for the desired signal from node j ($\text{SINR}(j, i)$) is given by

$$\text{SINR}(j, i) = \frac{P_j G_{ji}}{\sum_{k \neq j} P_k G_{ki} + P_{\text{thermal}}} \tag{4.1}$$

where P_{thermal} is the thermal noise. When the TP is fixed and common among all nodes, this equation is a function of only the channel gains. Node j's packet can be correctly received at node i if $\text{SINR}(j,i)$ is above a certain threshold (say, SINR_{th}) that reflects the QoS of the link.[3] Even if there is an interfering transmitter, say, v, that is within the transmission range of i, it may still be possible for i to correctly receive j's packet. A simplified analysis was given by Xu et al. [75], who assume that interference is attributed to only one node v. They show that under a path loss factor of 4, i can correctly receive the desired packet as long as v is at distance $1.78d$ or more from i, where d is the distance between i and j (assuming a common TP). Hence, in many cases, v can be allowed to transmit and cause interference at i, but not necessarily collide with i's reception of j's packet. Therefore, *interference and collision are not equivalent*. If high throughput is desired, then interference (hence, concurrent sessions in the same vicinity) should be allowed as long as collisions are prevented. The IEEE 802.11 approach equates interference with collision. The implications of this conservative approach are (1) it negatively impacts the channel utilization by not allowing concurrent transmissions to take place over the reserved floor and (2) the received power may be far more than necessary to achieve the required SINR_{th}, thus wasting energy and shortening the network lifetime. The discussion above is also valid when nodes can vary their TPs according to a certain protocol. Hence, there is a need for a solution, possibly a multilayer one, that allows concurrent transmissions to take place in the same vicinity and simultaneously conserves energy.

4.1.2   Tradeoffs in Selecting the Transmission Range

The selection of the "best" transmission range has been investigated extensively in the literature. It has been shown [24,25] that a higher network capacity can be achieved by transmitting packets to the nearest neighbor in the forward progress direction. The intuition behind this result is that halving the transmission range increases the number of hops by 2 but decreases the area of the reserved floor to one-fourth of its original value, hence allowing for more concurrent transmissions to take place in the neighborhood. In addition to improving network throughput, controlling the transmission range plays a significant role in reducing the energy

---

[3]Note that SINR_{th} already includes the effect of any employed forward error correction scheme.

required to deliver a packet over a multihop path of short per hop distances. On the other hand, the TP determines who can hear the signal, and so reducing it can adversely impact the connectivity of the network by reducing the number of active links and, potentially, partitioning the network. Thus, to maintain connectivity, power control should be carried out while accounting for its impact on the network topology. Furthermore, since route discovery in MANETs is often *reactive* (i.e., the path is acquired on demand), power control can be used to influence the decisions made at the routing layer by controlling the transmission power of the *route request* (RREQ) packets (discussed in more detail in Section 4.3.2).

The preceding discussion provides sufficient motivation to dynamically adjust the TP for data packets. However, there are many open questions at this point; perhaps the most interesting one is whether TPC is a network or a MAC layer issue. The interaction between the network and MAC layers is fundamental for power control in MANETs. On one hand, the power level determines who can hear the transmission, and hence, it directly impacts the selection of the next hop. This obviously is a network layer issue. On the other hand, the power level also determines the floor that the terminal reserves exclusively for its transmission through an access scheme. Obviously, this is a MAC layer issue. Hence, we have to introduce power control from the perspectives of both layers. Other important questions are

How can a terminal find an energy-efficient route to the destination?

What are the implications of adjusting the transmission powers of data and control packets?

How can multiple transmissions take place simultaneously in the same vicinity?

We address all of these questions in the subsequent sections.

4.2 ENERGY-ORIENTED POWER CONTROL APPROACHES

In this section, we present basic power control approaches that aim at reducing the energy consumption of nodes and prolonging the lifetime of the network. Throughput and delay are secondary objectives in such approaches.

4.2.1 TPC for Data Packets Only

One possible way to reduce energy consumption is for the communicating nodes to exchange their RTS/CTS packets at maximum power (P_{max}) but send their DATA/ACK packets at the minimum power (P_{min}) needed for reliable communication [23,34,52]. Determination of the value of P_{min} is based on the required QoS (i.e., the $SINR_{th}$), the interference level at the receiver, the antenna configuration (omnidirectional or directional), and the channel gain between the transmitter and the receiver. We refer to this basic protocol as SIMPLE. Note that SIMPLE and the IEEE 802.11 scheme have the same *forward progress rate* per hop; that is, the distance traversed by a packet in the direction of the destination is the same for both protocols. Thus, the two protocols achieve comparable throughputs.[4] However, energy consumption in SIMPLE is expectedly less. The problem with SIMPLE, however, is when a minimum-hop routing protocol (MHRP) (which is still the de facto routing approach in MANETs) is used at the network layer. In selecting the next hop (NH), a MHRP favors nodes in the direction of the destination that are *farthest* from the source node, but still within its maximum transmission range. When network density is high, the distance between the source node and the NH is very close to the maximum transmission range; thus, SIMPLE would be preserving very little energy. The problem lies in the poor selection of the NH (i.e., links are long), and so a more "intelligent" routing protocol that finds an energy-efficient route to the destination is required. In other words, *for SIMPLE to provide good energy saving, a power-aware protocol on top of SIMPLE is needed*, which is the topic of the next section.

4.2.2 Power-Aware Routing Protocols (PARPs)

The first generation of routing protocols for MANETs [30,49,50,59] are essentially MHRPs that do not consider power efficiency as the main goal. Singh et al. [63] first raised the power awareness issue in ad hoc routing and introduced new metrics for path selection, which include the energy consumed per packet, network connectivity duration (i.e., the time before network partitions), node power variance, cost per packet, and maximum node cost. PARPs discussed in the remainder of this section use one or more of these metrics in path selection [31].

The first wave of PARPs [10,42] was based on proactive shortest-path algorithms such as distributed Bellman–Ford. Instead of delay or hop count as the link weight, these protocols use energy-related metrics such as signal strength, battery level of each node, and power consumption per transmission. The link condition and power status of each node are obtained via a periodic route table exchange, as done in proactive routing protocols. Chang and Tassiulas [10] argue that the sole minimization of the total consumed energy per end-to-end packet delivery drains out the power of certain nodes in the network. Instead, energy consumption should be balanced among nodes to increase the network lifetime. Flow augmentation and redirection is incorporated in their routing scheme to split traffic. Krishnamachariy et al. [38] studied the issue of robustness to node failures in the context of energy-efficient networks. They showed that the energy cost of robustness obtained from multipath routing could be high. An alternative to routing through many paths is the use of higher TPs with fewer paths.

Proactive shortest-path algorithms are suitable mainly for networks with little (or no) mobility, such as sensor networks. Their applicability to highly mobile networks is questionable. The reason is that proactivity implies that each node must periodically exchange local routing and power information with neighboring nodes,

---

[4]In fact, SIMPLE achieves less throughput than the IEEE 802.11 because of the interference with the ACK reception at the source node [33].

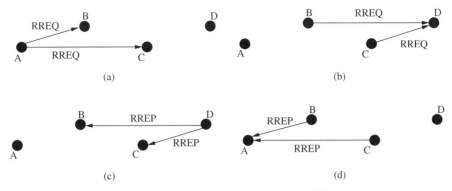

Figure 4.2   Route discovery process in DSR.

which incurs significant control overhead. Proactive routing schemes were shown to consume more power than on-demand routing protocols [76], as transmitting more control packets results in more energy consumption. Power-aware routing optimization (PARO) [23] is an example of an on-demand power-aware routing protocol. In PARO, one or more intermediate nodes elect to forward packets on behalf of source–destination pairs, thus reducing the aggregate TP consumption. However, as PARO's sole focus is on minimizing the TP consumed in the network, it does not account for balancing the energy consumption between nodes.

Careful modifications, however, have to be introduced to standard on-demand protocols (e.g., dynamic source routing (DSR) [30]) in order to incorporate energy awareness. In particular, the flooding techniques in the route discovery process of reactive protocols must be adjusted. To see why, consider Figure 4.2, where node A wants to find a route to node D. The most energy-efficient route is $A \rightarrow B \rightarrow C \rightarrow D$. In DSR route discovery process, node A broadcasts a route request (RREQ) packet to its neighbors (Fig. 4.2a). Assuming that this packet is heard by B and C, both nodes rebroadcast the RREQ packet (Fig. 4.2b). B's and C's broadcast packets are heard by all nodes. However, since C has already broadcasted the same request earlier, it does not rebroadcast the RREQ packet received from B, and vice versa. Node D now replies back with a route reply (RREP) packet (Fig. 4.2c), which is propagated back to node A by both B and C (Fig. 4.2d). Hence, the on-demand route discovered is either $A \rightarrow B \rightarrow D$ or $A \rightarrow C \rightarrow D$, and not the most energy-efficient route $A \rightarrow B \rightarrow C \rightarrow D$.

More recently, some work has been done to solve this problem [16,41]. Doshi et al. [16] proposed a DSR-based scheme in which each RREQ message includes the power at which that message was transmitted. Using this information and the received signal strength, the receiver of the RREQ calculates the minimum power required for the RREQ sender to successfully transmit a packet to that receiver. This information is inserted into the RREQ message and rebroadcasted by the receiver. The destination node inserts in the RREP message the power information for each hop in the path. Now, each node along the path uses the information contained in the RREQ and RREP messages to decide whether it lies on a lower energy path than the one advertised in the RREP. If so, the node sends to the source a "gratuitous" reply containing the lower energy path. This modified DSR protocol discovers routes that minimize the total TP per packet and is inline with the mechanism of PARO [23].

Maleki et al. [41] proposed a protocol that maximizes the lifetime of the network. In this protocol, every node other than the destination calculates its link cost (using battery life as the cost metric) and adds it to a path cost variable that is sent in the header of the RREQ packet. On receiving the RREQ packet, an intermediate node starts a timer and saves this path cost as a variable `min-cost`. If another RREQ packet with the same destination and sequence number arrives, its minimum-cost value is compared with the stored one. If the new packet has a lower path cost, it is forwarded, and `min-cost` is changed to this lower cost. When the destination receives the RREQ packet, it starts a timer and collects all RREQ packets with the same source and destination fields. Once the timer expires, the destination node chooses the minimum-cost route and sends a RREP back to the source.

An open problem that has not received much attention in the literature is the energy consumed in discovering a minimum-energy route, that is, the energy spent on RREQ and RREP packets. Since these packets are small, the energy spent on them may, at first, seem insignificant. Unfortunately, this is not the case; simulation results [29] show that for DSR, the overhead of the route discovery process can be up to 38% of the total received bytes. This high overhead is due primarily to the flooding nature of the route discovery process, which results in redundant broadcasts, contention, and collisions. These drawbacks are collectively referred to as the *broadcast storm problem* [47]. Another problem with several energy-aware protocols is related to the number of RREP messages. As explained before, the scheme presented by Doshi et al. [16] tries to produce energy-efficient paths by letting a node snoop on RREPs and send gratuitous replies once it finds out that it lies on a lower energy path than the one advertised in the RREP. In medium to dense networks, this approach results in sending a large number of gratuitous replies back to the source node (i.e., RREQ implosion). As we pointed out earlier [46], the severity of the problems mentioned above could be reduced by restricting the TP of the RREQ themselves.[5] This problem is still an open issue for research.

The main goal of energy-efficient routing protocols is to maximize the overall network lifetime. Compared to MHRPs, PARPs select paths that have longer hops and shorter distances per hop. Longer hop paths are usually more energy-efficient, as each node consumes less power when forwarding packets to nearby next hops. A careful design is required, however, as the aim is to build routes that require the least energy consumption for *reliable* end-to-end packet delivery (including recovery at the MAC or transport levels), and not simply minimum per hop energy consumption [7]. Moreover, as longer paths involve more forwarding nodes, the bandwidth allocation problem can be more complex.

---

[5]The operation details of our earlier work [46] are studied in Section 4.3.2.

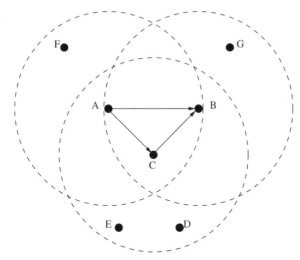

Figure 4.3   Drawbacks of the PARP/SIMPLE approach. Nodes E and D have to defer their transmissions when the data packets from A to B are routed via node C.

4.2.3   Limitations of the PARP/SIMPLE Approach

In the previous section, we have shown how a PARP/SIMPLE combination can significantly reduce the energy consumption of a MANET. This energy preservation, however, comes at the expense of a decrease in network throughput and an increase in packet delays. To illustrate, consider the example in Figure 4.3. Nodes A, B, and C are within each other's maximum transmission range. Node A wants to send packets to node B. According to a MHRP/802.11 solution, node A sends its packets directly to B. Thus, nodes E and D, who are unaware of the transmission $A \rightarrow B$, are able to communicate concurrently. On the other hand, according to a PARP/SIMPLE approach, data packets from A to B must be routed via node C, and thus nodes E and D have to defer their transmissions for two data packet transmission periods. More generally, all nodes within C's range but outside A's or B's ranges are not allowed to transmit, as they are first silenced by C's CTS to A, and then again by C's RTS to B. This example shows that a PARP/SIMPLE approach forces more nodes to defer their transmissions, resulting in lower network throughput compared with a MHRP/802.11 approach.

The total packet delay is also higher in the case of PARP/SIMPLE. The total delay consists of transmission periods and contention periods. Using a MHRP/802.11 scheme, delivering a packet from A to B involves one transmission period plus the duration it takes A to acquire the channel. Conversely, in the case of PARP/SIMPLE, the total delay consists of two packet transmission periods and two contention periods to acquire the channel (one for A and one for C). Since each RTS/CTS exchange reserves a fixed (maximum) floor, the *total* reserved floor in this case is greater, and so is the contention (and total) delay. The message that we are trying to convey here is that *the design of an energy-efficient protocol stack* (*network and MAC layers*) *should not occur at the cost of network throughput and delay performance.*

4.3   TPC: THE MAC PERSPECTIVE

The throughput degradation in PARP/SIMPLE has to do with the fixed-power *exclusive reservation* mechanism at the MAC layer. So it is natural to consider a medium access solution that allows for the adjustment of the reserved floor depending on the *data* TP. A power-controlled MAC protocol reserves different floors for different packet destinations. In such a protocol, both the channel bandwidth and the reserved floor constitute network resources that nodes contend for. For systems with a shared data channel (i.e., one node uses all the bandwidth for transmission), the floor becomes the single critical resource. This is in contrast to cellular systems and the IEEE 802.11 scheme, where the reserved floor is always fixed; in the former, the reserved floor is the whole cell, while in the latter it is the maximum transmission range. Note that in ad hoc networks, a node that reserves a larger floor uses more resources.

4.3.1   Topology Control Algorithms

We now present a family of protocols that use TPC as a means of controlling network topology (e.g., reducing node degree while maintaining a connected network). The size of the reserved floor in these protocols varies in time and among nodes, depending on the network topology. Rodoplu and Meng [58] proposed a distributed position-based topology control algorithm that consists of two phases. Phase 1 is used for link setup and configuration, and is done as follows. Each node broadcasts its position to its neighbors and uses the position information of its neighbors to build a sparse graph called the *enclosure graph*. In phase 2, nodes find the "optimal" links on the enclosure graph by applying the distributed Bellman–Ford shortest-path algorithm with power consumption as the cost metric. Each node i broadcasts its cost to its neighbors, where the cost of node i is defined as the minimum power necessary for i to establish a path to a destination. The protocol requires nodes to be equipped with GPS receivers. Ramanathan and Rosales-Hain [56] suggested a protocol that exploits global topological information provided by the proactive routing protocol to reduce the nodes' transmission powers such that the degree of each node is upper- and lower-bounded. The protocol, however, does not guarantee full network connectivity. In another study [70] a cone-based solution that guarantees network connectivity was proposed. Each node i gradually increases its TP until it finds at least one neighbor in every cone of angle $\alpha = 2\pi/3$ centered at i (a $5\pi/6$ angle was later proved to guarantee network connectivity). Node i starts the algorithm by broadcasting a "Hello" message at low TP and collecting replies. It gradually increases the TP to discover more neighbors and continuously caches the direction in which replies are received. It then checks

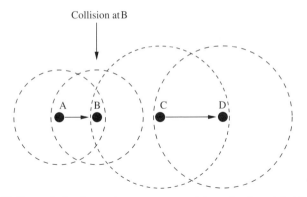

Figure 4.4   Challenge in implementing power control in a distributed fashion. Node C is unaware of the ongoing transmission $A \rightarrow B$, and hence it starts transmitting to node D at a power that destroys B's reception.

whether each cone of angle α contains a node. The protocol assumes the availability of directional information (angle of arrival), which requires extra hardware such as the use of more than one antenna. ElBatt et al. [18] proposed the use of a synchronized global signaling channel to build a global network topology database, where each node communicates only with its nearest N neighbors (N is a design parameter). The protocol, however, requires a signaling channel in which each node is assigned a dedicated slot. The broadcast incremental power (BIP) algorithm was developed by Wieselthier et al. [71]. BIP is a centralized energy-efficient broadcast–multicast algorithm that aims at producing a minimum-power tree, rooted at source nodes. A distributed version of BIP was later proposed [72].

One common deficiency of these protocols is that periodic or on-demand reconfiguration of the network topology is always needed if nodes are moving. This affects network resource availability and increases packet delays, especially at peak load times. Another critical drawback is their sole reliance on CSMA for accessing or reserving the shared wireless channel. It has been shown [68] that using CSMA alone for accessing the channel can significantly degrade network performance (throughput, delay, and power consumption) because of the well-known hidden terminal problem. Unfortunately, this problem cannot be overcome by simply using a standard RTS/CTS-like channel reservation approach, as explained in the example in Figure 4.4. Here, node A has just started a transmission to node B at a power level that is just enough to ensure coherent reception at B. Suppose that node B uses the same power level to communicate with A. Nodes C and D are outside the floors of A and B, so they do not hear the RTS/CTS exchange between A and B. For nodes C and D to be able to communicate, they have to use a power level that is reflected by the transmission floors in Figure 4.4 (the two circles centered at C and D). However, the transmission $A \rightarrow B$ transmission, causing a collision at B. In essence, the problem is caused by the asymmetry in the transmission floors (i.e., B can hear C's transmission to D but C cannot hear B's transmission to A).

4.3.2 Interference-Aware MAC Design

The topology control protocols discussed in Section 4.3.1 lack a proper channel reservation mechanism (e.g., RTS/CTS-like), which negatively impacts the achievable throughput under these protocols. To address this issue, more sophisticated MAC protocols are needed, in which information about an ongoing transmission is made known to all possible interferes. Before proceeding to discuss these protocols, we first explain two important design considerations that are fundamental to the operations of these protocols.

4.3.2.1 Timescale of Power Control

TPC schemes in MANETs are fundamentally different from those in cellular systems. In cellular systems, each time a new session is started or terminated, the powers of ongoing transmissions are renegotiated, whereas in MANETs, power is allocated only once at the start of the session; that is, the whole data packet is transmitted at one power level, regardless of what follows the start of that packet transmission.

The flexibility of the cellular approach allows it to provide more "optimal" solutions in the sense that more sessions can be admitted for the same amount of total power. However, the price of this flexibility is that the *entire* state of the system (power used by every node in the network) must be known whenever a new session is to be admitted. Moreover, it requires nodes with ongoing transmissions to be able to receive some power control information (i.e., renegotiation), which is very difficult, if not impossible, in single-channel, signal transceiver distributed systems, such as MANETs. Because of the overhead involved in distributing the entire system state of all nodes, and the infeasibility of the feedback channel model, the cellular approach for TPC cannot be used for MANETs.

4.3.2.2 Minimum versus Controlled Power

A fundamental design principle that is not well understood is whether a node should transmit at the minimum power required for reliable communication, or that the node should transmit at some controlled power (higher than the minimum) in order to optimize a certain criterion. Here, we advocate the second choice using intuition, an example, and through inspection of deployed wireless systems that are proven to be successful.

Suppose that the TPC scheme allocates the *minimum* transmission power ($P_{\text{min}}^{(ji)}$) required for node j to transmit to node i. Then $P_{\text{min}}^{(ji)}$ is given by

$$P_{\text{min}}^{(ji)} = \frac{\mu\left(P_{\text{thermal}} + P_{\text{MAI–current}}^{(i)}\right)}{G_{ji}} \tag{4.2}$$

where μ is SINR_{th} and $P_{\text{MAI–current}}^{(i)}$ is the *current* interference from all already ongoing (interfering) transmissions.

This $P_{\text{min}}^{(ji)}$, however, does not allow for any interference tolerance at node i, and hence all neighbors of node i will have to defer their transmissions during i's

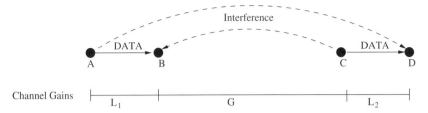

Figure 4.5 Example of a topology where the two interfering transmissions $A \rightarrow B$ and $C \rightarrow D$ can proceed simultaneously if A's and C's transmission powers are appropriately chosen.

ongoing reception (i.e., no simultaneous transmissions can take place in the neighborhood of i). Theoretically, the interference range is infinite, and practically, it is very large.[6] Clearly, a TPC scheme that allocates only $P_{\min}^{(ji)}$ is not attractive if high throughput is desirable.

Consider now a TPC scheme that allocates a power higher than $P_{\min}^{(ji)}$ such that nodes at some interfering distance from the receiver i can access the channel if their transmissions do not disturb i's reception. Clearly, such a scheme would significantly increase the throughput [43,46]. To understand this further, let us consider the line topology in Figure 4.5. Node A is transmitting to node B, and node C is transmitting to node D. An interference is induced from A to D and from C to B. The channel gains between the nodes are also shown in that figure. For node B to receive A's transmission reliably and for node D to receive C's transmission reliably, the following two conditions must hold:

$$\frac{P_A L_1}{P_{\text{thermal}} + P_C G} \geq \mu, \quad \frac{P_C L_2}{P_{\text{thermal}} + P_A L_1 G L_2} \geq \mu \qquad (4.3)$$

where L_1 is the channel gain between nodes A and B, G is the gain between nodes B and C, L_2 is the gain between nodes C and D, and P_i is the TP used by node i. Solving (4.3) for the minimum P_A and P_C, we get

$$\begin{aligned}
(P_A)_{\min} &= \frac{\mu P_{\text{thermal}}(L_2 + \mu G)}{L_1 L_2 (1 - \mu^2 G^2)} \\
(P_C)_{\min} &= \frac{\mu P_{\text{thermal}}(1 + \mu L_2 G)}{L_2 (1 - \mu^2 G^2)}
\end{aligned} \qquad (4.4)$$

As a numerical example, let $\mu = 6$, $L_1 = 0.6$, $L_2 = 0.5$, and $G = 0.1$. Then $P_A \simeq 34.4 P_{\text{thermal}}$ and $P_C \simeq 24.4 P_{\text{thermal}}$. Thus, although the two transmissions interfere with each other, it is possible to allow them to proceed simultaneously. The power-controlled dual channel (PCDC) [46] is an example of a protocol that achieves that; it requires the node that starts transmitting first, say, node A, to use more than $P_{\min}^{(AB)}$ for its TP. PCDC, however, requires a two-channel architecture.

---

[6]Qiao et al. [53] the auhors derived a finite value for the interference range in the case of minimum TP. However, the thermal noise power was not taken into account in that derivation.

Note that if A starts transmission first (i.e., $P_{\text{MAI-current}}^{(B)} = 0$) and uses the minimum required power according to Equation (4.2) namely, $P_{\min}^{(AB)} = 10 P_{\text{thermal}}$, then it will *not* be possible for C to transmit to D while A is sending to B, since C's transmission will disturb B's reception. This example clearly shows that transmitting at the minimum required power to overcome the current level of interference will severely impact the throughput by not allowing for any future concurrent transmissions in the vicinity of a receiving node. It should be emphasized, however, that increasing the power of a certain session above $P_{\min}^{(.)}$ is useful *only* if there is a protocol that allows for concurrent interference-limited transmissions, since otherwise, using more than $P_{\min}^{(.)}$ actually introduces more interference to other nearby nodes at no advantage, thus decreasing the network throughput. We also emphasize that the argument that increasing the TP above $P_{\min}^{(.)}$ is beneficial in terms of throughput does *not* contradict the analysis by Gupta and Kumar [24], who proved that using a smaller transmission range increases the network capacity. The intuition behind that result is that decreasing the *area of the reserved floor* allows for more concurrent transmissions to take place in the neighborhood. The argument in this chapter is inline with that; increasing the TP (not range) increases the interference margin at the receiver, thus, as in their study [24], allowing for more concurrent transmissions to take place in the neighborhood. The challenge, however, is to come up with a protocol for MANETs that allows for concurrent interference-limited transmissions.

We also note that in deployed cellular systems [48], the basestation instructs nodes to transmit at power higher than $P_{\min}^{(.)}$. This allows for some interference margin at the basestation, and thus, for new calls to be admitted. In fact, it has been proved [4] that to increase channel capacity, the TP of nodes must be increased by a common factor.

One can understand the intuition behind the discussion above by considering Equation (4.1). The reader can verify that increasing the power of all active nodes by a certain factor will actually improve the link SINR at all receivers, or, alternatively, increase the "free capacity" of the system, thus allowing for more nodes to access the channel.

4.3.3   Interference-Aware MAC Protocols

Figure 4.6 illustrates the intuition behind interference-aware MAC protocols. Node A intends to send its data to B. Before this transmission can take place, node B broadcasts some "collision avoidance information" to all possible interfering neighbors, which include C, D, and E. Unlike the RTS/CTS packets used in the 802.11 scheme, this "collision avoidance information" does not prevent interfering nodes from accessing the channel. Instead, it *bounds the transmission powers* of future packets generated by these nodes. Thus, in Figure 4.6, future transmitters (D and E in this example) can proceed only if the powers of their signals are not high enough to collide with the ongoing reception at node B.

To understand what this "collision avoidance information" is and how nodes can make use of it, let us consider Equation (4.1) again. Recall that a packet is correctly received if the SINR is above SINR_{th}. By allowing nearby nodes to transmit concurrently, the interference power at receiver i increases, and so $\text{SINR}(j,i)$

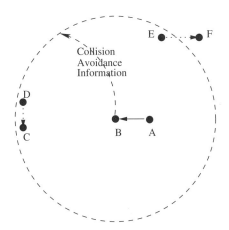

Figure 4.6 Broadcasting collision avoidance information in interference-aware MAC protocols.

decreases. Therefore, to be able to correctly receive the desired packet at node i, the TP at node j must be computed while taking into account *potential* future transmissions in the neighborhood of receiver i. This is achieved by incorporating an *interference margin* in the computation of $SINR(j,i)$. This margin represents the additional interference power that receiver i can tolerate while ensuring coherent reception of the upcoming packet from node j. Nodes at some interfering distance from i can now start new transmissions while the transmission $j \rightarrow i$ is taking place. The interference margin is incorporated by scaling up the TP at node j beyond what is minimally needed to overcome the current interference at node i. Because of the distributed nature of the TPC problem, it makes sense that the computation of the appropriate TP level is made by the *intended receiver*, which is more capable of determining the potential interferers in its neighborhood than the transmitter. Note that the power level is determined for *each* data packet separately (possibly via an RTS/CTS handshake), just before the transmission of that packet. This is in contrast to cellular networks in which the power is determined not only at the start of the transmission but also while the packet is being transmitted (e.g., the TP is updated every 125 μs in the IS-95 standard for cellular systems). This issue will be discussed in further detail shortly. Now, a node with a packet to transmit is allowed to proceed with its transmission if the TP will not disturb the ongoing receptions in the node's neighborhood *beyond the allowed interference margin(s)*. Allowing for concurrent transmissions increases network throughput and decreases contention delay.

Interference-aware MAC protocols differ from each other mainly in how they compute the "collision avoidance information" and how they distribute it to the neighboring nodes. Monks et al. [43] proposed the *power-controlled multiple access* (PCMA) protocol, in which each receiver sends busy-tone pulses to advertise its interference margin. The signal strength of the received pulses is used to bound the TP of the (interfering) neighboring nodes. A potential transmitter, say, j, first senses the busy-tone channel to determine an upper bound on its TP for all its control and data packets, adhering to the most sensitive receiver in its neighborhood. After that, node j sends its RTS at the determined upper bound and waits for a CTS. If the receiver, say, i, is within the RTS range of node j, and the power needed to send back the CTS is below the power bound at node i, node i sends back a CTS, allowing the transmission to begin. The simulation results due to monks et al. [43] show significant throughput gain (more than twice) over the 802.11 scheme. However, the choice of energy-efficient links is left to the upper layer (e.g., a PARP). Furthermore, the interference margin is fixed and it is not clear how it can be determined. Contention among busy tones is also not addressed. Finally, according to PCMA, a node may send many RTS packets without getting any reply, thus wasting the node's energy and the channel bandwidth.

The use of a separate control channel in conjunction with a busy-tone scheme has been proposed [74]. The sender transmits the data packets and the busy tones at reduced power, while the receiver transmits its busy tones at maximum power. A node estimates the channel gain from the busy tones and is allowed to transmit if its transmission is not expected to add more than a fixed interference to the ongoing receptions. The protocol is shown to achieve considerable throughput improvement over the original *dual busy-tone multiple access* (DBTMA) protocol [15]. The authors, however, make strong assumptions about the interference power. Specifically, they assume that the antenna is able to reject any interfering power that is less than the power of the "desired" signal (i.e., they assume perfect capture) and that there is no need for any interference margin. Also, the power consumption of the busy tones was not addressed. Furthermore, as in PCMA, the choice of energy-efficient links is left to the upper layer.

An issue that has been left unsolved in the two protocols described above is the contention between nodes broadcasting RREQ packets. These packets are broadcast by a source node to inquire about the path to a given destination. A neighboring node that receives the RREQ packet and has a path to the destination responds to the source using a *route reply* packet. Otherwise, the neighboring node rebroadcasts the RREQ packets to its own neighbors. It is easy to see that the first broadcast of a RREQ packet is likely to be followed by a high contention period during which several nodes attempt to rebroadcast this packet. This may result in many collisions between RREQ packets (the transmissions of which are typically unacknowledged), which delays the process of finding the destination and requires retransmitting these packets. The problem is known as the "broadcast storm problem" [47]. Furthermore, the transmission and reception powers of the routing packets themselves can be significant, given the large overhead of these packets.

The power controlled dual channel (PCDC) protocol proposed in [46] emphasizes the interplay between the MAC and network layers, whereby the MAC layer indirectly influences the selection of the next hop by properly adjusting the power of the RREQ packets. According to PCDC, the available bandwidth is divided into two frequency-separated channels for data and control. Each data packet is sent at a power level that accounts for a receiver-dependent interference margin. This margin allows for concurrent transmissions to take place in the neighborhood of the receiver, provided these transmissions do not individually interfere with the ongoing reception by more than a fraction of the total interference margin. The "collision avoidance information" is inserted into the CTS packet, which is sent at maximum power over the control channel, thus informing all possible interferers about the ensuing data packet and allowing for interference-limited simultaneous transmissions to take place in the neighborhood of a receiving node. Furthermore, each node continuously caches the estimated channel gain and angle of arrival of every signal it receives over the control channel, regardless of the intended destination of this signal. This information is used to construct an energy-efficient subset of neighboring nodes, called the *connectivity set* (CS). The intuition behind the algorithm is that the CS must contain only neighboring nodes with which direct communication requires less power than does indirect (two-hop) communication via any other node that is already in the CS. Let $P_{\text{conn}}^{(i)}$ denote the minimum power required for node i to reach the *farthest* node in its CS. Node i *uses this power level to broadcast its RREQ packets*. This results in two significant improvements:

1. Any simple MHRP can now be used to produce routes that are very power efficient and that increase network throughput (i.e., reduce the total reserved floor) — hence, no intelligence is needed at the network layer and no link information (e.g., power) has to be exchanged or included in the RREQ packets in order to find power-efficient routes (clearly, this reduces complexity and overhead).

2. Considering how RREQ packets are flooded throughout the network, significant improvements in throughput and power consumption can be achieved by limiting the broadcasting of these packets to nodes that are within the connectivity range $P_{\text{conn}}^{(i)}$.

It has been shown [46] that if the network is connected under a fixed-power strategy (i.e., RREQ packets are broadcasted using power P_{max}, then it must also be connected under a CS-based strategy. Figure 4.7 depicts an example of the resulting topologies under the two strategies.

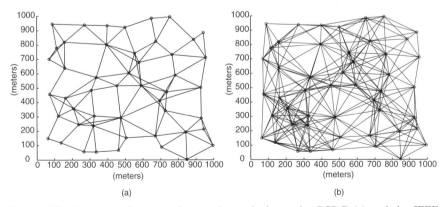

(a)

(b)

Figure 4.7 Instances of generated network topologies under PCDC (a) and the IEEE 802.11 (b) scheme.

PCDC was shown to achieve considerable throughput improvement over the 802.11 scheme *and* significant reduction in energy consumption. The authors, however, did not account for the processing and reception powers, which increase with the number of hops along the path (note that PCDC results in longer paths than the 802.11 scheme when both are implemented below a MHRP). Furthermore, there is an additional signaling overhead in PCDC due to the introduction of new fields in the RTS and CTS packets.

4.3.4 A Note on Mobility and Power Control

The interference-aware protocols discussed above rely on channel gain information to determine the appropriate transmission powers for data packets. The channel gain is often estimated from the transmission and reception powers of the RTS and CTS packets that precede a data packet. Therefore, for these protocols to function properly, it is important that the channel gain remain stationary for the period from the estimation time until the data packet is fully received. Since channel gain may change because of mobility, we need to consider the impact of mobility on the channel characteristics during the transmission period of a data packet.

In a multipath environment, multiple versions of the transmitted signal arrive at the receiver at slightly different times and combine to give a resultant signal that can vary widely in amplitude and phase. The spectral broadening caused by this variation is measured by the Doppler spread, which is a function of the relative velocity (v) of the mobile and the angle between the direction of motion and the directions of arrival of the multipath waves [57]. The variation can be equivalently measured in the time domain using the *coherence time* (T_c), which is basically a statistical measure of the time duration over which the channel can be assumed time-invariant. As a rule of thumb in modern communication systems, $T_c \approx 0.423/f_m$, where $f_m = v/\lambda$ is the maximum Doppler shift and λ is the wavelength of the carrier signal. Now, at a mobile speed of $v = 1$ m/s (3.6 km/h) and 2.4 GHz carrier frequency, $T_c \approx 52.89$ ms. This time reduces to 2.64 ms when $v = 20$ m/s (72 km/h). For a channel bandwidth of 2 Mbps (megabits per second), it takes 4 ms to transmit a 1000-byte packet. Note that the propagation delay and the turnaround time (the time it takes a node to switch from a receiving mode to a transmitting mode) are in the order of μs, and so can be ignored in the calculations. So the assumption about the channel stationarity is valid only when the packet transmission time is less than T_c. This places a restriction on the maximum mobile speed or, alternatively, the size of the data packet (for a given channel data rate). Power-controlled MAC protocols should take this tradeoff into account at the design stage.

4.3.5 Other TPC Approaches

So far we have investigated TPC in MAC perspectives. In this section, we describe two additional TPC approaches that adopt philosophies to the problem different from what has been discussed so far. The first one is clustering [39,66]. Kwon and Gerla [39] employ an elected cluster head (CH) to perform the function of a basestation in a cellular system. It uses closed-loop power control to adjust the transmission powers of nodes in the cluster. Communications between different clusters occur via *gateways*, which are nodes that belong to more than one cluster. This approach simplifies the forwarding function for most nodes, but at the expense of reducing network utilization since all communications have to go through the CHs. This can also lead to the creation of bottlenecks. A joint clustering/TPC protocol was proposed in a paper by Kawadia and Kumar [35], where clustering is implicit and is based on TP levels, rather than on addresses or geographic locations. No CHs or gateways are needed. Each node runs several routing layer agents that correspond to different power levels. These agents build their own routing tables by communicating with their peer routing agents at other nodes. Each node along the packet route determines the lowest-power routing table in which the destination is reachable. The routing overhead in this protocol grows in proportion to the number of routing agents, and can be significant even for simple mobility patterns (recall that for DSR, RREQ packets account for a large fraction of the total received bytes).

Another novel approach for TPC is based on joint scheduling and power control [17]. This approach consists of scheduling and power control phases. The purpose of the scheduling phase is to eliminate strong interference that cannot be overcome by TPC. It also renders the TPC problem similar to that of cellular systems. In the scheduling phase, the algorithm searches for the largest subset of nodes that satisfy "valid scenario constraints." A node satisfies such constraints if it does not transmit and receive simultaneously, does not receive from more than one neighbor at the same time, and is spatially separated from other interferers by a distance of at least D when receiving from a neighbor node. This D is set to the "frequency reuse distance" parameter used in cellular systems. In the TPC phase, the algorithm searches for the largest subset of users generated from the first phase that satisfy admissibility (SINR) constraints. The complexity of both phases is exponential in the number of nodes. Because the algorithm is invoked on a slot-by-slot basis, it is computationally expensive for real-time operation. ElBatt and Ephremides [17] proposed heuristics to reduce the computational burden. A simple heuristic for the scheduling phase is to examine the set of valid scenarios *sequentially* and defer transmissions accordingly. There is still a need for a centralized controller to execute the scheduling algorithm (i.e., the solution is not fully distributed). For the TPC phase, a cellular-like solution that involves deferring the user with the minimum SINR is examined in an attempt to lower the level of multiple access interference. It is assumed here that the SINR measurement at each receiver is known to *all* transmitters (e.g., via flooding).

4.4 COMPLEMENTARY APPROACHES AND OPTIMIZATIONS

In this section, we discuss three approaches that interact with TPC protocols to further enhance the throughout and energy consumption of MANETs.

4.4.1 Transmission Rate Control

Variable rate support is another optimization that TPC protocols have not yet considered. In fact, rate control and TPC are two sides of one coin; if the channel gain is high, then it is possible to reduce the power and/or increase the rate. The reason for this is that the quality of a reception is adequately measured by the *effective bit energy-to-noise spectral density ratio* at the detector, denoted by E_b/N_0, where $E_b \overset{\text{def}}{=} P/R_b$, P is the reception power, and R_b is the data rate. Hence, under the same modulation and channel coding schemes, decreasing the signal power by a factor of λ is exactly equivalent [in terms of reception quality, i.e., bit error rate (BER)] to increasing the data rate by λ. If the modulation and/or the channel coding schemes are changed, this factor could be different, but the general trend is the same.

Although rate increase and power decrease are equivalent to some extent, there are several reasons that could make the former a more attractive approach than the latter:

1. The efficiency of the power amplifier (percentage of the the TP to the overall power consumption of the amplifier) in the transmitter circuitry increases with the TP [60]. Hence, operating at rate $2R_0$ bps and power $2P_0$ watts is more energy-efficient than operating at rate R_0 bps and power P_0 watts, since the power amplifier efficiency is higher in the former. Furthermore, it is difficult to design an efficient power amplifier that has a wide range of power levels.

2. The total energy consumption (E_{total}) of sending and receiving a packet of size l at rate R bps consists of (a) the energy consumed in processing the packet [power consumed by DACs, mixers, amplifiers, voltage controlled oscillators (VCOs), synthesizers, etc.) at the transmitter side (lP^T_{proc}/R), (b) the energy consumed in the power amplifier (lP_t/R), where P_t is the transmitted signal power, and (c) the energy consumed in processing the packet at the receiver side (lP^R_{proc}/R). Accordingly, $E_{\text{total}} = l(P_t + P^T_{\text{proc}} + P^R_{\text{proc}})/R$. Real-life measurements indicate that not only both P^T_{proc} and P^R_{proc} are nonnegligible, but also, they are hardly affected by R [44].

These results are important and can be interpreted as follows. Let E_b/N_0 be the required signal quality to send a packet of length l over a certain link, say, i. In the first scenario, we use rate $R = R_0$ and a transmission power $P_t = P_0$ to achieve E_b/N_0. In the second scenario, we use rate $R = 2R_0$ and hence, we require $P_t = 2P_0$ in order to achieve the same E_b/N_0. Then, given our discussion above, the second scenario is actually more energy-efficient than the first one. Specifically, E_{total} in the first scenario is

$$100 * \frac{P^T_{\text{proc}} + P^R_{\text{proc}}}{2P_t + P^T_{\text{proc}} + P^R_{\text{proc}}}$$

of E_{total} in the second scenario. This increase can go up to 50% in current wireless network interface cards [2]. Clearly, this shows that there is potential for significant energy saving by controlling the transmission rate. Finally, note that if the transmission rate is increased, both the sender and the receiver will spend less time in sending/receiving the packet. This allows for longer sleep time, and hence, preserves even more energy.

One drawback of increasing the transmission rate is that the spreading gain (the ratio of the used Fourier bandwidth over the data rate) of the communication system is decreased. The less the spreading gain, the less immune is the signal against interference from other nodes and devices that use the same spectrum. Nonetheless, the tradeoff is definitely worth investigating, and more research is needed to evaluate the benefits and drawbacks of rate control in MANETs. Furthermore, the performance achieved through TPC can be further improved by allowing for dynamic adjustment of the information rate [20]. The mechanics of such an approach are yet to be explored.

4.4.2 Directional Antennas

So far, we have assumed that each node is equipped with an omnidirectional antenna, where the distribution of the TP is equal in all directions. One possible approach to increase the throughput and reduce the energy consumption is to employ directional antennas (DAs). The use of these antennas allows the transmitter to focus its TP in the receiver direction, thus achieving better range or, alternatively, saving power. Moreover, DAs allow for more efficient use of the channel,which increases the network throughput. Because of these advantages, directional antennas are deployed in IS-95 and third-generation cellular systems [40]. For instance, partitioning the cell into three 120° sectors using DAs increases the capacity (maximum number of users) by a factor of ∼3. Furthermore, it can provide a power gain of 18 dBi, which translates into increased coverage or energy saving.

The use of DAs for MANETs is fraught with challenges. One challenge is related to the size of a typical DA. At 2.4 GHz, a six-element circular array with interelement spacing of 0.4 wavelength has a radius of nearly 4.8 cm, making this DA system rather bulky for small mobile nodes such as handheld devices and laptops. This is one of the reasons (besides cost) why *mobile* nodes in cellular networks do not currently employ DAs, and usage of DAs have been limited to basestations.

Another challenge is that MAC protocols for MANETs with omnidirectional antennas (e.g., IEEE 802.11) are not suitable for usage with DAs. These protocols are designed according to assumptions that are not valid for DAs; for instance, nodes have equal reception sensitivity and radiate equal power in all directions, and therefore, nodes that are able to hear the RTS or the CTS packet are the ones that can cause collisions. When directional antennas are used, the radiated power and reception sensitivity between any two nodes are a function of the angular orientation of these nodes. Thus, the scheme of using equal power for RTS/CTS and data packets no longer prevents potential interferers from transmitting. Several

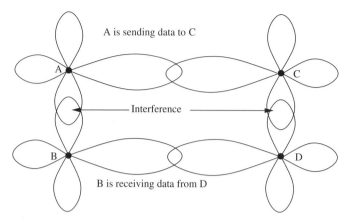

Figure 4.8   Minor lobe interference problem in proposed MAC protocols for DAs.

protocols have proposed for DAs in MANETs [6,8,13,26,36,37,54,67,77], which improve the spatial reuse by allowing simultaneous data transmissions. These transmissions are permitted provided nodes that intend to transmit must point the main lobe of their DAs away from nodes with ongoing receptions. The nodes keep track of the prohibited directions by various mechanisms, for instance, by setting the directional network allocation vector (DNAV) as in two other studies [6,67] or by location tables as in a study by Korakis et al. [37]. Choudhary et al. [13] extended the range of directional antennas, where a *multihop* RTS mechanism is used to beamform two faraway nodes in each other's direction before data transmission. Huang et al. [26] extended the busy-tone concept, originally designed for MANETs with omnidirectional antennas [15], to the case of directional antennas. In another paper [8], a time-slotted scheduling channel access scheme was developed for multibeam adaptive array antennas (MANETs).

In all of these proposals, transmitters are prevented from pointing their *main* lobes toward nodes that are currently receiving. However, all practical DAs have *minor* lobes with significant radiation power. For example, for a six-element circular array, minor lobes could have a peak gain of 10 dBi, where the power radiated in the peak minor-lobe direction is 10 times greater than an isotropic antenna (6 times greater compared with a typical omnidirectional antenna with a gain of 2.2 dB [2]). Thus, a receiver in the direction of a minor lobe of a transmitter will experience considerable interference, which may lead to packet collisions. This problem is illustrated in Figure 4.8. In this figure, node *A* is sending data to node *C* after an RTS/CTS exchange. Node *D* has received a CTS from node *C*. According to many proposed protocols for DAs [6,37,67], node *D* is free to transmit as long as it does not beamform in the direction of *C*. Meanwhile, node *D* sends an RTS to *B* since it is not located in the direction of *C*. Thereafter, *D* starts sending data packets to *B*, causing interference at *C* from its minor-lobe radiation. Likewise, node *B*, which is located in the direction of *A*'s minor lobe, experiences interference that could cause a collision. At high loads, this channel access problem, which is exclusive to DAs, is exacerbated, leading to a high probability of collisions and adversely affecting all the protocols mentioned above.

At the network layer, Spyropoulos and Raghavendra [65] proposed a scheme to conserve energy and increase network lifetime based on the use of directional antennas. This scheme first builds "minimum energy consumed per packet" routes using Dijkstra-like algorithms, and then schedules nodes transmissions by executing a series of maximum weight matchings. The scheme is shown to be energy-efficient compared with shortest-path routing using omnidirectional antennas. However, since each node is assumed to have a single-beam directional antenna, the sender and the receiver must redirect their antenna beams toward each other before transmission and reception can take place. Moreover, it is preferred that each node participate in only one data session at a time, as redirecting antenna requires large energy consumption. These restrictions factor into large delay, and hence the scheme is not adequate for time-sensitive data transmission.

4.4.3 TPC for CDMA-Based Ad Hoc Networks

Power control for CDMA-based MANETs is another interesting topic that has not received enough attention. Because of its demonstrated superior performance (compared with TDMA and FDMA), CDMA has been chosen as the access technology of choice in cellular systems, including the more recently adopted 3G systems. It is therefore natural to consider the use of CDMA in MANETs. Interestingly, the IEEE 802.11 standard uses spread-spectrum techniques at the physical layer, but only to mitigate the interference of the unlicensed, heavily used, 2.4-GHz Industrial, Scientific, and Medical (ISM) radio band.[7] More specifically, in the 802.11 protocol, all transmitted signals are spread using a *common* pseudo-random-noise (PN) code, precluding the possibility of multiple concurrent transmissions in the vicinity of a receiver.

The use of CDMA in MANETs is not straightforward because of the time-asynchronous nature of ad hoc systems [45], which makes it impossible to design PN that are orthogonal for all time offsets [51]. This results in nonnegligible cross-correlations between different PN codes, thus inducing multiaccess interference (MAI). The *near–far* problem is a severe consequence of MAI, whereby a receiver who is trying to detect the signal of, say, transmitter i may be much closer in distance to, say, transmitter j than i. When all transmission powers are equal, the signal from j will arrive at the receiver in question with a sufficiently larger power than that of i's signal, causing incorrect decoding of i's signal.

The majority of CDMA-based MAC protocols for MANETs that have been proposed in the literature [19,27,28,64] have overlooked the near–far problem, and assumed a synchronous orthogonal CDMA system, which we now know to be impractical. As we pointed out in another study [45], the near–far problem can cause a significant reduction in network throughput, and hence cannot be overlooked when designing CDMA-based MAC protocols for MANETs.

---

[7]The ISM band is also used by the HomeRF wireless networking system, cordless analog and digital phones, microwave ovens, and some medical equipment.

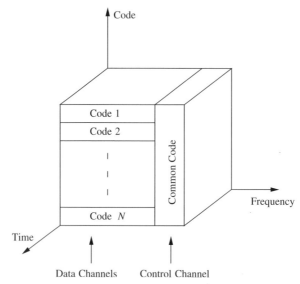

Figure 4.9 Data and control codes in CA-CDMA.

According to the CA-CDMA protocol that we proposed earlier [45], the near–far problem in MANETs requires a combined channel access/TPC solution. The authors proposed the architecture in Figure 4.9, where two *frequency* channels, one for data and the for control, are used. A common spreading code is used by all nodes over the control channel, while *several* node-specific codes can be used over the data channel.

On receiving the RTS packet, the intended receiver node i decides on the data packet TP, depending on the planned "loading" of the network. Node i can then calculate the amount of additional interference power that it can tolerate from future transmissions without impacting its future reception. Node i then inserts this information in the CTS packet and sends this packet at maximum power over the control channel. Neighbors of node i use this information and the estimated channel gain between them and receiver i to decide on the maximum power that they can use for their future transmissions without disturbing i's reception. We [45] have solved a few of the challenging issues in ad hoc CDMA system, but definitely more work is still needed to better understand the capacity of a CDMA-based MANETs, the optimal design of TPC for such networks, the interoperability with the existing IEEE 802.11 standard, and many other issues.

4.5 POWER SAVING MODES

In this section, we survey some of the well-known power saving mode (PSM) approaches, including that of the IEEE 802.11 standard. According to the IEEE 802.11 PSM [1], time is divided into beacon intervals and nodes start and finish

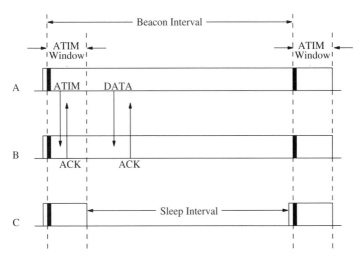

Figure 4.10 Power-saving mode in the IEEE 802.11 standard.

each beacon interval at about the same time. It is assumed that all nodes are *fully connected and synchronized* using periodic beacon transmissions. Figure 4.10 illustrates the IEEE 802.11 PSM. At the start of each beacon interval, there exists an interval called the *announcement traffic indication message* (ATIM) window, where every node should be in the awake state. If node *A* has buffered packets to node *B*, it sends an ATIM packet to node *B* during this interval. If *B* receives this packet, it replies back with an acknowledgment. Both *A* and *B* will then stay awake for the rest of that entire beacon interval. If a node has not sent or received any ATIM packet during the ATIM window (e.g., node *C* in Fig. 4.10), it enters the sleep mode until the next beacon time.

Several enhancements to the IEEE 802.11 PSM scheme were proposed in the literature. Cano and Manzoni [9] proposed a power conserving algorithm that allows a node to enter the sleep state if it overhears an RTS or a CTS packet between some other nodes for the duration of the data packet (the RTS and CTS packets specify duration of the ensuing data packet transmission). However, as pointed out elsewhere [32], this approach is not always suitable because of the time and energy costs associated with a packet-by-packet sleep-to-active transition.

In PAMAS [62] each node uses two separate channels for control and data packets. Nodes exchange *probe* messages over the control channel in order to determine when to power on and off. This scheme has the disadvantage of requiring two channels for communication. Chiasserini and Rao [12] proposed a scheme that allows mobile hosts to select their sleep patterns according to their battery status and the target QoS. However, a special hardware called *remote activated switch* (RAS) is required to receive wakeup signals when the mobile host has entered a sleep state.

The PSM issue has also been studied from a network layer viewpoint. When the node density is high and there exist redundant routes between the source and the receiver, intermediate nodes of secondary routes can be put to sleep. Using geographic location information, the geographi adaptive fidelity (GAF) algorithm [76] divides the network space into virtual grids. To balance the load on the nodes, GAF uses application and system information to decide which nodes should be put to sleep, which ones should remain active, and when to make the switch. Each grid has at least one active node at any given time. The main drawback of GAF is its requirement of geographic location information. SPAN [11] addresses this problem and instead uses periodic local broadcast messages. Each node periodically decides whether to go in the sleep mode or to remain active and participate as one of the "coordinators." A node becomes a coordinator when two neighboring nodes cannot directly communicate with each other and there is no other coordinator to forward packets between them. The role between coordinators and sleeping nodes are rotated so that nodes do not drain out their power. This class of algorithms works well only when node density is relatively high. The network can become isolated as a result of some nodes being powered off. Moreover, when packets are destined to a node that is turned off, other nodes need to buffer these packets. A novel approach for combining PSM and TPC has been proposed [61]. However, this approach requires an access point and cannot be used for distributed MANETs.

Arguably, the main challenge for PSM protocols in MANETs is achieving clock synchronization. Recall that the 802.11 PSM assumes a fully connected synchro-nized network. The lack of synchronization complicates the problem since a host has to predict when other hosts will be awake. To address this challenge, Tseng et al. [69] proposed several asynchronous PSM mechanisms. In that work, the authors enforce nodes to send more beacon packets than in the IEEE 802.11 scheme, allowing for more accurate neighborhood information. Furthermore, they ensure that the wakeup period of any two neighbors will overlap, for instance, by making the awake period equal to at least as half of the beacon interval.

Another challenge for PSM protocols is the fixed size of the ATIM window. It has been shown [73] that any fixed ATIM window size cannot perform well in all situations, when throughput and energy consumption are considered. If the ATIM window is too large, there would be less time for the actual data transmission; if it is too small, there may not be enough time available to announce the buffered data packet by transmitting ATIM frames. This problem was addressed by Jung and Vai-dya [32], who proposed an adaptive scheme that dynamically adjusts the size of the ATIM window on the basis of the backlog and some overheard information. Furthermore, the authors proposed allowing nodes to enter the sleep mode after completing their transmissions and receptions that were explicitly announced dur-ing the ATIM window. However, this is done only when the time left until the next ATIM window is not small, in order to avoid the time and energy costs associated with a sleep-to-active transition.

4.6 SUMMARY AND OPEN ISSUES

Transmission power control has a great potential to improve the throughput perfor-mance of a MANET and simultaneously decrease energy consumption. In this chapter, we surveyed several TPC approaches. Some of these approaches (e.g., PARP/SIMPLE) are successful in achieving the second goal, but sometimes at the expense of a reduction (or at least, no improvement) in the throughput performance. By locally broadcasting "collision avoidance information," some protocols are able to achieve both goals of TPC simultaneously. These protocols, however, are designed on the basis of assumptions (e.g., channel stationarity and reciprocity) that are valid only for certain ranges of speeds and packet sizes. Furthermore, they generally require additional hardware support (e.g., duplexers). The key message in the design of efficient TPC schemes is to account for the interplay between the routing (network), MAC layer, and the physical layer.

Many interesting open problems remain to be addressed. Interference-aware TPC schemes are quite promising, but their feasibility and design assumptions need to be evaluated. For instance, the protocols in a few other studies [43,46,74] assume that the channel gain is the same for the control (or busy tone) and data channels, and that nodes can transmit on one channel and simultaneously receive on the other. For the first assumption to hold, the control channel must be within the coherence bandwidth of the data channel, which places an upper bound on the allowable frequency separation between the two channels. However, for the second assumption to hold, there must be some minimal channel spacing between the two channels that are used for simultaneously transmitting and receiving from the same node. Typically, 5% of the nominal RF frequency is needed to keep the price and complexity of air interface reasonable [57]. However, spacing the control (or busy-tone) and data channel by this much frequency spectrum would make the first assumption invalid. Ideally, one would like to have a *single-channel* TPC solution that preserves energy while increasing spatial reuse.

Interoperability with existing standards and hardware is another important issue. Currently, most wireless devices implement the IEEE 802.11b standard. TPC schemes proposed in the literature (e.g., interference-aware protocols) are often not backward-compatible with the IEEE 802.11 standard, which makes it difficult to deploy such schemes in real networks. The convergence of these TPC algorithms is yet to be determined; one possible approach for this is based on noncooperative game theory proposed by Altman and Altman [3]. Another important issue is the incorporation of a sleep mode in the design of TPC protocols. A significant amount of energy is consumed by unintended receivers. In many cases, it makes sense to turn off the radio interfaces of some of these receivers to prolong their battery lives. The effect of this on the TPC design has not been explored.

Increasing the data rate versus decreasing the TP is another interesting issue. Research should also focus on the energy consumption of the various stages in the total energy consumption and not only on the transmitted signal power. Moreover, the interaction with PSM modes is also crucial. Directional antennas has been proposed as a means of increasing network capacity under a fixed-power strategy. The use of TPC in MANETs with directional antennas can provide significant energy saving. However, the access problem is now more difficult because of the resurfacing of various problems such as the effect of minor lobes and deafness, which need to be addressed. Furthermore, TPC plays a significant role in solving the near–far problem in CDMA-based MANETs.

ACKNOWLEDGMENTS

The work of M. Krunz was supported by the National Science Foundation under grants CCR 9979310 and ANI 0095626, and by the Center for Low Power Electronics (CLPE) at the University of Arizona. CLPE is supported by NSF (grant EEC-9523338), the State of Arizona, and a consortium of industrial partners.

REFERENCES

1. International Standard ISO/IEC 8802-11; ANSI/IEEE Std 802.11, 1999 ed., Part 11, *Wireless LAN Medium Access Control* (*MAC*) *and Physical Layer* (*PHY*) *Specifications.*

2. The Cisco Aironet 350 Series of wireless LAN, http://www.cisco.com/warp/public/cc/pd/witc/ao350ap/prodlit/a350c ds.pdf.

3. E. Altman and Z. Altman, S-modular games and power control in wireless networks, *IEEE Trans. Autom. Control* **48**:839–842 (May 2003).

4. D. Ayyagari and A. Ephramides, Power control for link quality protection in cellular DS-CDMA networks with integrated (packet and circuit) services, *Proc. ACM MobiCom Conf.*, 1999, pp. 96–101.

5. N. Bambos, Toward power-sensitive networks architecture in wireless communications: Concepts, issues, and design aspects, *IEEE Pers. Commun. Mag.* **5**:50–59 (June 1998).

6. S. Bandyopadhyay, K. Hasuike, S. Horisawa, and S. Tawara, An adaptive MAC protocol for wireless ad hoc community network (WACNet) using electronically steerable passive array radiator antenna, *Proc. IEEE GlobeCom Conf.*, 2001, pp. 2896–2900.

7. S. Banerjee and A. Misra, Minimum energy paths for reliable communication in multi-hop wireless networks, *Proc. ACM MobiHoc Conf.*, June 2002, pp. 146–156.

8. L. Bao and J. Garcia-Luna-Aceves, Transmission scheduling in in adhoc networks with directional antennas, *Proc. ACM MobiCom Conf.*, 2002.

9. J. C. Cano and P. Manzoni, Evaluating the energy-consumption reduction in a MANET by dynamically switching-off network interfaces, *Proc. 6th IEEE Symp. Computers and Communications*, July 2001.

10. J. H. Chang and L. Tassiulas, Energy conserving routing in wireless ad-hoc networks, *Proc. IEEE INFOCOM Conf.*, March 2000, pp. 22–31.

11. B. Chen, K. Jamieson, H. Balakrishnan, and R. Morris, Span: An energy-efficient coordination algorithm for topology maintenance in ad hoc wireless networks, *Proc. ACM MobiCom Conf.*, July 2001, pp. 85–96.

12. C. F. Chiasserini and R. R. Rao, A distributed power management policy for wireless ad hoc networks, *Proc. IEEE Wireless Communications and Networking Conf.*, September 2000, pp. 1209–1213.

13. R. R. Choudhury, X. Yang, R. Ramanathan, and N. H. Vaidya, Using directional antennas for medium access control in ad hoc networks, *Proc. ACM MobiCom Conf.*, 2002, pp. 59–70.

14. M. S. Corson, J. P. Macker, and G. H. Cirincione, Internet-based mobile ad hoc networking, *IEEE Internet Comput.* **3**(4):63–70 (July/Aug. 1999).

15. J. Deng and Z. Haas, Dual busy tone multiple access (DBTMA): A new medium access control for packet radio networks, *Proc. IEEE ICUPC*, Oct. 1998, pp. 973–977.

16. S. Doshi, S. Bhandare, and T. X. Brown, An on-demand minimum energy routing protocol for a wireless ad hoc network, *ACM SIGMOBILE Mobile Comput. Commun. Rev.* **6**:50–66 (July 2002).

17. T. ElBatt and A. Ephremides, Joint scheduling and power control for wireless ad-hoc networks, *Proc. IEEE InfoCom Conf.*, 2002, pp. 976–984.

18. T. A. ElBatt, S. V. Krishnamurthy, D. Connors, and S. Dao, Power management for throughput enhancement in wireless ad-hoc networks, *Proc. IEEE ICC Conf.*, 2000, pp. 1506–1513.

19. J. Garcia-Luna-Aceves and J. Raju, Distributed assignment of codes for multihop packet-radio networks, *Proc. IEEE MilCom Conf.*, 1997, pp. 450–454.

20. A. Goldsmith and P. Varaiya, Increasing spectral efficiency through power control, *Proc. IEEE ICC Conf.*, 1993, pp. 600–604.

21. A. J. Goldsmith and S. B. Wicker, Design challenges for energy-constrained ad hoc wireless networks, *IEEE Wireless Commun.* **9**:8–27 (Aug. 2002).

22. J. Gomez, A. Campbell, M. Naghshineh, and C. Bisdikian, A distributed contention control mechanism for power saving in random access ad-hoc networks, *Proc. IEEE Inte. Workshop on Mobile Multimedia Commun.*, 1999, pp. 114–123.

23. J. Gomez, A. T. Campbell, M. Naghshineh, and C. Bisdikian, PARO: Supporting dynamic power controlled routing in wireless ad hoc networks, *ACM/Kluwer J. Wireless Networks* **9**(5):443–460 (2003).

24. P. Gupta and P. R. Kumar, The capacity of wireless networks, *IEEE Trans. Inform. Theory* **46**(2):388–404 (March 2000).

25. T.-C. Hou and V. O. K. Li, Transmission range control in multiple packet radio networks, *IEEE Trans. Commun.* **34**(1):38–44 (Jan 1986).

26. Z. Huang, C. Shen, C. Srisathapornphat, and C. Jaikaeo, A busy tone based directinal MAC protocol for ad hoc networks, *Proc. IEEE MilCom Conf.*, 2002.

27. K.-W. Hung and T.-S. Yum, The coded tone sense protocol for multihop spread-spectrum packet radio networks, *Proc. IEEE GlobeCom Conf.*, 1989, pp. 712–716.

28. M. Joa-Ng and I.-T. Lu, Spread spectrum medium access protocol with collision avoidance in mobile ad-hoc wireless network, *Proc. IEEE InfoCom Conf.*, 1999, pp. 776–783.

29. P. Johansson, T. Larsson, N. Hedman, B. Mielczarek, and M. Degermark, Scenario-based performance analysis of routing protocols for mobile ad-hoc networks, *Proc. ACM MobiCom Conf.*, 1999, pp. 195–206.

30. D. Johnson and D. Maltz, Dynamic source routing in ad hoc wireless networks, in T. Imielinski and H. Korth, eds., *Mobile Computing*, Kluwer, Publishing Company, 1996, Chapter 5, pp. 153–181.

31. C. E. Jones, K. M. Sivalingam, P. Agrawal, and J. C. Chen, A survey of energy efficient network protocols for wireless networks, *ACM/Kluwer J. Wireless Networks* **7**(4):343–358 (2001).

32. E.-S. Jung and N. H. Vaidya, An energy efficient MAC protocol for wireless LANs, *Proc. IEEE InfoCom Conf.*, 2002, pp. 1756–1764.

33. E.-S. Jung and N. H. Vaidya, A power control MAC protocol for ad hoc networks, *Proc. ACM MobiCom Conf.*, 2002, pp. 36–47.

34. P. Karn, MACA — a new channel access method for packet radio, *Proc. 9th ARRL Computer Networking Conf.*, 1990, pp. 134–140.

35. V. Kawadia and P. R. Kumar, Power control and clustering in ad hoc networks, *Proc. IEEE InfoCom Conf.*, 2003, pp. 459–469.

36. Y.-B. Ko, V. Shankarkumar, and N. H. Vaidya, Medium access control protocols using directional antennas in ad hoc networks, *Proc. IEEE InfoCom Conf.*, 2000, pp. 13–21.

37. T. Korakis, G. Jakllari, and L. Tassiulas, A MAC protocol for full exploitation of directional antennas in ad-hoc wireless networks, *Proc. ACM MobiHoc Conf.*, 2003, pp. 95–105.

38. B. Krishnamachariy, Y. Mourtada, and S. Wicker, The energy-robustness tradeoff for routing in wireless sensor networks, *Proc. IEEE ICC Conf.*, 2003, pp. 1833–1837.

39. T. J. Kwon and M. Gerla, Clustering with power control, *Proc. IEEE MilCom Conf.*, 1999, pp. 1424–1428.

40. J. C. Liberti, Jr. and T. S. Rappaport, *Smart Antennas for Wireless Communication: IS-95 and Third Generation CDMA Applications*, Prentice-Hall, 1999.

41. M. Maleki, K. Dantu, and M. Pedram, Power-aware source routing protocol for mobile ad hoc networks, *Proc. ACM Int. Symp. Low Power Electronics and Design*, Aug. 2002, pp. 72–75.

42. A. Michail and A. Ephremides, Algorithms for routing session traffic in wireless ad-hoc networks with energy and bandwidth limitations, *Proc. IEEE PIMRC*, Oct. 2001.

43. J. Monks, V. Bharghavan, and W.-M. Hwu, A power controlled multiple access protocol for wireless packet networks, *Proc. IEEE InfoCom Conf.*, 2001, pp. 219–228.

44. J. Monks, J.-P. Ebert, A. Wolisz, and W.-M. Hwu, A study of the energy saving and capacity improvement potential of power control in multi-hop wireless networks, *Proc. IEEE LCN Conf.*, Nov. 2001, pp. 550–559.

45. A. Muqattash and M. Krunz, CDMA-based MAC protocol for wireless ad hoc networks, *Proc. ACM MobiHoc Conf.*, June 2003.

46. A. Muqattash and M. Krunz, Power controlled dual channel (PCDC) medium access protocol for wireless ad hoc networks, *Proc. IEEE InfoCom Conf.*, 2003, pp. 470–480.

47. S.-Y. Ni, Y.-C. Tseng, Y.-S. Chen, and J.-P. Sheu, The broadcast storm problem in a mobile ad hoc network, *Proc. ACM MobiCom Conf.*, 1999, pp. 151–162.

48. T. Ojanperä and R. Prasad, *Wideband CDMA for Third Generation Mobile Communications*, Artech House, 1998.

49. C. Perkins and E. Royer, Ad-hoc on-demand distance vector routing, *Proc. 2nd IEEE Workshop on Mobile Computing Systems and Applications (IEEE WMCSA'99)*, Feb. 1999, pp. 90–100.

50. C. E. Perkins and P. Bhagwat, Highly dynamic destination-sequenced distance-vector routing (DSDV) for mobile computers, *Proc. ACM SigComm Conf.*, London, Sept. 1994, pp. 234–244.

51. J. G. Proakis, *Digital Communications*, McGraw-Hill, 2001.

52. M. B. Pursley, H. B. Russell, and J. S. Wysocarski, Energy-efficient transmission and routing protocols for wireless multiple-hop networks and spread spectrum radios, *Proc. EUROCOMM*, 2000, pp. 1–5.

53. D. Qiao, S. Choi, A. Jain, and K. G. Shin, Miser: An optimal low-energy transmission strategy for IEEE 802.11a/h, *Proc. 9th Annual Int. Conf. Mobile Computing and Networking*, 2003, pp. 161–175.

54. R. Ramanathan, On the performance of ad hoc networks with beam forming antennas, *Proc. IEEE GlobeCom Conf.*, 2001, pp. 95–105.

55. R. Ramanathan and J. Redi, A brief overview of ad hoc networks: Challenges and directions, *IEEE Commun. Mag.*, 20–22 (May 2002).

56. R. Ramanathan and R. Rosales-Hain, Topology control of multihop wireless networks using transmit power adjustment, *Proc. IEEE InfoCom Conf.*, 2000, pp. 404–413.

57. T. Rappaport, *Wireless Communications: Principles and Practice*, Prentice-Hall, 2002.

58. V. Rodoplu and T. Meng, Minimum energy mobile wireless networks, *IEEE J. Select. Areas Commun.*, **17**(8):1333–1344 (Aug. 1999).

59. E. M. Royer and C.-K. Toh, A review of current routing protocols for ad hoc mobile wireless networks, *IEEE Pers. Commun. Mag.*, **6**(2):46–55 (April 1999).

60. J. F. Sevic, Statistical characterization of RF power amplifier efficiency for CDMA wireless communication systems, *Proc. IEEE Wireless Communications Conf.*, Aug. 1997, pp. 110–113.

61. T. Simunic, H. Vikalo, P. Glynn, and G. D. Micheli, Energy efficient design of protable wireless systems, *Proc. Int. Symp. Low Power Electronics and Design*, 2000, pp. 49–54.

62. S. Singh and C. S. Raghavendra, PAMAS — power aware multi-access protocol with signalling for ad hoc networks, *ACM SIGCOMM Comput. Commun. Rev.* **28**(3):5–26 (1998).

63. S. Singh, M. Woo, and C. S. Raghavendra, Power aware routing in mobile ad hoc networks, *Proc. ACM MobiCom Conf.*, 1998, pp. 181–190.

64. E. Sousa and J. A. Silvester, Spreading code protocols for distributed spread-spectrum packet radio networks, *IEEE Trans. Commun.* **36**(3):272–281 (March 1988).

65. A. Spyropoulos and C. Raghavendra, Energy efficient communications in ad hoc networks using directional antennas, *Proc. IEEE InfoCom Conf.*, April 2003.

66. M. E. Steenstrup, Self-organizing network control structures: Local algorithms for forming global hierarchies, *Proc. IEEE MilCom Conf.*, Oct. 2001, pp. 952–956.

67. M. Takai, J. Martin, A. Ren, and R. Bagrodia, Directional virtual carrier sensing for directional antennas in mobile ad hoc networks, *Proc. ACM MobiHoc Conf.*, 2002, pp. 59–70.

68. F. A. Tobagi and L. Kleinrock, Packet switching in radio channels: Part II — the hidden terminal problem in carrier sense multiple-access and the busy-tone solution, *IEEE Trans. Commun.* **23**(12):1417–1433 (Dec. 1975).

69. Y.-C. Tseng, C.-S. Hsu, and T.-Y. Hsieh, Power-saving protocols for IEEE 802.11-based multihop ad hoc networks, *Proc. IEEE InfoCom Conf.*, June 2002, pp. 200–209.

70. R. Wattenhofer, L. Li, P. Bahl, and Y.-M. Wang, Distributed topology control for power efficient operation in multihop wireless ad hoc networks, *Proc. IEEE InfoCom Conf.*, 2001, pp. 1388–1397.

71. J. E. Wieselthier, G. D. Nguyen, and A. Ephremides, On the construction of energy-efficient broadcast and multicast trees in wireless networks, *Proc. IEEE InfoCom Conf.*, March 2000, pp. 585–594.

72. J. E. Wieselthier, G. D. Nguyen, and A. Ephremides, Distributed algorithms for energy-efficient broadcasting in ad hoc networks, *Proc. IEEE MilCom Conf.*, Oct. 2002.

73. H. Woesner, J.-P. Ebert, M. Schlager, and A. Wolisz, Power-saving mechanisms in emerging standards for wireless LANs: The MAC level perspective, *IEEE Pers. Commun.*, 40–48 (1998).

74. S.-L. Wu, Y.-C. Tseng, and J.-P. Sheu, Intelligent medium access for mobile ad hoc networks with busy tones and power control, *IEEE J. Select. Areas Commun.* **18**(9):1647–1657 (2000).

75. K. Xu, M. Gerla, and S. Bae, How effective is the IEEE 802.11 RTS/CTS handshake in ad hoc networks? *Proc. IEEE GlobeCom Conf.*, Nov. 2002, pp. 72–76.

76. Y. Xu, J. Heidemann, and D. Estrin, Geography-informed energy conservation for ad hoc routing, *Proc. ACM MobiCom Conf.*, July 2001, pp. 70–84.

77. S. Yi, Y. Pei, and S. Kalyanaraman, On the capacity improvement of ad hoc wireless networks using directional antennas, *Proc. ACM MobiHoc Conf.*, 2003, pp. 108–116.

78. R. Zheng, J. C. Hou, and L. Sha, Asynchronous wakeup for ad hoc networks, *Proc. ACM MobiHoc Conf.*, June 2003, pp. 35–45.

79. R. Zheng and R. Kravets, On-demand power management for ad hoc networks, *Proc. IEEE InfoCom Conf.*, April 2003.

CHAPTER 5

Routing Algorithms for Energy-Efficient Reliable Packet Delivery in Multihop Wireless Networks

SUMAN BANERJEE

Department of Computer Sciences, University of Wisconsin, Madison

ARCHAN MISRA

IBM T. J. Watson Research Center, Hawthorne, New York

5.1 INTRODUCTION

Lowering energy consumption is a key goal in many multihop wireless networking environments, especially when the individual nodes of the network are battery-powered. This requirement has become increasingly important for new generations of mobile computing devices (such as PDAs, laptops, and cellular phones) because the energy density achievable in batteries has grown only at a linear rate, while processing power and storage capacity have both grown exponentially. As a consequence of these technological trends, many wireless-enabled devices are now primarily energy-constrained; while they possess the ability to run many sophisticated multimedia networked applications, their operational lifetime between recharges is often very small (sometimes less than 1 hr). In addition, the energy consumed in communication by the radio interfaces is often higher than, or at least comparable to, the computational energy consumed by the processor.

Various energy-aware routing protocols have thus been proposed to lower the communication energy overhead in such multihop wireless networks. In contrast to conventional wired routing protocols that try to utilize the minimum-hop route (one that minimizes the number of unique links), these protocols [2,19,20] typically aim to utilize the most energy-efficient route. These protocols exploit the fact that

Mobile, Wireless, and Sensor Networks: Technology, Applications, and Future Directions
Edited by Rajeev Shorey, Akkihebbal L. Ananda, Mun Choon Chan, and Wei Tsang Ooi
Copyright © 2006 John Wiley & Sons, Inc.

the transmission power needed on a wireless link is a nonlinear function of the link distance, and assume that the individual nodes can *adapt* their transmission power levels. As a consequence of this, it turns out that choosing a route with a large number of short-distance hops often turns out to consume significantly less energy than an alternative one with a few long-distance hops. (Of course, if the radios all used an identical transmission power independent of the link distance, and if all the wireless links are error-free, then conventional minimum-hop routing (e.g., RIP [11] and OSPF [14]) is also the most energy-efficient.)

For wireless links, a signal transmitted with power P_t over a link with distance D is attenuated and is received with power

$$P_r \propto \frac{P_t}{D^{K(D)}} \quad K(D) \geq 2 \qquad (5.1)$$

where $K(D)$ depends on the propagation medium, antenna characteristics,[1] and channel parameters, such as the radiofrequency. Since most wireless receivers are able to correctly decode the received signal as long as its power is above a certain fixed threshold,[2] energy-efficient algorithms typically set the transmission power to be proportional to $D^{K(D)}$. If the link cost in a routing algorithm is then assigned proportional to this tranmission power, a minimum-cost path will then correspond to a route that consumes the lowest *cumulative energy* for a single-packet transmission. A number of energy-efficient routing schemes, such as PAMAS [20] and PARO [8], utilize this approach to choose minimum-energy paths.

In this chapter, we present modifications to this basic approach for computing the minimum-energy path for *reliable packet delivery*. In practical wireless networks with nonnegligible link loss rates, packet retransmission or forward-error correction codes are employed to ensure reliable end-to-end delivery over the entire wireless path. Moreover, higher-layer protocols such as TCP or SCTP employ additional source-initiated retransmission mechanisms to ensure end-to-end reliability. Accordingly, we show that for reliable energy-efficient communication, the routing algorithm must *consider not only the distance of each link but its quality (in terms of its error rate) as well*. Intuitively, the cost of choosing a particular link is defined not simply in terms of the basic transmission power but also the overall transmission energy (including possible retransmissions) needed to ensure eventual error-free delivery. This is specially important in practical multihop wireless environments, where packet loss rates can be as high as 15–25 %.

Besides presenting the algorithmic modifications needed to compute a minimum-energy path for reliable communication, we shall also consider the

---

[1]Note that P_r represents the average power level of the received signal; the instantaneous received signal strength will vary around this mean value, due to additional effects such a fading or noise. In many cases, K is typically around 2 for short distances and omnidirectional antennas and around 4 for longer distances.

[2]More accurately, a receiver is able to correctly receive a transmission as long as the *ratio* of the signal's power to the cumulative power to all interfering signal and noise lies above a threshold. For our purposes, we model each link a independent, assuming that the links are either non interfering (e.g., have different frequencies or orthogonal CDMA codes) or use a MAC protocol to avoid contention.

challenges with implementing this algorithm in practical multihop ad hoc networks. In particular, conventional routing protocols are "proactive" and compute paths for each (source–destination) pair irrespective of whether those paths are needed or used. This requires the periodic exchange or flooding of routing messages, which can itself consume significant energy, especially when the traffic flows are sparsely distributed. To avoid these overheads, a family of "reactive" routing protocols has been proposed specifically for wireless networks. These protocols (e.g., AODV [17] and DSR [9]) compute routes on demand, when they are needed for a specific traffic flow. Using AODV as a representative protocol, we shall explain the enhancements needed to compute minimum-energy reliable paths with a reactive protocol.

5.1.1   The Underlying Wireless Network Model

To study the impact of link error rate on the energy required to ensure reliable packet delivery, we shall use two fundamentally distinct operating models:

1. *End-to-end retransmission* (EER), where the individual links do not provide link layer retransmission and reliable packet transfer is achieved only via retransmission initiated by the source node
2. *Hop-by-hop retransmission* (HHR), where each individual link provides reliable forwarding to the next hop using localized packet retransmission

To capture the potential effect of retransmissions on the overall energy consumption, we define a new link cost metric that incorporates both the link distance and the link error rate. Minimum-cost route computation based on this metric leads to energy-efficient paths for reliable communication for both EER and HHR scenarios. However, we shall show that such a link cost can be exactly defined only for the HHR scenario; for the EER framework, we can devise only an *approximate* cost function. By using simulation studies, we shall also demonstrate how the choice of parameters in the approximate EER cost formulation represents a tradeoff between energy efficiency and the achieved throughput. While the use of link quality in the definition of a link cost has been previously suggested as a routing metric to reduce queuing delays and loss rates, its implicit effect on energy efficiency of packet delivery has not been studied before. By incorporating the link error rates in the link cost, we show that 30–70% energy savings can often be achieved under realistic operating conditions.

Under both the EER and HHR models, the choice of links with relatively high error rates can significantly increase the overall energy overhead, due to the increase in the number of retransmissions needed to ensure reliable packet delivery. Of course, most practical multihop wireless networks follow the HHR model to counteract the effect of low-quality individual links. Note that the choice of a poor-quality link increases the energy overhead even when each node uses a constant transmission power. However, analysis of the effect of link error rates is more interesting for the variable-power case; we shall show that the choice between a path with many short-range hops and another with fewer long-range hops involves a *tradeoff between the reduction in the transmission energy for a single packet and the potential increase in the frequency of retransmissions.* As the fixed-power model is a trivially special case of the variable-power model, the analytical and performance results presented in this chapter are restricted to the more general variable-power model.

5.1.2 Roadmap

The rest of this chapter is organized as follows. Section 5.2 provides a survey of related work on energy-efficient routing. Section 5.3 formulates the reliable transmission energy problem as a function of the number of hops, and the error rates of each hop, both the EER and HHR cases and analyzes its effect on the optimum number of hops in the variable-power scenario. It also demonstrates the agreement between our idealized energy computation and real TCP behavior. Section 5.4 shows how to form link costs that lead to the selection of minimum-energy paths. In Section 5.5 we present the results of simulation studies to study the performance of our minimum-cost algorithms for the variable-power model. In Section 5.6, we explain how the computation of minimum-energy paths can be achieved using the ad hoc on-demand distance vector (AODV) protocol, a well-known on-demand ad hoc routing protocol. Section 5.7 then presents the results of simulation studies on the performance of our proposed energy-aware extensions to AODV. Finally, Section 5.8 summarizes the main conclusions of our research, and presents a set of open issues and research challenges.

5.2 RELATED WORK

Many energy-aware routing protocols aim to choose a route between a given (source–destination) pair that minimizes the cumulative transmission energy (the sum of the transmitter power levels over all the consituent links). PAMAS [20] is one energy-aware MAC/routing protocol, which proposes to set the link cost equal to the transmission power; the minimum-cost path is then equivalent to the one that uses the smallest cumulative energy. In the variable-power case, where nodes adjust their power on the basis of the link distance, such a formulation often selects a path with a large number of hops. A link cost that includes the receiver power as well has been presented [19]. By using a modified form of the Bellman–Ford algorithm, this approach results in the selection of paths with a smaller number of hops than in the power-aware multiaccess protocol with signaling (PAMAS). The power-aware route optimization (PARO) algorithm [8] has also been proposed as a distributed route computation technique for variable-power scenarios, and aims to generate a path with a larger number of short-distance hops. According to the PARO protocol, a candidate intermediary node monitors an ongoing direct communication between two nodes and evaluates the potential for power savings by inserting itself in the forwarding path — in effect, replacing the direct hop between the two nodes by two smaller hops through itself.

Alternative metrics, besides the minimum cumulative transmission energy, have also been considered for selection of energy-efficient routes in wireless environments. Indeed, selecting minimum-energy paths can sometimes unfairly penalize a subset of the nodes; for instance, if several minimum-energy routes have a common node in the path, the battery of that node will be exhausted quickly. Researchers have thus used an alternate objective function — maximizing the network lifetime — that considers both the energy consumption of a particular path and the remaining battery capacity of nodes on that path. The key idea is to distribute the energy expenditure across all the constituent nodes, selecting a less energy-efficient path if it helps extend the lifetime of a node nearing battery exhaustion. For example, Singh et al. [21] use node "capacity" as a routing metric, where the capacity of each node was a decreasing function of the residual battery capacity. A minimum-cost path selection algorithm then helps to steer routes away from paths where many of the intermediate nodes are facing battery exhaustion. Similarly, the MMBCR and CMMBCR algorithms [23] use a max–min route selection strategy, choosing a path that has the largest capacity value for its most critical ("bottleneck") node, where the bottleneck node for any given path is the one that has the least *residual* battery capacity. In an earlier study [13] we extended this approach to variable-power scenarios, by defining a combined node–link metric that normalizes the residual battery capacity of a node by the transmission power on an associated link. While we shall focus purely on computing the minimum-energy path in this chapter, we note that our techniques can be easily adapted to such battery-aware algorithms.

The proposed proactive routing protocols for wireless ad hoc environments (e.g., AODV [17], DSR [9]) contain special features to reduce the signaling overheads and convergence problems caused by node mobility and link failures. While some features of such protocols are implementation-specific, they generally aim to compute minimum-delay paths, and thus often choose minimum-hop paths rather than minimum-energy routes. We shall present the modifications needed to compute minimum-energy paths using such on-demand protocols.

Most prior work on the effect of link quality on packet transmissions has focused on the problem of intelligent *link scheduling* [4,18,25] rather than energy-efficient routing. More recently, Gass et al. [6] have proposed a transmission power adaptation scheme to control the link quality of individual frequency hopping wireless links. In contrast, our work explicitly formulates the overall transmision energy in terms of the link error rates and the associated retransmission probabilities, and uses this formulation to efficiently compute minimum-energy paths for reliable communication. Since our mathematical models assume that each node "knows" the packet error rate on its outgoing links, we shall also explain how this rate can be practically computed using AODV control packets.

5.3 ENERGY COST ANALYSIS AND MINIMUM-ENERGY PATHS

In this section, we demonstrate how the error rate associated with a link affects (1) the overall probability of reliable delivery and consequently (2) the energy associated with the reliable transmission of a single packet. For any particular link $\langle i,j \rangle$ between a transmitting node i and a receiving node j, let $T_{i,j}$ denote the transmission power and $p_{i,j}$ represent the packet error probability. Assuming that all packets are of a constant size, the energy involved in a packet transmission $E_{i,j}$ is simply a fixed multiple of $T_{i,j}$.

Any signal transmitted over a wireless medium experiences two different effects: attenuation due to the medium, and interference with ambient noise at the receiver. Due to the characteristics of the wireless medium, the transmitted signal suffers an attenuation proportional to $D^{K(D)}$, where D is the distance between the receiver and the transmitter. The ambient noise at the receiver is independent of the distance between the source and distance, and depends purely on the operating conditions at the receiver. The bit error rate associated with a particular link is essentially a function of the ratio of this received signal power to the ambient noise. In the constant-power scenario, $T_{i,j}$ is independent of the characteristics of the link $\langle i,j \rangle$ and is a constant. In this case, a receiver located farther away from a transmitter will suffer greater signal attenuation [proportional to $D^{K(D)}$] and will, accordingly, be subject to a larger bit error rate. In the variable-power scenario, a transmitter node adjusts $T_{i,j}$ to ensure that the strength of the (attenuated) signal received by the receiver is *independent of D* and is above a certain threshold level *Th*. Accordingly, the optimal transmission power associated with a link of distance D in the variable-power scenario is given by

$$T_{\text{opt}} = Th \times \gamma \times D^{K(D)} \tag{5.2}$$

where γ is a proportionality constant and $K(D)$ is the coefficient of attenuation ($K \geq 2$). Since *Th* is typically a technology-specific constant, we can see that the optimal transmission energy over such a link varies as follows:

$$E_{\text{opt}}(D) \propto D^{K(D)} \tag{5.3}$$

If links are considered error-free, then minimum-hop paths are the most energy-efficient for the fixed-power case. Similarly, in the absence of transmission errors, paths with a large number of small hops are typically more energy-efficient in the variable-power case. However, in the presence of link errors, neither of these choices necessarily gives the most energy-efficient path. We now analyze the interesting consequences of this behavior for the variable-power scenario (for both the EER and HHR cases).

5.3.1   Optimal Minimum-Energy Paths in EER Case

In the EER case, a transmission error on any link leads to an end-to-end retransmission over the path. Given the variable-power formulation of E_{opt} in Equation (5.3), it is easy to see why placing an intermediate node along the straight line between two adjacent nodes (breaking up a link of distance D into two shorter links of distance D_1 and D_2 such that $D_1 + D_2 = D$) always reduces the total E_{opt}. In fact,

PARO [7] works using precisely such an estimation. From a reliable transmission energy perspective, such a comparison is inadequate since it does not include the effect on the overall probability of error-free reception.

To understand the energy tradeoff involved in choosing a path with multiple short hops over one with a single long hop, consider communication between a sender (S) and a receiver (R) separated by a distance D. Let N represent the total number of hops between S and R, so that $N - 1$ represents the number of forwarding nodes between the endpoints. For notational ease, let these nodes be indexed as $i : i = \{2, \ldots, N\}$, with node i referring to the $(i$ - $1)$th intermediate hop in the forwarding path; also, node 1 refers to S and node $N + 1$ refers to R. Also, assume that $K(D)$ is a constant for the given link distances, so that $K(D)$ may be replaced by a constant K. In this case, the total optimal energy spent in simply transmitting a packet once (without considering whether the packet was reliably received) from the sender to the receiver over the $N - 1$ forwarding nodes is

$$E_{\text{total}} = \sum_{i=1}^{N} E_{\text{opt}}^{i,i+1} \tag{5.4}$$

or, using Equation (5.3), we obtain

$$E_{\text{total}} = \sum_{i=1}^{N} \alpha D_{i,i+1}^{K} \tag{5.5}$$

where $D_{i,j}$ refers to the distance between nodes i and j and α is a proportionality constant. To understand the transmission energy characteristics associated with the choice of $N - 1$ intermediate nodes, we compute the lowest possible value of E_{total} for any given layout of $N - 1$. Using very simple symmetry arguments, it is easy to see that the minimum transmission energy case occurs when each of the hops are of equal length D/N. In that case, E_{total} is given by

$$E_{\text{total}} = \sum_{i=1}^{N} \alpha \frac{D^K}{N^K} = \frac{\alpha D^K}{N^{K-1}} \tag{5.6}$$

For computing the energy spent in *reliable delivery*, we now consider how the choice of N affects the probability of transmission errors and the consequent need for retransmissions. Clearly, increasing the number of intermediate hops increases the likelihood of transmission errors over the entire path.

Assuming that each of the N links has an independent packet error rate of p_{link}, the probability of a transmission error over the entire path, denoted by p, is given by

$$p = 1 - (1 - p_{\text{link}})^N \tag{5.7}$$

The number of transmissions (including retransmissions) necessary to ensure the successful transfer of a packet between S and D is then a geometrically distributed random variable X, such that

$$\text{Prob}\{X = k\} = p^{k-1} \times (1 - p), \ \forall \ k$$

The *mean* number of individual packet transmissions for the successful transfer of a single packet is thus $1/(1 - p)$. Since each such transmission uses total energy E_{total} given by Equation (5.6), the total expected energy required in the reliable transmission of a single packet is given by

$$E_{\text{total rel}}^{\text{EER}} = \alpha \frac{D^K}{N^{K-1}} \cdot \frac{1}{1 - p} = \frac{\alpha D^K}{N^{K-1}(1 - p_{\text{link}})^N} \tag{5.8}$$

This equation clearly demonstrates the effect of increasing N on the total energy necessary; while the term N^{K-1} in the denominator increases with N, the error-related term $(1 - p_{\text{link}})^N$ decreases with N. By treating N as a continuous variable and differentiating, it follows that the optimal value of the number of hops N_{opt} is given by

$$N_{\text{opt}} \approx \frac{(K - 1)}{-log(1 - p_{\text{link}})}$$

Thus a larger value of p_{link} corresponds to a smaller value for the optimal number of intermediate forwarding nodes. Also, the optimal value for N increases linearly with the attenuation coefficient K. There is thus clearly an optimal value of N; *while lower values of N do not exploit the potential reduction in the transmission energy, higher values of N cause the overhead of retransmissions to dominate the total energy budget.*

To study these tradeoffs graphically, we plot $E_{\text{total rel}}^{\text{EER}}$ against varying N (for different values of p_{link}) in Figure 5.1. For this graph, α and D (which are really arbitrary scaling constants) in the analysis are kept at 1 and 10, respectively, and $K = 2$. The graph shows that for low values of the link error rates, the probability of transmission errors is relatively insignificant; accordingly, the presence of multiple short-range hop nodes leads to a significant reduction in the total energy consumption. However, when the error rates are higher than $\sim 10\,\%$, the optimal value of N is fairly small; in such scenarios, any potential power savings due to the introduction of an intermediate node are negated by a sharp increase in the number of transmissions necessary due to a larger effective path error rate. *In contrast to earlier work, our analysis shows that a path with multiple shorter hops is thus not always more beneficial than one with a smaller number of long-distance hops.*

5.3.1.1   *Energy Costs for TCP Flows*

Our formulation [Eq. (5.8)] provides the total energy consumed per packet using an ideal retransmission mechanism. TCP's flow control and error recovery algorithms could potentially lead to different values for the energy consumption, since TCP behavior during loss-related transients can lead to unnecessary retransmissions. While the effective TCP throughput (or goodput) as a function of the end-to-end loss probability has been derived in several analyses [5,10], we are interested in

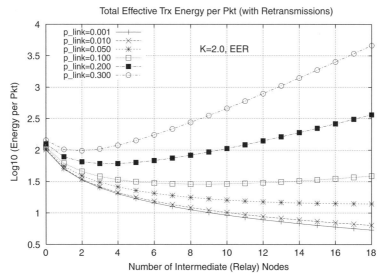

Figure 5.1   Total energy costs versus number of forwarding nodes (EER).

the total number of packet transmissions (including retransmissions) for a TCP flow subject to a variable packet loss rate. We thus use simulation studies using the $ns - 2$ simulator[3] to measure the energy requirements for reliable TCP transmissions. Figure 5.2 plots the energy consumed by a persistent TCP flow, as well as

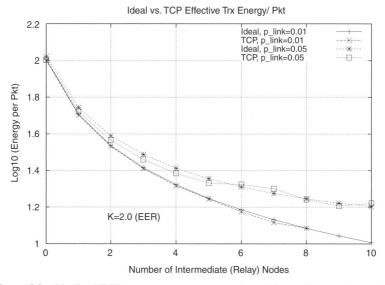

Figure 5.2   Idealized/TCP energy costs versus number of forwarding nodes (EER).

[3]Available at http://www.isi.edu/nsnam/ns.

the ideal values computed using Equation (5.8), for varying N and for $p_{\text{link}} = \{0.01, 0.05\}$. We observe good agreement between our analytical predictions and TCP-driven simulation results. This verifies the practical utility of our analytical model.

5.3.2 Optimal Minimum-Energy Paths in HHR Case

In the case of the HHR model, a transmission error on a specific link implies the need for retransmissions on that link alone. This is a better model for multihop wireless networking environments, since wireless link layers typically employ link layer retransmissions. In this case, the link layer retransmissions on a specific link ensure that the transmission energy spent on the other links in the path is *independent* of the error rate of that link. For our analysis, we do not bound the maximum number of permitted retransmissions; a transmitter continues to retransmit a packet until the receiving node acknowledges error-free reception. (Clearly, practical systems would typically employ a maximum number of retransmission attempts to bound the forwarding latency.) Since our primary focus is on energy-efficient routing, we also do not explicitly consider the effect of such retransmissions on the overall forwarding latency of the path in this paper.

Since the number of transmissions on each link is now *independent of the other links* and is geometrically distributed, the total energy cost for the HHR case is

$$E_{\text{total rel}}^{\text{HHR}} = \sum_{i=1}^{N} \alpha \frac{D_{i,i+1}^k}{1 - p_{i,i+1}} \tag{5.9}$$

In the case of N intermediate nodes, where each hop is of distance D/N and has a link packet error rate of p_{link}, this reduces to

$$E_{\text{total rel}}^{\text{HHR}} = \alpha \frac{D^K}{N^{K-1} * (1 - p_{\text{link}})} \tag{5.10}$$

Figure 5.3 plots the total energy for the HHR case, for $K = 2$ and different values of N and p_{link}. In this case, it is easy to see that the *total energy required always decreases with increasing N, following* the $1/N^{K-1}$ *asymptote*. The logarithmic scale for the energy cost compresses the differences in the value of $P_{\text{total rel}}^{\text{HHR}}$ for different p_{link}. If all links have the same error rate, it would therefore be beneficial to substitute a single hop with multiple shorter hops.

In Figure 5.3 we can also observe the effect of increasing value of p_{link} for a fixed N. As expected, a higher link error rate leads to larger number of retransmissions and a higher energy consumption. It is important to note that the effect of increasing link error rates is more significant in the EER case — in Figure 5.1, for $N = 10$, increasing the loss probability from 0.1 to 0.2 increases the energy consumption 10-fold.

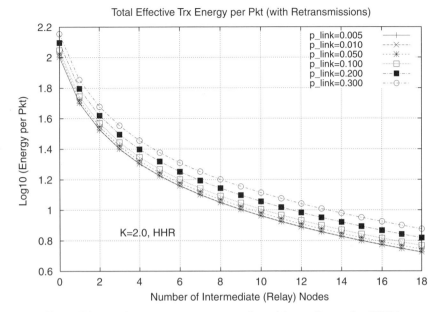

Figure 5.3   Total energy costs versus number of forwarding nodes (HHR).

A comparison of the energy consumption in the EER and HHR cases for identical values of N and K shows that the energy consumption in the EER case is at least an order of magnitude larger, for even moderate values of the link error rate. By avoiding the end-to-end retransmissions, the HHR approach can significantly lower the total energy consumption. These analyses reinforce the requirements of link layer retransmissions in any radio technology used in multihop, ad hoc wireless networks.

5.4   ASSIGNING LINK COSTS FOR MINIMUM-ENERGY RELIABLE PATHS

In contrast to traditional Internet routing protocols, energy-aware routing protocols typically compute the shortest-cost path, where the cost associated with each link is some function of the transmission (and/or reception) energy associated with the corresponding nodes. To adapt such minimum-cost route determination algorithms (such as the Dijkstra or the Bellman–Ford algorithm) for energy-efficient reliable routing, the link cost must now be a function of not just the associated transmission energy but the link error rates as well. Using such a metric would allow the routing algorithm to select links that present the optimal tradeoff between low transmission power and low link error rates. As we shall shortly see, defining such a link cost is possible only in the HHR case; approximations are needed to define suitable cost metrics in the EER scenario.

Consider a graph with the set of vertices representing the communication nodes and a link $\{i,j\}$ representing the direct hop between nodes i and j. For generality, assume an asymmetric case where $\{i,j\}$ is not the same as $\{j,i\}$; moreover, $\{i,j\}$ refers to the link used by node i to transmit to node j. A link is assumed to exist between node pair $\{i,j\}$ as long as node j lies within the transmission range of node i. This transmission range is uniquely defined for the constant-power case; for the variable-power case, this range is really the *maximum permissible range* corresponding to the maximum transmission power of a sender. Let $E_{i,j}$ be the energy associated with the transmission of a packet over link $l_{i,j}$, and $p_{i,j}$ be the link packet error probability associated with that link. (In the fixed-power scenario, $E_{i,j}$ is independent of the link characteristics; in the variable-power scenario, $E_{i,j}$ is a function of the distance between nodes i and j.) Now, the routing algorithm's job is to compute the shortest path from a source to the destination that minimizes the sum of the energy costs over each constituent link.

5.4.1 Hop-by-Hop Retransmission (HHR)

Consider a path P from a source node S (indexed as node 1) to node D (indexed as node $N+1$) that consists of $N-1$ intermediate nodes indexed as $2, \ldots, N$. Then, choosing path P for communication between S and D implies that the total energy cost is given by

$$E_P = \sum_{i=1}^{N} \frac{E_{i,i+1}}{1 - p_{i,i+1}} \tag{5.11}$$

Choosing a minimum-cost path from node 1 to node $N+1$ is thus equivalent to choosing the path P that minimizes Equation (5.11). It is thus easy to see that the corresponding link cost for link $L_{i,j}$, denoted by $C_{i,j}$, is given by

$$C_{i,j} = \frac{E_{i,j}}{1 - p_{i,j}} \tag{5.12}$$

5.4.2 End-to-End Retransmission (EER)

In the absence of hop-by-hop retransmissions, the total energy cost along a path contains a multiplicative term involving the packet error probabilities of the individual constituent links. In fact, assuming that transmission errors on a link do not stop downstream nodes from relaying the packet, the total transmission energy can be expressed as follows:

$$E_P = \frac{\sum_{i=1}^{N} E_{i,i+1}}{\prod_{i=1}^{N}(1 - p_{i,i+1})} \tag{5.13}$$

Given this form, the total cost of the path cannot be expressed as a linear sum of individual link costs,[4] thereby making the exact formulation inappropriate for traditional minimum-cost path computation algorithms. We therefore concentrate on alternative formulations of the link cost, which allow us to use conventional distributed shortest-cost algorithms to compute "approximate" minimum-energy routes.

A study of Equation (5.13) shows that using a link with a high p can be very detrimental in the EER case; an error-prone link effectively drives up the energy cost for all the nodes in the path. Therefore, a useful heuristic function for link cost should have a superlinear increase with increase in link error rate; by making the link cost for error-prone links prohibitively high, we can ensure that such links are usually excluded during shortest-cost path computations.

In particular, for a path consisting of k identical links (i.e., have the same link error rate and link transmission cost), Equation (5.13) will reduce to

$$E_P = \frac{kE}{(1-p)^k} \tag{5.14}$$

where p is the link error rate and E is the transmission cost across each of these links. This leads us to propose a heuristic cost function for a link, as follows

$$C_{i,j}^{\text{approx}} = \frac{E_{i,j}}{(1-p_{i,j})^L} \tag{5.15}$$

where $L = 2, 3, \ldots$, and is chosen to be identical for all links.[5] Clearly, if the exact pathlength is known and all nodes on the path have identical link error rates and transmission costs, L should be chosen equal to that pathlength. However, in accordance with current routing schemes, we require that a link should associate only a single link cost with itself, irrespective of the lengths of specific routing paths that pass through it. Therefore, we need to fix the value of L, independent of the different paths that cross a given link. If better knowledge of the network paths is available, *then L should be chosen to be the average pathlength of this network*. Higher values of L impose progressively stiffer penalties on links with nonzero error probabilities. Given this formulation of the link cost, the minimum-cost path computation effectively computes the path with the minimum "approximate" energy cost given by

$$E_P \sim \sum_{i=1}^{N} \frac{E_{i,i+1}}{(1-p_{i,i+1})^L} \tag{5.16}$$

As with our theoretical studies in Section 5.3, the analysis here does not directly apply to TCP-based reliable transport, since TCP's loss recovery mechanism can

---

[4]We do not consider solutions that require each node or link to separately advertise two different metrics. It is possible to define optimal energy efficient paths if nodes distributed two separate metrics: (1) $E_{i,j}$ and (2) $\log(1-p_{i,j})$. The cumulative values $\sum E_{i,j}$ and $\sum \log(1-p_{i,j})$ can be used by nodes to compute such optimal paths.

[5]There should be an L factor in the numerator too [as in Equation (5.14), but since this is identical for all links, it can be ignored].

lead to additional transients. In the next section, we shall use simulation-based studies to study the performance of our suggested modifications to the link cost metric in typical ad hoc topologies.

5.5   PERFORMANCE EVALUATION: MINIMUM ENERGY PATHS

The analysis of the previous section provides a foundation for devising energy-conscious protocols for reliable data transfer. In this section, we report on simulation-based studies that examine the performance of our suggested techniques for computing energy-efficient reliable communication paths. We performed our simulations in the ns-2 simulator. We experimented with different types of traffic sources:

1. For studies using the EER framework, we used TCP flows implementing the NewReno version of congestion control.
2. For studies using the HHR framework, we used both UDP and TCP flows. In UDP flows, packets are inserted by the source at regular intervals.

To study the performance of our suggested schemes, we implemented and observed three separate routing algorithms:

1. The *minimum-hop routing algorithm*, where the cost of all links is identical and independent of both the transmission energy and the error rate.
2. The *energy-aware* (EA) *routing algorithm*, where the cost associated with each link is the energy required to transmit a single packet (without retransmission considerations) across that link.
3. Our *retransmission-energy-aware* (RA) algorithm, where the link cost includes the packet error rates, and thus considers the impact of retransmissions necessary for reliable packet transfer. For the HHR scenario, we use the link cost of Equation (5.12); for the EER model, we use the "approximate" link cost of Equation (5.15) with $L = 2$. In Section 5.5.3.2, we also study the effect of varying the L parameter.

For our experiments, we used different topologies having up to 100 nodes randomly distributed on a square region, to study the effects of various schemes on energy requirements and throughputs achieved. In this section, we discuss in detail results from one representative topology, where 49 nodes were distributed over a 70×70-unit grid, equispaced 10 units apart (Fig. 5.4). The maximum transmission radius of a node is 45 units, which implies that nodes have between 14 and 48 neighbors in this topology.

All the routing techniques were then run on these static topologies to derive the least-cost paths to each destination node. To simulate the offered traffic load typical of such ad hoc wireless topologies, each corner node on the grid has 3 active flows, providing a total of 12 flows. Since our objective was to study the transmission

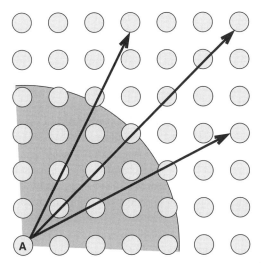

Figure 5.4   The 49-node topology. The shaded region marks the maximum transmission range for the corner node, *A* There are three flows from each of the 4 corner nodes, for a total of 12 flows.

energies alone, we did not consider other factors such as link congestion and buffer overflow. Thus, each link had an infinitely larger transmit buffer; the link bandwidths for all links (point to point) was set to 11 Mbps. Each simulation was run for a fixed duration.

5.5.1 Modeling Link Errors

The relation between the bit error rate (p_b) over a wireless channel and the received power level P_r is a function of the modulation scheme. However, in general, several modulation schemes exhibit the following generic relationship between p_b and P_r

$$p_b \propto \text{erfc}\left(\sqrt{\frac{\text{constant} \times P_r}{N \times f}} \right)$$

where N is the noise spectral density (noise power per hertz), f is the raw channel bit rate, and erfc(x) is defined as the complementary function of erf(x) and is given by

$$\text{erfc}(x) = 1 - \frac{2}{\sqrt{\pi}} \int_0^x e^{-t^2} dt$$

As specific examples, the bit error rate is given by $p_b = \text{erfc}(\sqrt{P_r/2N\,f})$ for coherent OOK (ON–OFF keying), by

$$p_b = (M - 1) \times \text{erfc} \sqrt{\frac{P_r \times \log_2(M)}{2\,N\,f}}$$

for M-ary FSK (frequency shift keying), and by

$$p_b = 0.5 \times \text{erfc}\sqrt{\frac{P_r}{Nf}} \qquad (5.17)$$

for binary phase shift keying (BPSK).

Since we are not interested in the details of a specific modulation scheme but merely want to study the general dependence of the error rate on the received power, we make the following assumptions:

1. The packet error rate p equals $S.p_b$, where p_b is the bit error rate and S is the packet size. This is an accurate approximation for small error rates p_b; thus, we assume that the packet error rate increases or decreases in direct proportion to p_b.

2. The received signal power is inversely proportional to D^K, where D is the link distance and K is the attenuation constant. Thus P_r can be replaced by P_t/D^K, where P_t is the transmitter power. We choose BPSK as our representative candidate and hence use Equation (5.17) to derive the bit error rate.

We report results for the variable power scenario, where all the nodes in the network dynamically able adjust their transmission power across the links. (Detailed results, including those for the fixed-power scenario, are available [1].) Each node chooses the transmission power level for a link so that the signal reaches the destination node with the *same constant* received power. Since we assume that the attenuation of signal strength is given by Equation (5.1), the energy requirements for transmitting across links of different lengths are as given by Equation (5.3).

Since all nodes now receive signals with the same power, the bit error rate, given by Equation (5.17), is now dependent only on the distance-independent receiver noise component. Accordingly, if we assume that the noise levels at different receivers are independent of one another, it follows that the bit error rates of different links are essentially random and do not depend on the link distance. We simply need to model the random ambient noise at each receiver. We chose the maximum error rate for a link due to ambient noise (p_{ambient}) for the different experiments in this case. We then chose the actual error rate for any particular link uniformly at random from the interval $(0, p_{\text{ambient}})$.

5.5.2 Metrics

To study the energy efficiency of the routing protocols, we observed two different metrics:

1. *Normalized Energy.* We first compute the average energy per data packet by dividing the total energy expenditure (over all the nodes in the network) by

the total number of unique packets received at any destination (sequence number for TCP and packets for UDP). We define the normalized energy of a scheme as the *ratio of the average energy per data packet for that scheme to the average energy per data packet required by the minimum-hop routing scheme*. Since the minimum-hop routing scheme clearly consumes the maximal energy, the normalized energy parameter provides an easy representation of the percentage energy savings achieved by the other (EA and RA) routing algorithms.

2. *Effective Reliable Throughput.* This metric counts the number of packets reliably transmitted from the source to the destination, over the simulated duration. Since all the plots show results of runs of different schemes over the same time duration, we do not actually divide this packet count by the simulation duration. Different routing schemes will differ in the total number of packets that the underlying flows are able to transfer over an identical time interval.

5.5.3 Simulation Results

We first present results for the HHR model, followed by the EER case.

5.5.3.1 HHR Model
In this model, each link implements its own localized retransmission algorithm to ensure reliable delivery to the next node on the path.

HHR with UDP Figure 5.5 shows the the total energy consumption for the routing schemes under link layer retransmissions (HHR case). We experimented with a

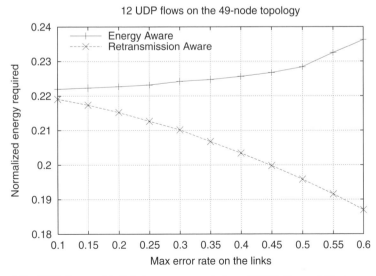

Figure 5.5 UDP flows with link layer retransmissions (HHR) for variable-transmission-power scenario.

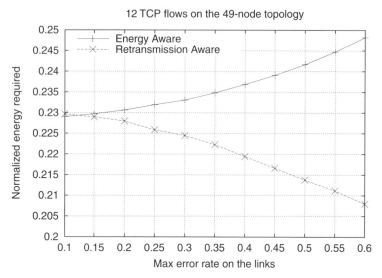

Figure 5.6 Energy required for TCP flows with link layer retransmissions (HHR) for variable-transmission-power scenario.

range of channel error rates to obtain these results. Both EA and RA schemes are a significant improvement over the minimum-hop routing scheme, as expected. However, with increasing channel error rates, the difference between the normalized energy required per reliable packet transmission for the RA and the EA schemes diverges. At some of the high channel error rates ($p_{ambient} = 0.5$), the energy requirements of the RA scheme are about 25 % lower than those of the EA scheme. Note that this error rate is only the maximum error rate for the link. The link error rates of individual links are typically much smaller. Also, it is only the normalized energy for the RA scheme that decreases. The absolute energy required obviously increases with an increasing value of p_{max}.

HHR with TCP In Figure 5.6, we observe the same metric for TCP flows. As before, the energy requirements of the RA scheme are much lower than those of the EA scheme. Additionally, we can observe (Fig. 5.7) that the number of data packets transmitted reliably for the RA scheme is much higher than that of the EA scheme, even though the RA scheme uses much lower energy per sequence number transmitted. This is so because the RA scheme chooses a path with lower error rates; thus the number of link layer retransmissions seen for TCP flows using the RA scheme is lower, and hence the roundtrip time delays are lower. The throughput T of a TCP flow, with roundtrip delay τ and loss rate p varies as follows [12]:

$$T(\tau, p) \sim \frac{1}{\tau} \times \frac{1}{\sqrt{p}} \qquad (5.18)$$

The RA scheme has smaller values of both p and τ and so has a higher throughput.

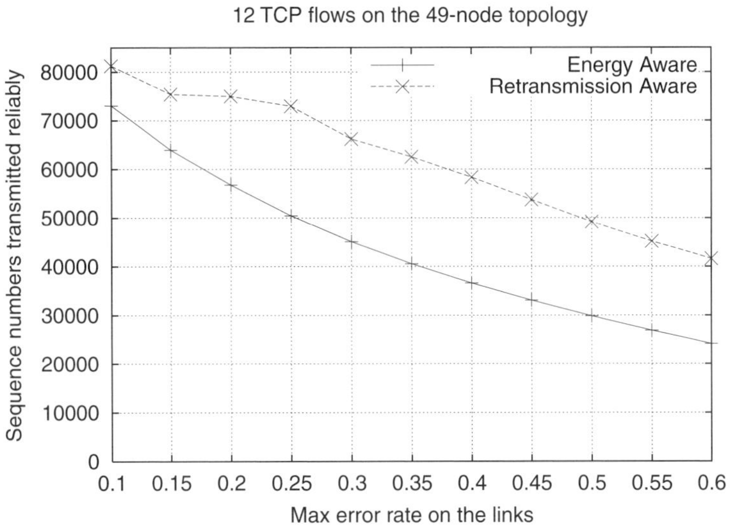

12 TCP flows on the 49-node topology

Figure 5.7   Reliable packet transmission for TCP flows with link layer retransmissions (HHR) for variable-transmission-power scenario.

5.5.3.2 EER Model
We now provide the results of our experiments under the EER scheme.

EER with TCP   For the EER case, it was often difficult to simulate links with high error rates — even with a small number of hops, each TCP packet is lost with a high probability and no data ever reaches their destinations. The energy savings achieved by the RA algorithm is more pronounced when no link layer retransmission mechanisms are present. For some of the higher link error rates simulated in this environment (e.g., $p_{max} = 0.22$), the energy savings of the RA scheme was nearly 65% of the EA scheme, as can be seen in Figure 5.8. Again, it is interesting to observe the data packets transmitted reliably by the EA and the RA schemes, simulated over the same duration (Fig. 5.9). The RA scheme transmits nearly an order of magnitude more TCP sequence numbers than does the EA scheme, even for relatively small maximum error rates (p_{max} between 0.1 and 0.14). While the total TCP goodput approaches zero for both schemes, as the link error rates increase, the rate of decrease in the TCP goodput is much higher for the EA scheme than the RA scheme.

Varying L   In Figure 5.10, we varied the L-parameter of Equation (5.15) for a specific error rate on the links (i.e., $p_{max} = 0.175$). The number of reliably transmitted packets increased monotonically with the value of L. However, the curve in the figure has a minimum "energy per reliably transmitted packet," corresponding to $L = 5$, in this example.[6] Varying the L value from this optimal value leads to poorer

---

[6]Finer measurements with many more L values would yield the exact L that minimizes this curve.

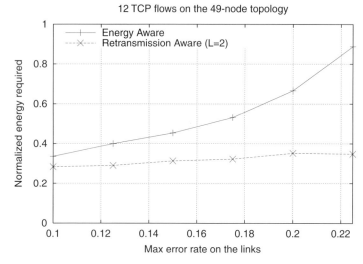

Figure 5.8   TCP flows with no link layer retransmissions (EER) for variable-transmission-power scenario.

energy efficiency (higher energy/packet). There is thus clearly a tradeoff between the achieved throughput, and the effective energy expended. To achieve a higher throughput, it is necessary to prefer fewer hops, as well as links with low error rates (higher error rate links will cause higher delays, due to retransmissions). This plot illustrates the following important point: *It is possible to tune the L parameter to choose an appropriate operating point that captures the tradeoff between (1) the*

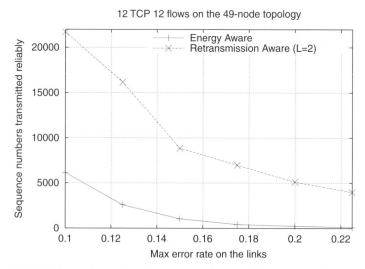

Figure 5.9   TCP flows with no link layer retransmissions (EER) for variable-transmission-power scenario.

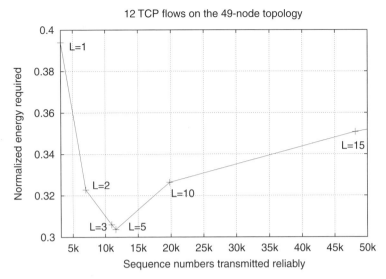

Figure 5.10   Varying the *L* parameter to trade off normalized energy and number of reliably transmitted sequence numbers.

achieved TCP throughput, and (2) the effective energy expended per sequence number received reliably. Of course, the right choice of *L* is expected to be topology-dependent. We leave the problem of developing an adaptive algorithm for optimizing *L* in a specific network as an open problem for future research.

5.6 ADAPTATIONS FOR ON-DEMAND ROUTING PROTOCOLS

We now describe how the proposed technique for calculating minimum energy reliable paths can be applied to on-demand (reactive) routing protocols. On-demand routing protocols, as the name suggests, calculate paths *on-demand.* In these protocols, link costs are not periodically distributed to all other nodes in the network; rather, routes are computed only when needed by particular sessions. Hence, it is comparatively more difficult to directly employ metric-based shortest-path computation algorithms to obtain minimum-energy routes. The problem becomes significantly harder for mobile networks since the link error rates (channel conditions) also change with node mobility. In the work presented here, we have experimented with the ad hoc on-demand distance vector routing protocol (AODV) [17]. This chapter describes our experience in developing a minimum-energy end-to-end reliable path computation mechanism for AODV. It should, however, become obvious from our description that our technique can be generalized to alternative on-demand routing protocols (e.g., DSR [9] and TORA [16]). Through our experimentation, we perform a detailed study of the AODV protocol and our energy-efficient variants, under various noise and node mobility conditions. As part of this study, we have identified some specific configurations where an on-demand protocol that does not consider noise characteristics can result in significantly lower throughput, even under conditions of low or moderate channel noise.

5.6.1 Estimating Link Error Rate

In order to implement our proposed mechanism, it is sufficient for each node to estimate only the error rate on its incoming wireless links from its neighboring nodes. In this section we discuss two possible mechanisms that allow each node to estimate the bit error rate p_b of its incoming links.

5.6.1.1 Estimation Using Radio Signal-to-Noise Ratio

As shown in Section 5.5.1, The bit error rate p_b of a wireless channel depends on the received power level P_r, of the signal. Most wireless interface cards typically measure the signal-to-noise ratio (SNR) for each received packet. SNR is a measure of the received signal strength relative to the background noise and is often expressed in decibels as

$$\text{SNR} = 10 \, \log \frac{P_r}{N} \tag{5.19}$$

From the SNR value measured by the wireless interface card we can calculate the ratio P_r/N [Eq. (5.19)]. Substituting this in the equation for bit error [e.g., Eq. (5.17) for BPSK modulation], we can estimate the bit error rate experienced by each received packet.

This SNR-based error rate estimation technique is useful primarily in free-space environments where such error models are applicable. Consequently this is not applicable for indoor environments, where signal path characteristics depend more on the location and properties of physical obstacles on the signal paths. For such environments we use an alternative technique that is based on empirical observations of link error characteristics, which we describe next.

5.6.1.2 Estimation Using Link Layer Probes

In this empirical technique, we estimate the bit error rate of the incoming links by using link layer probe packets. Each node periodically broadcasts a probe packet within its local neighborhood. Each such packet has a local sequence number that is incremented with each broadcast. Each neighbor of this node receives only a subset of these probes. The remaining ones are lost because of channel errors. We define the time period between successive correctly received probes as an *epoch*. Correct reception of a probe terminates an epoch. Each node stores the sequence number of the last correctly received probe from each of its neighbors. On the reception of the next (*i*th) probe from a node, the receiving node can calculate s_i, the number of probes lost in the last epoch. The total number of probes broadcast in this epoch is $s_i + 1$. Note that the packet error rate (p) for a probe packet of length `packet_size` bits (assuming independent bit errors) is given by

$$p = 1 - (1 - p_b)^{\texttt{packet\_size}} \tag{5.20}$$

The packet error rate for probes over the last epoch can also be calculated as follows:

$$p = \frac{s_i}{s_i + 1} \tag{5.21}$$

Therefore, the receiving node can compute the incoming link BER of the last epoch as follows:

$$p_b = 1 - \exp\frac{\log(1 - \frac{s_i}{s_i+1})}{\texttt{packet\_size}} \tag{5.22}$$

Unicast packets in wireless environments sometimes use channel contention mechanisms, such as the RTS/CTS technique employed for data packets in IEEE 802.11. Such a contention mechanism is not employed for broadcast packets. As a consequence, broadcast packets are more prone to losses due to collision than are unicast packets. Therefore our probe-based bit error rate estimation technique can potentially overestimate the actual bit error rate experienced by the data packets. This overestimation is further magnified in the highly loaded areas of the network.

Although the estimated bit error rate for the incoming links could be higher than the actual one, using the probe-based mechanism is still applicable for the following reasons:

1. The bit error rate is overestimated in parts of the wireless network with high traffic load. Since our route computation technique is biased against high bit error rates, routes will naturally avoid these areas of high traffic load. This will lead to an even distribution of traffic load in the network, increasing network longevity, and decreasing contention.

2. The criteria of selecting optimum route in our algorithm is based on the relative costs of the routes, not the actual costs. As the traffic load on the network gets evenly balanced the different links, the bit error rate will be equally overestimated for all the links. This implies that the relative ordering of the link costs is largely unaffected by overestimation of the bit error rate of the links. Consequently, the proposed scheme is still able to choose the appropriate energy efficient reliable routes using the probe-based techniques.

5.6.1.3 *Estimation for Variable-Power Case*

To continuously monitor and update the bit error rate, we make use of packets that are guaranteed to be periodically exchanged between neighboring nodes. In the AODV protocol, each node periodically broadcasts a "Hello" packet to detect its local neighborhood, Therefore, we leverage this packet for the bit error rate estimation as follows. In the SNR-based approach, we measure the SNR value and infer the bit error rate for each received "Hello" packet. In the probe-based approach, we treat each "Hello" packet as a probe.

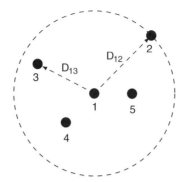

Figure 5.11   Calculating the bit error rate for the variable power case.

For the fixed-transmission-power case, bit error rate estimation can be performed using "Hello" packets exactly as described in Sections 5.6.1 and 5.6.1. However, both these techniques need to be modified for the variable power scenario. As shown in Section 5.5.1, the bit error rate of an incoming link depends on the signal power level of the received packet. For the fixed-transmission-power case, both "Hello" and data packets are transmitted with the same constant power by all nodes. Therefore for a specific pair of transmitting and receiving nodes, the bit error rate estimated for the "Hello" packet [Eq. (5.22)] is equally applicable to the data packets. However, the same is not true for the variable-power-transmission case.

In the variable-power case, the transmission power used for a given data packet is formulated in Equation (5.2) and depends on the distance of the link. However, "Hello" packets are broadcast to all possible neighbors and are transmitted with the fixed maximum transmission power ($P_{t,\max}$). For example, in Figure 5.11, node 1 would transmit a "Hello" with power $P_{t,\max} = P_{Th}.\gamma.D_{12}^k$, where D_{12} represents the maximum transmission range of node 1. It will, however, transmit a data packet to node 3 with the power $P_t = P_{Th}.\gamma.D_{13}^k$. Clearly, $P_{t,\max} > P_t$. Therefore node 3 receives the "Hello" packet at a higher power level than does the data packet. Clearly, packets received at a higher power (e.g., "Hello" packets) will experience lower bit error rate than will the packets received at lower power (e.g., data packets). Therefore a suitable adjustment is required to estimate the bit error rate for data packets in the variable-transmission-power case.

In the variable-power case, a node chooses the transmission power level such that the power level of the received data packet at the receiver is P_{Th}. So for the SNR-based technique, we can estimate the bit error rate for data packets by substituting P_{Th} for P_r in Equation (5.17). The estimated bit error rate for data packets on an incoming link is calculated as $p_b = 0.5 \operatorname{erfc}(\sqrt{P_{Th}/Nf})$, where the noise, N can be calculated using the SNR and P_r values measured by the wireless interface card, using Equation (5.19). We needed to apply a related but different correction scheme to estimate the bit error rate for data packets when using the probe-based technique. We omit the details of this scheme because of space constraints.

For both the SNR-based and probe-based schemes, each node continuously updates its estimate of the bit error rate using an exponentially weighted moving average of the sampled bit error rate values. As in all such averaging techniques, the estimate can be biased toward newer samples depending on the rate at which the noise conditions on the link change. In general, the link error characteristics change with increasing node mobility, and so the estimation can be increasingly biased toward newer samples, with increasing node mobility.

In free-space environments, the SNR-based technique allows faster convergence (i.e., a few "Hello" packets are sufficient) to estimate the actual bit error rate value. The probe-based technique needs a large number of "Hello" packets to accurately estimate the bit error rate. Clearly the computation of energy efficient routes depends on the accuracy of the link error estimation. The SNR-based technique can provide very accurate estimations of bit error rate in free-space environments. It is, however, not applicable in indoor environments, where the probe-based technique provides a reasonable estimate of the actual link error rates. Additionally, node mobility affects the estimation accuracy of both these schemes. We study the effects of node mobility on link error estimates and energy efficient route computation in Section 5.7.

5.6.2 AODV and Its Proposed Modifications

The *ad hoc on demand distance vector* (AODV) routing protocol is an on-demand routing protocol designed for ad hoc mobile networks. AODV not only builds routes only when necessary but also maintains such routes only as long as data packets actively use the route. AODV uses sequence numbers to ensure the freshness of routes.

AODV builds routes using a route request–route reply query cycle. When a source node desires a route to a destination for which it does not already have a route, it broadcasts a route request (RREQ) packet across the network. Nodes receiving this packet update their information for the source node and set up backward pointers to the source node in the route tables. In addition to the source node's IP address, current sequence number, and broadcast ID, the RREQ also contains the most recent sequence number for the destination of which the source node is aware. A node receiving the RREQ may send a route reply (RREP) if it is either the destination or if it has a route to the destination with corresponding sequence number greater than or equal to that contained in the RREQ. If this is the case, it unicasts a RREP back to the source. Otherwise, it rebroadcasts the RREQ. Nodes keep track of the RREQ's source IP address and broadcast ID. If they receive a RREQ that they have already processed, they discard the RREQ and do not forward it.

As the RREP propagates back to the source, nodes set up forwarding pointers to the destination. Once the source node receives the RREP, it may begin to forward data packets to the destination. If the source later receives a RREP containing a greater sequence number or contains the same sequence number with a smaller hop count, it may update its routing information for that destination and begin using the better route.

As long as the route remains active, it will continue to be maintained. A route is considered active as long as there are data packets periodically traveling from the source to the destination along that path. Once the source stops sending data packets, the links will time out and eventually be deleted from the intermediate node routing tables. If a link break occurs while the route is active, the node upstream of the break propagates a route error message (RERR) to the source node to inform it of the now unreachable destination(s). On receiving such an RERR, the source node will reinitiate route discovery, if it is still interested in a route to that destination node. A detailed description of the AODV protocol can be found in a paper by Perkins and Royer [17].

We now describe the set of modifications to the AODV protocol that are required to select energy-efficient paths for reliable data transfer. Our proposed modifications adhere to the on-demand philosophy, where paths are still computed on demand and as long as an existing path is valid, we do not actively change the path. Clearly, other alternate designs are possible where even small changes in link error rates can be used to trigger exploration of better (i.e., more energy-efficient) paths. However, we view such a design as a deviation from the on-demand nature. Therefore, our proposed (energy-efficient) route computations are invoked for the same set of events as basic AODV — either in response to a new route query or to repair the failure of an existing route.

5.6.2.1 AODV Messages and Structures

To perform energy-efficient route computation for reliable data transfer, we need to exchange information about energy costs and loss probabilities for links that lie on the candidate paths. This information exchange is achieved by adding additional fields to existing AODV messages (RREQ and RREP), as decribed below, and does not require the specification of any new message type:

- *RREQ Message.* The information passed and accumulated through the RREQ messages is used by the destination node to judge which candidate path has the minimum cost. The new fields in RREQ are

    C_{req} — stores the average energy cost to transmit a data packet from the source to the current node along the path traversed by the RREQ message.

    E_{req} — used only for the EER case. It stores the summation of energy consumed to transmit data packet over the links traversed by RREQ starting from the source node to the current node. In calculating this field we assume no link error rates, which means that the packet is transmitted only once per link. This field is calculated as $E_{req} = \sum_{\forall l} E_l$, where $\forall l$ denotes the set of links traversed by RREQ.

    Q_{req} — also used only in the EER case. It stores the probability of transmitting a data packet successfully over the links that the RREQ traversed starting from the source node to the current node as $Q_{req} = \prod_{\forall l} (1 - p_l)$, where $\forall l$ denotes the set of links traversed by RREQ and p_l is calculated as in Equation (5.20).

- *RREP Message.* The information passed through the RREP messages is used by each node along the reply path, to compute the cost of the partial route starting from the current node to the destination node. Some of this information is also stored in the *routing table* node to be used later by other RREQ messages looking for the same destination. The new fields in RREP are

C_{rep} — stores the average energy cost to transmit a *data* packet over the links traversed by RREP starting from the current node to the destination node. Like the C_{req} field, its interpretation is different for the HHR and the EER cases.

E_{rep} — Analogous to E_{req} (for all links from the current node to the destination node). It is used only in the EER case.

Q_{rep} — analogous to Q_{rep} (for all links from the current node to the destination node). It is used only in the EER case.

$P_{t,rep}$ — the transmission power level that the recipient of the RREP message should use in forwarding data packets toward the destination.

$Bcast_{rep}$ — the RREQ message ID that uniquely identifies the broadcast RREQ message that led to the generation of this RREP message.

- *Broadcast ID Table.* Each node maintains an entry in the broadcast ID table for each route request query and does not further forward a RREQ already seen by the node. The new fields added to the table are

  H_{bid} — the number of hops traversed by the RREQ starting from the source node to the current node.

  C_{bid} — stores the value of the C_{req} field in the received RREQ message.

  E_{bid} — stores the value of the E_{req} field in the received RREQ message.

  Q_{bid} — stores the value of the Q_{req} field in the RREQ message.

  $Prev_{bid}$ — stores the ID of the node from which the current node received the RREQ message. This entry is updated for each received RREQ message forwarded by the current node. In case the received RREQ is dropped, this field is not updated.

- *Route Table.* A node maintains an entry in the route table for each destination to which it knows a route. The new fields in this table are

  C_{rt} — stores the value of the C_{req} field in the RREQ message or the C_{rep} field in the RREP message received by the current node. If the C_{rep} value is stored in this field, on receiving future RREQ messages for this destination node, it can be used as an estimate of the cumulative downstream cost from this node to the destination node.

  E_{rt} — stores the value of the E_{req} field in the RREQ message or the E_{rep} field in the RREP message received by the current node.

  Q_{rt} — stores the value of the Q_{req} field in the RREQ message or the Q_{rep} field in the RREP message received by the current node.

  $P_{t,rt}$ — stores the value of $P_{t,rep}$ field in the RREP message.

5.6.3 Route Discovery

Route discovery consists of two phases: route request phase and route reply phase. We now describe our modifications to these two phases.

5.6.3.1 Route Request Phase

The source node triggers the route discovery by initializing a RREQ message with $C_{req} = 0$, $E_{req} = 1$, and $Q_{req} = 1$ (the latter two are valid for the EER case). Other fields are initialized as in the original algorithm. RREQ messages are transmitted at the node's maximum power level in order to reach all legitimate one-hop neighbors. When an intermediate node n_i receives RREQ message from a previous node n_{i-1}, it updates the fields in the broadcast ID table entry corresponding to this route request message. If appropriate, it also forwards the RREQ message downstream after updating the message fields.

In order to apply the updates, node n_i calculates the energy (E_l) consumed by node n_{i-1} in a single transmission attempt of a data packet over the link $l = \langle i - 1, i \rangle$. For the fixed-power case, the transmission power P_t is a globally known constant. In the variable-power case, the control messages, such as "Hello" and RREQ messages, are sent with a fixed maximum transmission power $P_{t,max}$, which is globally known. Data packets are sent with transmission power P_t such that the received power of the data packet at node n_i is just above the threshold, equal to P_{Th}. Therefore, node n_i can calculate the transmit power to be used by node n_{i-1} for data packets as

$$p_t = P_{Th} * \frac{P_{t,max}}{P_{r,max}} \tag{5.23}$$

where $P_{r,max}$ is the power level at which the "Hello" and RREQ messages from n_{i-1} are received at n_i. E_l is a fixed multiple of Pt.

Subsequently, node n_i updates fields in the RREQ message as follows:

• HHR case:

$$C_{req} = C_{req} + \frac{E_l}{1 - p_l} \tag{5.24}$$

• EER case:

$$E_{req} = E_{req} + E_l \tag{5.25}$$

$$Q_{req} = Q_{req} * (1 - p_l) \tag{5.26}$$

$$C_{req} = \frac{E_{req}}{1 - Q_{req}} \tag{5.27}$$

The packet error rate (p_l) is calculated by node n_i using Equation (5.20) and the bit error rate estimate that is stored in the *neighbor list*.

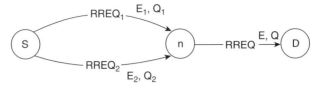

Figure 5.12   Calculating the energy cost for the EER case.

Node n_i examines the broadcast identification number[7] (Bid_{req}) stored in the RREQ message to check whether it has seen any previous RREQ message belongs to the same route request phase. If this is the first instance for this RREQ, node n_i adds a new entry in its *broadcast ID table* and initializes its values as $H_{\text{bid}} = H_{\text{req}}, C_{\text{bid}} = C_{\text{req}}, E_{\text{bid}} = E_{\text{req}}, Q_{\text{bid}} = Q_{\text{req}}$, and $\text{Prev}_{\text{bid}} = n_{i-1}$ where H_{req} is the number of hops traversed by the RREQ messages stored within the RREQ message. Otherwise a previous RREQ message has been seen by the node n_i. In this case it compares the updated cost value in the RREQ message with that stored in the *broadcast ID table* entry. In the HHR case, if the Boolean expression

$$(C_{\text{req}} < C_{\text{bid}}) \quad \text{OR} \quad (C_{\text{req}} = C_{\text{bid}} \quad \text{AND} \quad H_{\text{req}} < H_{\text{bid}}) \tag{5.28}$$

is true, then this RREQ message is forwarded further. Otherwise the currently best known route has lower cost than the new route discovered by this RREQ message, and so is discarded.

For a correct formulation in the EER case, the same comparison rule [expression (5.28)] used in the HHR case does not apply. This is because the cost function is not linear in the EER case. Consider Figure 5.12, in which node n receives two RREQ messages through two different paths from the source. The end-to-end energy costs for the two paths are $(E_1 + E)/(1 - Q_1.Q)$ and $(E_2 + E)/(1 - Q_2.Q)$, respectively. The node n should choose the path defined by $RREQ_1$ if and only if

$$\frac{E_1 + E}{1 - Q_1.Q} < \frac{E_2 + E}{1 - Q_2.Q} \tag{5.29}$$

However, at node n, information on E and Q is not available, and so this inequality cannot be evaluated. Therefore, to optimally compute energy-efficient routes in the EER case, each separate RREQ message needs to be forwarded toward the destination. To do this, we also need a separate entry in the broadcast ID table for each such message. This can potentially lead to an exponential growth in the size of the broadcast table, and hence is not practical. Therefore, in practice we propose the same forwarding mechanism as used in the HHR case and maintain only a single entry in the broadcast ID table for each route request. This implies that the paths chosen in the EER case are not optimal. The quality of the chosen paths can be improved by increasing the state maintained in the broadcast ID table and forwarding correspondingly more eligible RREQ messages toward the destination.

---

[7]The broadcast identification uniquely identifies all the RREQ messages belonging to the same route request phase.

5.6.3.2 *Route Reply Phase*

In AODV, the route reply (RREP) message can be generated by either the destination or by an intermediate node that is aware of *any* path to the destination. In our modified version of AODV, generation of the RREP message is based on the cost of the candidate paths. If the destination node receives a set of RREQ messages from different paths, it chooses the path with the lowest cost among these alternatives and generates a RREP message along this path. Since the destination node receives multiple RREQ messages, it has two choices: (1) immediately reply with a RREP message for each better (i.e., more energy-efficient) route discovered by a new RREQ message, or (2) wait for a small timeout to allow all RREQ messages to discover routes, and then send a single RREP response for the best discovered route. Clearly, the former approach will allow the destination node to select the optimum route at the expenses of transmitting multiple RREP messages, The latter approach results in just a single transmission of RREP message at the expense of higher route setup latency. For the results of this chapter, we choose to implement the first approach of sending multiple RREP messages.

When the destination node receives the first RREQ, it updates its corresponding broadcast ID table entry to reflect the cost of the route traversed by this RREQ message as described in the previous section. For each duplicate RREQ message received by an alternate path from the source, the destination node compares the cost traversed by this new RREQ with the one stored locally in its broadcast ID table using expression (5.28). If the expression is valid, then the cost of the path traversed by the new RREQ is lower. In such a case, the node updates its local information and responds with a RREP message for the path traversed by this new RREQ. Otherwise, it ignores the RREQ message. An intermediate node that receives a RREQ message for a destination can also generate a RREP message if it has a well-known route to the destination[8] and the cost of the partial path traversed by this RREQ is less than the cost stored locally.

The node generating the RREP message copies the RREQ id to the $Bcast_{rep}$ in the RREP message. For the variable-power case, it also calculates the transmission power to be used by the previous hop node to transmit the data packets. This value is computed using Equation (5.23) and is put in the $P_{t,rep}$ field of the RREP message. The different fields of the RREP message are computed as

- HHR case:

$$C_{rep} = \frac{E_l}{1 - p_l} + C_{rt}$$

- EER case:

$$E_{rep} = E_l + E_{rt}$$
$$Q_{rep} = (1 - p_l) * Q_{rt}$$
$$C_{rep} = \frac{E_{rep}}{1 - Q_{rep}}$$

---

[8]By "well known" we mean that the cost of the route from the current node to the destination is known.

where p_l is the packet error rate of the previous hop for data packets. If the node generating the RREP message is the destination itself, then $C_{rt} = 0, E_{rt} = 0$, and $Q_{rt} = 1$. The node forwards the RREP message to node $Prev_{bid}$ stored in the corresponding broadcast ID table entry to this RREQ.

When a node receives a RREP message for the first time, it creates an entry in the route table corresponding to this RREP. It initializes the fields of this entry as $C_{rt} = C_{rep}, E_{rt} = E_{rep}, Q_{rt} = Q_{rep}$, and $P_{t,rt} = P_{t,rep}$. If such an entry already exists, the node compares the cost values as described in the route request phase. If the new path has lower cost, then the route table entries are updated and the entries in the RREP message are appropriately updated and forwarded.

To update the RREP message, the node calculates E_l and p_l values corresponding to the link between this node and the node $Prev_{bid}$ stored in corresponding entry of broadcast ID table. The node updates the $P_{t,rep}$ field using Equation (5.23). It also updates the other fields of the RREP message (i.e., C_{rep}, E_{rep}, and Q_{rep}) in the same way as done for a RREQ message as described by Equations (5.24)–(5.27).

As described above, the node may forward multiple RREP messages in response to better routes found by successive RREQ messages that indicate progressively lower-cost routes.

5.7 PERFORMANCE EVALUATION: EXTENSIONS FOR ON-DEMAND PROTOCOLS

We now report our simulation-based studies on the performance of the AODV protocol, both with and without our energy-aware modifications. We use the same simulation environment as described in Section 5.5.

These simulations model various scenarios of channel noise, interference between nodes due to channel contention, node mobility, and their effects on performance. A full description of these experiments can be found in a study by Nadeem et al. [15]. In this chapter we report a snapshot of these results on the same 49-node topology of Figure 5.4 using UDP flows. Each UDP packet was 1000 bytes long, and the simulations were each run for 250 s.

We primarily compare the performance of the retransmission-aware (RA) variant of the AODV protocol to the energy-aware (EA) variant. For the sake of completeness, we also include the performance of the basic minimum hop or shortest-delay (SD) AODV protocol. For the RA variant we experimented with both techniques for link error estimation, namely, SNR-based, called "RA (SNR)," and probe-based, called "RA (probe)."

For different experiments we varied the noise at different points on the topologies. We partitioned the entire square region into small square grids (50 × 50 units each). Each of these small square regions was assigned a single noise level. Note that the bit error rate of a wireless link depends on the noise level and regions with higher noise have higher bit error rates for the corresponding wireless links. The noise for the different small square grids was chosen to vary between two configurable parameters, N_{min} and N_{max}, corresponding to minimum and maximum noise,

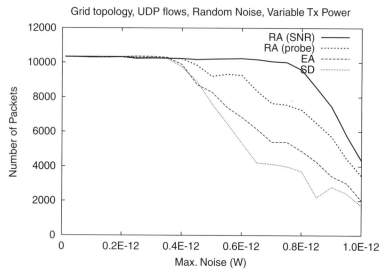

Figure 5.13   Effective reliable throughput (grid topology, variable transmission power in random-noise environment).

respectively. We experimented with different noise distributions over the entire region. In this chapter, we focus only on a random noise environment. In this scenario, we chose N_{min} to be 0.0 W and varied N_{max} to range between 0.0 and 1.0×10^{-11} W in different experiments. The bit error rate at a wireless receiver is given by Equation (5.17).

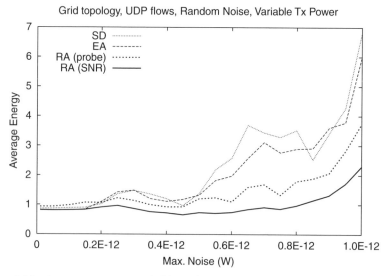

Figure 5.14   Average energy costs (grid topology, variable transmission power in random-noise environment).

In Figures 5.13 and 5.14 we plot the effective reliable throughput and average energy costs for these random-noise environments on a grid topology. The receiver power threshold P_{Th} was chosen to be 1.0×10^{-9} W. The transmission power for a given link was chosen such that the receiving node receives the packet with this power.

Our results show that the other schemes are as good as the RA scheme only in zero-noise environments. For all other cases, the RA scheme shows significant performance improvement, with the performance gain becoming larger with increasing levels of noise.

Between the two link error estimation techniques, the SNR-based scheme performs better, since we model a free-space environment. However, the RA scheme that uses probe-based link error estimation also provides significant performance benefits over EA and SD schemes.

5.8 CONCLUSIONS

In this chapter, we have shown why the *effective total transmission energy*, which includes the energy spent in potential retransmissions, is the proper metric for reliable, energy-efficient communications. The energy efficiency of a candidate route is thus critically dependent on the packet error rate of the underlying links, since they directly affect the energy wasted in retransmissions. Our analysis of the interplay between error rates, number of hops, and transmission power levels reveals several key results:

1. Even if all links have identical error rates, it is not always true that splitting a long-distance (high-power) hop into multiple short-distance (low-power) hops results in overall energy savings.

2. Any routing algorithm must evaluate a candidate link (and the path) on the basis of both its power requirements and its error rate.

3. Link-layer retransmission support (HHR) is almost mandatory for a wireless, ad hoc network, since it can reduce the effective energy consumption by at least an order of magnitude.

4. The advantage of using our proposed retransmission aware routing scheme is significant irrespective of whether fixed or variable transmission power is used by the nodes to transmit across links.

In the model used in this chapter, we have considered the energy consumption only for packet transmission $E_{i,j}$ in formulating our link cost $C_{i,j}$ in Equations (5.12) and (5.15). Different studies (e.g., see the article by Stemm and Katz [22]) have shown that the energy expended for packet reception is sometimes comparable to the energy consumed for packet transmission for some wireless technologies. This packet reception cost can be easily accommodated in our energy cost formulation. For example, in the HHR framework, we can simply modify Equation (5.12) as $C_{i,j} = (E_{i,j} + R_{i,j})/(1 - p_{i,j})$, where $R_{i,j}$ is the energy consumed for packet reception on link $\langle i,j \rangle$.

Our current work on energy-efficient routing assumes that all the nodes in the network are always available to route all packets. In reality, since nodes consume power even in idle mode, significant overall energy savings can be achieved by turning off an appropriate subset of the nodes without losing connectivity or network capacity. There has been much work on topology control algorithms (e.g., SPAN [3] and GAF [24]), based on the notion of connected dominating sets, that reduce energy consumption precisely by periodically putting some nodes to sleep. These protocols, however, have so far focused on reducing the ambient energy cost without sacrificing the available throughput. Clearly, the two approaches of minimum-energy routing and topology control may be combined to further reduce the energy overhead. For example, it may be better to keep more than the minimally necessary set of nodes awake, if this facilitates the selection of paths (for the active flows) with significantly lower reliable communication costs. Combining these two approaches constitutes an area of potential future research.

REFERENCES

1. S. Banerjee and A. Misra, Minimum energy paths for reliable communication in multi-hop wireless networks, *Proc. MobiHoc Conf.*, June 2002.

2. J.-H. Chang and L. Tassiulas, Energy conserving routing in wireless ad-hoc networks, *Proc. InfoCom, Conf.*, March 2000.

3. B. Chen, K. Jamieson, H. Balakrishnan, and R. Morris. Span: An energy-efficient coordination algorithm for topology maintenance in ad hoc wireless networks, *ACM Wireless Networks J.* **8**(5) (Sept. 2002).

4. A. El Gamal, C. Nair, B. Prabhakar, E. Uysal-Biyikoglu, and S. Zahedi, Energy-efficient scheduling of packet transmissions over wireless networks, *Proc. IEEE InfoCom Conf.*, June 2002.

5. S. Floyd, Connections with multiple congested gateways in packet-switched networks Part 1: One-way traffic, *Compu. Commun. Rev.* **21**(5) (Oct. 1991).

6. J. Gass Jr., M. Pursley, H. Russell, and J. Wysocarski, An adaptive-transmission protocol for frequency-hop wireless communication networks, *Wireless Networks* **7**(5):487–495 (Sept. 2001).

7. J. Gomez and A. Campbell, Power-aware routing optimization for wireless ad hoc networks, *Proc. High Speed Networks Workshop (HSN)*, June 2001.

8. J. Gomez, A. Campbell, M. Naghshineh, and C. Bisdikian, Conserving transmission power in wireless ad hoc networks, *Proc. Int. Conf. Networking Protocols*, Nov. 2001.

9. D. Johnson and D. Maltz, Dynamic source routing in ad hoc wireless networks, *Mobile Comput. 153–181* (1996).

10. T. Lakshman, U. Madhow, and B. Suter, Window-based error recovery and flow control with a slow acknowledgment channel: A study of TCP/IP performance, *Proc. InfoCom Conf.*, April 1997.

11. G. Malkin, RIP version 2, RFC 2453, IETF, Nov. 1998.

12. M. Matthis, J. Semke, J. Madhavi, and T. Ott, The macrosocopic behavior of the TCP congestion avoidance algorithm, *Comput. Commun. Rev.* **27**(3) (July 1997).

13. A. Misra and S. Banerjee, MRPC: Maximizing network lifetime for reliable routing in wireless environments, *Proc. IEEE Wireless Communications and Networking Conf. (WCNC)*, March 2002.

14. J. Moy, OSPF version 2, RFC 2328, IETF, April 1998.

15. T. Nadeem, S. Banerjee, A. Misra, and A. Agrawala, Energy-efficient reliable paths for on-demand routing protocols, *Proc. 6th IFIP IEEE Int. Conf. Mobile and Wireless Communication Networks*, Oct. 2004.

16. V. Park and M. Corson, A highly adaptive distributed routing algorithm for mobile wireless networks, *Proc. InfoCom, Conf.*, April 1997.

17. C. Perkins and E. Royer, Ad-hoc on-demand distance vector routing, *Proc. 2nd IEEE Workshop on Mobile Computing Systems and Applications*, Feb. 1999.

18. B. Prabhakar, E. Uysal-Biyikoglu, and A. El Gamal, Energy-efficient transmission over a wireless link via lazy packet scheduling, *Proc. IEEE InfoCom, Conf.*, April 2001.

19. K. Scott and N. Bamboos, Routing and channel assignment for low power transmission in PCS, *Proc. ICUPC*, Oct. 1996.

20. S. Singh and C. Raghavendra, Pamas-power aware multi-access protocol with signaling for ad hoc networks, *ACM Commun. Rev.* (July 1998).

21. S. Singh, M. Woo, and C. Raghavendra, Power-aware routing in mobile ad-hoc networks, *Proc. MobiCom Conf.*, Oct. 1998.

22. M. Stemm and R. Katz, Measuring and reducing energy consumption of network interfaces in hand-held devices, *IEICE Trans. Fund. Electron. Commun. Comput. Sci.* (special issue on mobile computing), **80**(8) (Aug. 1997).

23. C. Toh, H. Cobb, and D. Scott, Performance evaluation of battery-life-aware routing schemes for wireless ad hoc networks, *Proc. ICC*, June 2001.

24. Y. Xu, J. Heidemann, and D. Estrin, Geography-informed energy conservation for ad hoc routing, *Proc. ACM MobiCom Conf.*, 2001.

25. M. Zorzi and R. Rao, Error control and energy consumption in communications for nomadic computing, *IEEE Trans. Comput.* **46**(3) (March 1997).

PART II

RECENT ADVANCES AND RESEARCH IN SENSOR NETWORKS

Sensor networks have attracted a lot of attention lately. These wireless networks consist of highly distributed nodes with energy and resource constraints. Driven by advances in microelectromechanical system (MEMS) microsensors, wireless networking, and embedded processing, ad hoc networks of sensors are becoming increasingly available for commercial and military applications such as environmental monitoring (e.g., traffic, habitat, security), industrial sensing and diagnostics (e.g., factory, appliances), critical infrastructure protection (e.g., power grids, water distribution, waste disposal), and situational awareness for battlefield applications. From the engineering and computing perspective, sensor networks offer a rich source of problems that include sensor tasking and control, tracking and localization, probabilistic reasoning, sensor data fusion, distributed databases, and communication protocols and theory that address network coverage, connectivity, and capacity, as well as system/software architecture and design methodologies. Moreover, in all of these issues there is a a need to consider many interdependent requirements such as efficiency–cost tradeoffs, robustness, self-organization, fault tolerance, scalability, and network longevity.

Chapters 6–9 take a systemic approach to address important and related issues of detection, tracking, and coverage in wireless sensor networks from different perspectives. Chapters 10 and 11 cover the areas of storage management and security in sensor networks.

In Chapter 6, Yu and Ephremides investigate the tradeoff between energy consumption and detection accuracy in sensor networks. The authors use three models to study the tradeoff. In the centralized model, the observed data are sent to the control node with no loss of information and the decision is based on all observations received. In the distributed model, each sensor node makes a local decision and transmits the binary decision to the central node. In the quantized model, observations are quantized into bits of data and sent to the control node. Detection

Mobile, Wireless, and Sensor Networks: Technology, Applications, and Future Directions
Edited by Rajeev Shorey, Akkihebbal L. Ananda, Mun Choon Chan, and Wei Tsang Ooi
Copyright © 2006 John Wiley & Sons, Inc.

performance is measured with respect to false alarm, detection probability, and the overall probability of error.

Numerical results obtained show that for low computation cost and high transmission cost, the distributed model performs the best. Therefore, this model consumes the lowest energy for the same overall probability of error. On the other hand, the centralized model is best for high computation cost and low transmission cost. Finally, the authors investigate the issue of robustness under two attack models: node destruction and observation deletion. Results show that the distributed model is the most robust against both the attacks while the centralized option is the most vulnerable.

Mobile target tracking (MTT) is a classic problem that has more recently been revisited in new sensor networks settings. While traditional MTT is based on powerful sensor nodes, more recent work on MTT differs by (1) using two to three orders of magnitude more sensors, (2) using sensors with limited sensing and processing capabilities, (3) using sensors that tend to be closer to target and can be quickly deployed, and (4) achieving detection redundancy with simultaneous detection by multiple sensors. These new sensor environments dictate a need for increased coordination and algorithms that are lightweight and power-efficient.

In Chapter 7, Gupta et al. present the problem of target tracking in three sections. First, two types of distributed tracking schemes are described. In the first scheme, sensors dynamically optimize the information utility of data for a given cost of computation. In the second scheme, sensors are assumed to be simple and can detect only one bit of information, for example, whether the target is within range. Usually, the detection capability of this scheme is limited to only direction and path of the target and the accuracy depends on the density of sensor.

In Section 7.2, protocols that support collaborative tracking are discussed. The tasks needed for collaboration include group management, state maintenance, and leader election. Use of a tracking tree has also been proposed, and the operations required are construction, expansion and pruning, and reconfiguration.

In Section 7.3, deployment strategies, in particular placement of sensors, are presented. The optimal solution can be approached using the concept of covering coding. Finally, the tradeoff between coverage and energy consumption, and the concept of heterogeneous sensor networks are also discussed.

The *field gathering sensor network* is a network of sensor nodes deployed for spatial and temporal measurements of a given set of parameters. In Chapter 8, Duarte-Melo and Liu investigate performance limits of lifetime and throughput in a sensor network. In the model used, sensors are deployed in a two-dimensional field, with a single collector.

Multiple approaches are cosidered starting with a formulation based on a fluid flow model and using a linear programming technique. The formulation can be modified to model data aggregation and limits on transmission range due to power constraints. However, this initial formulation is limited in that the results obtained are specific to a particular network layout or precise location of each node in the network. A more general model can be constructed by considering the case where the deployment probability distribution of node placement is known. With this assumption, the problem can be solved by discretizing the density function and again using the linear programming formulation. Numerical experiments show that over randomly generated topologies, the approximation is good even if the network is not very dense.

In Chapter 9, Ghosh and Das discuss the importance of coverage and connectivity in sensor networks for ensuring efficient resource management. The main focus is on optimal coverage of a sensor field while maintaining the global network connectivity at the same time.

The authors present the mathematical model for communication, sensing, and coverage in sensor networks and the graph-theoretic background for sensor connectivity issues. Then the coverage algorithms based on exposure path and sensor deployment strategies are introduced. Each method is explained in terms of its applicability, complexity, and other issues that are specific to a sensor network or its underlying application. The survey presented in that chapter provides pointers to many research issues relating to sensor coverage, as well as a concise tabulation of various characteristics of all sensor deployment algorithms.

Storage management is an area of sensor network research that is beginning to attract attention. The need for storage management arises primarily in the class of sensor networks where information collected by the sensors is not relayed to observers in real time. In such storage-bound networks, the sensors are limited not obly in terms of available energy but also in terms of storage.

In Chapter 10, Tilak et al. consider the problem of storage management for sensor networks where data are stored in the network. Two classes of applications are discussed where such storage is needed: (1) *offline scientific monitoring*, where data are collected offline and periodically gathered by an observer for later playback and analysis; and (2) *augmented reality applications*, where data are stored in the network and used to answer dynamically generated queries from multiple observers. The authors have identified the goals, challenges, and design considerations present in storage-bound sensor networks. The three components in storage management that are described in detail are the system support for storage, collaborative storage, and indexing and retrieval.

In Chapter 11, Anjum and Sarkar present a survey that addresses security in sensor networks. The assumptions in sensor networks are that energy and computation are the fundamental constraints, the nodes are densely deployed and are prone to failure, and there is no global identification. The main types of attack considered are tampering with a sensor device, jamming of wireless signals, and link layer attacks on the MAC protocols. One defensive mechanism is the use of data encryption/authentication using very lightweight encryption. In order to use encryption, a suitable key management scheme is needed. Such a scheme should use symmetric-key, does not need prior knowledge of neighbors, and cannot completely trust the neighbors. Other issues discussed in Chapter 11 are intrusion detection, secure routing protocols and secure computation, and aggregation of collected data.

■■■■■ **CHAPTER 6**

Detection, Energy, and Robustness in Wireless Sensor Networks

LIGE YU and ANTHONY EPHREMIDES

Department of Electrical and Computer Engineering, University of Maryland, College Park

6.1 INTRODUCTION

Wireless sensor networks are composed of sensor nodes that must cooperate in performing specific functions. In particular, with the ability of nodes to sense, process data, and communicate, they are well suited to perform event detection, which is clearly a prominent application of wireless sensor networks. Hence the distributed, or decentralized, detection of wireless sensor networks has been studied quite extensively since the late 1980s [1–11].

For a wireless sensor network performing a distributed detection function, most of the previous work has focused on developing the optimal decision rules or investigating the statistical properties for different scenarios. For example, the structure of an optimal sensor configuration was studied for the scenario where the sensor network is constrained by the capacity of the wireless channel over which the sensors are transmitting [1], the performance of a parallel distributed detection system was investigated where the number of sensors is assumed to tend to infinity [3], optimum distributed detection system design has been studied [4] for cases with statistically dependent observations from sensor to sensor, another study [7] focused on a wireless sensor network with a large number of sensors based on a specific signal attenuation model, and investigated the problem of designing an optimum local decision rule, and Shi et al. [10] and Zhang et al. [11] have studied the problem of binary hypothesis testing using binary decisions from independent and identically distributed sensors and developed the optimal fusion rules.

On the other hand, energy efficiency has always been a key issue for sensor networks as sensor nodes must rely on small, nonrenewable batteries. Raghunathan

Mobile, Wireless, and Sensor Networks: Technology, Applications, and Future Directions
Edited by Rajeev Shorey, Akkihebbal L. Ananda, Mun Choon Chan, and Wei Tsang Ooi
Copyright © 2006 John Wiley & Sons, Inc.

145

et al. [12] summarize several energy optimization and management techniques at different levels, in order to enhance the energy awareness of wireless sensor networks. Meanwhile a lot of related work has been done to improve the energy efficiency of sensor networks [13–17], but focusing mostly on clustering mechanisms [13,14], routing algorithms [16], energy dissipation schemes [14,17], sleeping schedules [15], and so on, where energy is usually traded for detection latency [15,16], network density [15,16], or computation complexity [14,17].

However, the energy concern in the detection problem of wireless sensor networks has not been adequately explored. Additionally, in a detection system the wireless sensor networks have to be robust in resisting various kinds of attack. Robustness therefore is another key issue for the wireless sensor networks from the viewpoint of security. In this chapter we investigate the three important issues, detection, energy, and robustness, in the detection scenario of wireless sensor networks. Specifically, we demonstrate a tradeoff between detection accuracy and energy consumption.

In a distributed detection process, sensor nodes are deployed randomly in the field and are responsible for collecting data from the surrounding environment. The observed data are processed locally if needed before they are transmitted to a control center with some routing scheme. A final decision is made at the control center on the basis of all the data sent from the sensor nodes. Various options for data processing are possible and result in different patterns of data transmission. Maniezzo et al. [17] investigated how energy consumption is affected by the tradeoff between local processing and data transmission. Also, it is clear that detection accuracy depends on the aggregated information contained in the data available to the control center. Therefore a connection between detection and energy can be established naturally through balancing local processing and data transmission. Energy efficiency is traded for detection performance in this way.

For a wireless sensor network performing a detection function, the observation data are usually spatially correlated across nodes [4] and temporally correlated at each single node. A routing scheme is necessary for data transmission from sensor nodes to the control center due to the limited power of nodes as well as the unexpected complexity of the hostile terrain [13,14,16]. Noise also needs to be considered as it may interfere with the data transmission, and the Gaussian noise case has also been studied [4,10]. However, as the first step in this direction, we attempt to obtain a beginning and basic result. Therefore we investigate a simplified wireless sensor network model where the abovementioned considerations are disregarded. Thus, we assume that each node independently observes, processes data, and transmits the processed data directly to the control center, in an error-free communication channel. The observations at each node and across nodes are independently and identically distributed (i.i.d.) conditioned on a certain hypothesis. Furthermore we start from the special case of binary hypothesis testing. By ignoring the spatial and temporal correlations, the routing issue, and so on, we simplify the problem to a basic level where the detection scheme would become simple and straightforward, and the detection accuracy as well as energy consumption can be computed by closed-form expressions. However, we should be aware that the simplified model is faraway from the realistic world; thus we plan to develop the model with more complicated considerations and investigate the new scenarios in future work.

On the basis of the simplified wireless sensor network model, we propose three operating options with different schemes for local processing and data transmission, known as the *centralized* option, the *distributed* option, and the *quantized* option. To be specific, the *centralized* option transmits all the information contained in the observed data to the control center, which results in a simple binary hypothesis testing problem. The optimal solution is given by the maximum a posteriori detector [18]. On the other hand, for the *distributed* option each sensor node makes its own decision by a local decision rule. The one-bit decisions are transmitted to the control center, where a final decision is made. The *quantized* option does some local processing at sensor nodes and transmits the resulted data to the control center, which contains partial information of the original observed data. For the distributed option and the quantized option, the global optimal detection schemes can always be obtained by exhaustive search, although it is not practical because of computation complexity. Therefore we adopt the identical local detector because of its asymptotic optimality [1,3]. Thus we develop the desired decision rule for each operating option where tremendous computations are avoided.

Having developed the decision rules, we focus on the detection mission. We compare the detection performance of each option for different values of system parameters. Then we establish an energy consumption model, where energy is assumed to be charged for data processing and data transmission, as introduced by Maniezzo et al. [17]. For our simplified wireless sensor network model, we assume sensor nodes to be homogeneous [13] in that they all adopt the identical detectors and communication systems. Meanwhile as the routing components are disregarded, the data transmission occurs only between sensor nodes and the control center. Therefore the energy consumption would depend only on the number of data processing operations and the number of bits in transmission, given all the other system parameters as fixed. We evaluate the "detection versus energy" performance by varying the values of system parameters for each operating option. Generally, detection accuracy is improved when more energy is consumed. However, the three options have different performances regarding the tradeoff between detection and energy, depending on the system parameters.

Finally we discuss the robustness issue of the wireless sensor networks. Specifically, we consider two forms of attack of node destruction and observation deletion for each operating option. For the observation deletion attack, the number of observations to each sensor node is not necessarily identical as before. Therefore the optimal decision rule of each option is reconsidered and modified. The comparison shows that the distributed option is the most robust option against both types of attack while the centralized option is the weakest one.

The remainder of this chapter is organized as follows. The simplified wireless sensor network model is described in Section 6.2 along with three operating options. In Section 6.3 we analyze and develop the optimal decision rule for each option. Numerical results of the detection performance are shown in Section 6.4, and energy efficiency is investigated in Section 6.5, as well as the tradeoff

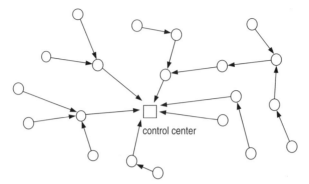

Figure 6.1 Typical wireless sensor network.

between detection accuracy and energy consumption. Section 6.6 discusses the robustness issue. Finally we conclude this chapter in Section 6.7.

6.2 MODEL DESCRIPTION

6.2.1 A Typical Wireless Sensor Network

A typical wireless sensor network consists of a number of sensor nodes and a control center. To perform a detection function, each sensor node collects observation data from the surrounding environment, does some processing locally if needed, and then routes the processed data to the control center. The control center is responsible for making a final decision based on all the data it receives from the sensor nodes. Figure 6.1 shows the structure of a typical wireless sensor network for detection.

6.2.2 Simplified Wireless Sensor Network Model

For a wireless sensor network to perform a detection function, routing usually is needed to transmit data from faraway nodes to the control center; spatial and temporal correlations exist among measurements across or at sensor nodes; and noise interference must be considered as well. However, to focus our attention on the key issues of detection and energy, we start with a simple model where such considerations are disregarded. Our assumptions for the simplified wireless sensor network model include

- No cooperations among sensor nodes — each sensor node independently observes, processes, and transmits data.
- No spatial or temporal correlation among measurements — observations are independent across sensor nodes, and at each single node.
- No routing — each sensor node sends data directly to the control center.
- No noise or any other interference — data are transmitted over an error-free communication channel.

The simplified wireless sensor network model is shown on Figure 6.2.

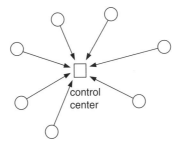

Figure 6.2 Simplified model.

Furthermore we start our investigation from the binary hypothesis testing, which has been widely studied before [1–11]. Let H indicate whether an event occurs $(H = H_1)$ or does not occur $(H = H_0)$, with the prior probabilities $P[H = H_1] = p$ and $P[H = H_0] = 1 - p$ $(0 < p < 1)$. We have K sensor nodes, $\{S_1, S_2, \ldots, S_K\}$, randomly deployed on the field; each node makes T binary observations; thus $Y_i(j)$ is the jth observation at S_i, $Y_i(j) = 0$ or 1, $i = 1, 2, \ldots, K$; $j = 1, 2, \ldots, T$. Observations on each sensor node and across sensor nodes are assumed to be independently and identically distributed (i.i.d.), conditioned on H_0 or H_1. Observations have the identical conditional *pmf* of $P[Y_i(j) = 1|H_0] = p_0$ and $P[Y_i(j) = 1|H_1] = p_1$, with $0 < p_0 < p_1 < 1$. The observation data can be processed locally at each sensor node if needed. The processed data are transmitted to the control center, where a final decision $\hat{H}$ is made. Our objective is to minimize the overall probability of error $(P[\hat{H} \neq H])$ at the control center.

6.2.3 Three Operating Options

For the simplified wireless sensor network model we propose three operating options with different schemes for local processing and data transmission:

1. *Centralized Option.* At each sensor node, the observation data are transmitted to the control center without any loss of information. The control center bases its final decision on the comprehensive collection of information.
2. *Distributed Option.* Each sensor node makes a local decision ($\hat{H}_i$ for S_i) and transmits a binary quantity b_i to the control center indicating its decision:

$$b_i = \begin{cases} 1 & \text{if} \quad \hat{H}_i = \hat{H}_1 \\ 0 & \text{if} \quad \hat{H}_i = \hat{H}_0 \end{cases}$$

   The final decision at the control center is based on the K binary quantities $\{b_1, b_2, \ldots, b_K\}$.
3. *Quantized Option.* Instead of sending all the information or sending a one-bit decision, each sensor node processes the observation data locally and sends a

quantized M-bit quantity (q_i for S_i, $q_i \in \{0, 1, \ldots, 2^M - 1\}$, $1 \leq M \leq T$) to the control center, and the control center makes the final decision based on the basis of the K quantized quantities $\{q_1, q_2, \ldots, q_K\}$.

With the operating options well defined, we are ready to analyze and develop the optimal decision rules for them.

6.3 ANALYSIS

As we have assumed, the observations are conditional i.i.d. random variables. Therefore the order of the observations does not matter to the detection. Since each observation is a binary random variable, we conclude that the number of 1s out of T observations (call it n_i at S_i) at each sensor node is a sufficient statistic for this node to make a decision.

This is verified when we look at the *likelihood ratio* at S_i:

$$LR_i = \frac{P[Y_i(1) \cdots Y_i(T)|H_1]}{P[Y_i(1) \cdots Y_i(T)|H_0]} = \frac{\binom{T}{n_i} p_1^{n_i} (1 - p_1)^{T-n_i}}{\binom{T}{n_i} p_0^{n_i} (1 - p_0)^{T-n_i}} = \left[\frac{p_1(1 - p_0)}{p_0(1 - p_1)}\right]^{n_i} \left(\frac{1 - p_1}{1 - p_0}\right)^{T}$$

$$(6.1)$$

which is determined only by n_i for given T, p_0, and p_1. Hence $\{n_1, n_2, \ldots, n_K\}$ form a sufficient statistic for the control center to make the final decision.

We derive the optimal decision rules for three operating options and develop the closed-form expressions of the corresponding detection performances as follows.

6.3.1 Centralized Option

For the centralized option, the minimum-probability-of-error decision rule is known as the maximum a posteriori detector for the binary hypothesis testing [18]; that is, we choose $\hat{H} = H_1$ if

$$P[H_1|n] \geq P[H_0|n] \tag{6.2}$$

where $n = \sum_{i=1}^{K} n_i$ represents the total number of 1s at the control center. From Bayes rule, that yields

$$\frac{P[n|H_1]}{P[n|H_0]} \geq \frac{P[H_0]}{P[H_1]} = \frac{1 - p}{p} \tag{6.3}$$

Since all the observations are conditional i.i.d. binary random variables, we have

$$P[n|H_1] = \binom{KT}{n} p_1^n (1 - p_1)^{KT-n}$$
$$P[n|H_0] = \binom{KT}{n} p_0^n (1 - p_0)^{KT-n}$$

Then Equation (6.3) becomes

$$\frac{\binom{KT}{n}p_1^n(1-p_1)^{KT-n}}{\binom{KT}{n}p_0^n(1-p_0)^{KT-n}} \geq \frac{1-p}{p}$$

which yields

$$n \geq \frac{\ln\dfrac{1-p}{p} + KT\ln\dfrac{1-p_0}{1-p_1}}{\ln\dfrac{p_1(1-p_0)}{p_0(1-p_1)}} = \gamma_c \quad \text{(threshold)} \tag{6.4}$$

Finally the desired decision rule at the control center is given by

$$\hat{H} = \begin{cases} H_1 & \text{if} \quad n \geq \gamma_c \\ H_0 & \text{if} \quad n < \gamma_c \end{cases} \tag{6.5}$$

This is the optimal decision rule for the centralized option in the sense that it achieves the minimal probability of error.

The corresponding detection performances in terms of false alarm (P_f) and detection probability (P_d) are given by

$$P_f = P[\hat{H} = H_1|H_0] = P[n \geq \gamma_c|H_0] = \sum_{n=\lceil \gamma_c \rceil}^{KT} \binom{KT}{n}p_0^n(1-p_0)^{KT-n} \tag{6.6}$$

$$P_d = P[\hat{H} = H_1|H_1] = P[n \geq \gamma_c|H_1] = \sum_{n=\lceil \gamma_c \rceil}^{KT} \binom{KT}{n}p_1^n(1-p_1)^{KT-n} \tag{6.7}$$

Furthermore, the overall probability of error can be computed in the same way for three options, given by

$$P_e = P[\hat{H} \neq H] = p(1 - P_d) + (1-p)P_f \tag{6.8}$$

6.3.2 Distributed Option

For the distributed option we consider the local decision rule at the sensor nodes and the final decision rule at the control center, respectively.

1. *Local Decision Rule*. As we have specified before, each sensor node applies a local decision rule to make a binary decision based on the T observations. A question yields naturally whether we should have an identical local decision rule for all the sensor nodes. Generally, an identical local decision rule does not result in an optimum system from a global point of view. However, it is still a suboptimal scheme if not the optimal one, which has been observed by some previous work. Irving and Tsitsiklis [9] showed that for the binary

hypothesis detection, no optimality is lost with identical local detectors in a two-sensor system; Chen and Papamarcou [3] showed that identical local detectors are asymptotically optimum when the number of sensors tends to infinity. Although their models are somehow different from ours, we will simply apply the identical local decision rule to our approach because it reduces the computation complexity dramatically. Furthermore, we assume that each sensor node does not have any information about other nodes, which means that the identical local decision rule would depend only on $\{T, p, p_0, p_1\}$, while the number of sensor nodes K is considered as global information and not available for decisionmaking of sensor nodes. Eventually the problem is simplified to a similar case for the centralized option, where the only difference is the number of observations changes from KT to T. Thus, for this binary hypothesis testing, from Equations (6.4) and (6.5), the optimal local decision rule for node S_i is given by

$$\hat{H}_i = \begin{cases} H_1 & \text{if} \quad n_i \geq \gamma_d \\ H_0 & \text{if} \quad n_i < \gamma_d \end{cases} \tag{6.9}$$

where the identical threshold for all the sensor nodes is

$$\gamma_d = \frac{\ln\dfrac{1-p}{p} + T\ln\dfrac{1-p_0}{1-p_1}}{\ln\dfrac{p_1(1-p_0)}{p_0(1-p_1)}} \tag{6.10}$$

2. *Final Decision Rule.* At the control center, a final decision is made on the basis of the K one-bit decisions $\{b_1, b_2, \ldots, b_K\}$, which are also i.i.d. binary random variables conditioned on a certain hypothesis H_0 or H_1. Therefore the number of 1s out of the K binary quantities $b = \sum_{i=1}^{K} b_i$ is a sufficient statistic. The optimal final decision rule at the control center is to choose $\hat{H} = H_1$ if

$$P[H_1|b] \geq P[H_0|b] \tag{6.11}$$

From Bayes rule, that is

$$\frac{P[b|H_1]}{P[b|H_0]} > \frac{P[H_0]}{P[H_1]} = \frac{1-p}{p} \tag{6.12}$$

Furthermore, let P_D and P_F respectively represent the detection probability and false alarm for the local decision at each node, which are given by

$$P_D = P[b_i = 1|H_1] = P[n_i \geq \gamma_d|H_1] = \sum_{n_i = \lceil \gamma_d \rceil}^{T} \binom{T}{n_i} p_1^{n_i} (1-p_1)^{T-n_i} \tag{6.13}$$

$$P_F = P[b_i = 1|H_0] = P[n_i \geq \gamma_d|H_0] = \sum_{n_i = \lceil \gamma_d \rceil}^{T} \binom{T}{n_i} p_0^{n_i} (1-p_0)^{T-n_i} \tag{6.14}$$

Similarly we have

$$P[b|H_1] = P\left[\sum_{i=1}^{K} b_i = b|H_1\right] = \binom{K}{b}P_D^b(1 - P_D)^{K-b}$$

$$P[b|H_0] = P\left[\sum_{i=1}^{K} b_i = b|H_0\right] = \binom{K}{b}P_F^b(1 - P_F)^{K-b}$$

Thus Equation (6.12) yields the final decision rule as

$$\hat{H} = \begin{cases} H_1 & \text{if} \quad b \geq \gamma_D \\ H_0 & \text{if} \quad b < \gamma_D \end{cases} \qquad (6.15)$$

where the threshold is given by

$$\gamma_D = \frac{\ln\dfrac{1-p}{p} + K\ln\dfrac{1 - P_F}{1 - P_D}}{\ln\dfrac{P_D(1 - P_F)}{P_F(1 - P_D)}} \qquad (6.16)$$

The overall false alarm and detection probability are given by

$$P_f = P[b \geq \gamma_D|H_0] = \sum_{b=\lceil \gamma_D \rceil}^{K} \binom{K}{b}P_F^b(1 - P_F)^{K-b} \qquad (6.17)$$

$$P_d = P[b \geq \gamma_D|H_1] = \sum_{b=\lceil \gamma_D \rceil}^{K} \binom{K}{b}P_D^b(1 - P_D)^{K-b} \qquad (6.18)$$

6.3.3 Quantized Option

For the quantized option, we develop the optimal quantization algorithm as well as the suboptimal quantization algorithm for different application scenarios.

1. *Optimal Quantization Algorithm.* Since the number of 1s out of T observations $\{n_1, n_2, \ldots, n_K\}$ form a sufficient statistic, it is sufficient to quantize n_i into an M-bit quantity q_i at S_i and send it to the control center. Therefore, the quantization algorithm is a mapping:

$$n_i \in \{0, 1, \ldots, T\} \longrightarrow q_i \in \{0, 1, \ldots, 2^M - 1\}; i = 1, \ldots, K$$

Obviously we have $1 \leq 2^M - 1 \leq T$, which yields $1 \leq M \leq \log_2(T + 1)$. Then

- Similar to the distributed option, we assume that all sensor nodes apply identical quantization algorithms. This is reasonable because the system parameters are the same for all the nodes; also, the assumption significantly simplifies the problem.

- Suppose that we already have some fixed quantization algorithm at the sensor nodes; there is always an optimal decision rule at the control center to decide $\hat{H}(q_1 \cdots q_K) = H_0$ or H_1. That is the binary hypothesis testing based on the K quantized quantities $\{q_1, \ldots, q_K\}$. So we choose $\hat{H} = H_1$ if

$$P[H_1|q_1 \ldots q_K] \geq P[H_0|q_1 \ldots q_K] \tag{6.19}$$

In much the same way as before, we have the optimal decision rule at the control center given by

$$
\hat{H} =
\begin{cases}
H_1 & \text{if} \quad \dfrac{P[q_1 q_2 \ldots q_K | H_1]}{P[q_1 q_2 \ldots q_K | H_0]} \geq \dfrac{1-p}{p} \\[4mm]
H_0 & \text{if} \quad \dfrac{P[q_1 q_2 \ldots q_K | H_1]}{P[q_1 q_2 \ldots q_K | H_0]} < \dfrac{1-p}{p}
\end{cases}
\tag{6.20}
$$

For a fixed set of quantized quantities $\{q_1, q_2, \ldots, q_K\}$, let $N_i \subseteq \{0, 1, 2, \ldots, T\}$ denote the set that $x \in N_i \Longleftrightarrow x$ is mapped to $q_i; i = 1, 2, \ldots, K$. Thus the probabilities can be computed as

$$P[q_1 \cdots q_K | H_1] = \sum_{n_1 \in N_1} \cdots \sum_{n_K \in N_K} \binom{T}{n_1} \cdots \binom{T}{n_K} p_1^{n_1 + \cdots + n_K} (1 - p_1)^{KT - n_1 - \cdots - n_K} \tag{6.21}$$

$$P[q_1 \cdots q_K | H_0] = \sum_{n_1 \in N_1} \cdots \sum_{n_K \in N_K} \binom{T}{n_1} \cdots \binom{T}{n_K} p_0^{n_1 + \cdots + n_K} (1 - p_0)^{KT - n_1 - \cdots - n_K} \tag{6.22}$$

Furthermore, let N represent the sufficient statistic set $\{n_1 n_2 \cdots n_K\}$; then the overall false alarm and detection probability can be expressed as

$$P_f = \sum_{N : \hat{H}(N) = H_1} \binom{T}{n_1} \cdots \binom{T}{n_K} p_0^{n_1 + \cdots + n_K} (1 - p_0)^{KT - n_1 - \cdots - n_K} \tag{6.23}$$

$$P_d = \sum_{N : \hat{H}(N) = H_1} \binom{T}{n_1} \cdots \binom{T}{n_K} p_1^{n_1 + \cdots + n_K} (1 - p_1)^{KT - n_1 - \cdots - n_K} \tag{6.24}$$

The optimal quantization algorithm can be obtained by exhaustive search. Specifically, we compute and then compare the probability of error with the optimal decision rule applied at the control center for each possible quantization algorithm that is applied at the sensor nodes; the one producing the minimal probability of error is the desired optimal quantization algorithm. However, the exhaustive search is not practical because the computation complexity would be too high for large K and T. Hence we develop the suboptimal quantization algorithm to somehow reduce the computation burden by avoiding the nonscalable computations.

2. *Suboptimal Quantization Algorithm.* The suboptimal quantization algorithm is inspired by the observed properties of the optimal quantization algorithm that was performed on selected examples for small values of K and T. It

would vary for different p, p_0, and p_1, and it does not depend on K, while the optimal quantization algorithm does. For example, as we demonstrate in the following sections, for $p = 0.5, p_0 = 0.2$, and $p_1 = 0.7$, by exhaustive search the optimal quantization algorithm is found to be a set of thresholds $0 = I_o(1) < I_o(2) < \cdots < I_o(2^M) < I_o(2^M + 1) = T + 1$, while our sub-optimal quantization algorithm determines a set of thresholds $I_s(1) < I_s(2) < \cdots < I_s(2^M + 1)$ in the following way:

$$I_s(1) = 0, \; I_s(2^M + 1) = T + 1$$

For $k = 2, 3, \ldots, 2^M$

$$I_s(k) = \begin{cases} k - 1, & \text{if } \; \lceil \gamma_d \rceil \le 2^{M-1} \le \dfrac{T+1}{2} \\[2mm] T - 2^M + k, & \text{if } \; \lceil \gamma_d \rceil \ge T + 1 - 2^{M-1} \ge \dfrac{T+1}{2} \\[2mm] \lceil \gamma_d \rceil - 1 - 2^{M-1} + k, & \text{else} \end{cases}$$

And the quantization is given by

$$n_i \in [I_s(k), I_s(k+1)) \Longrightarrow q_i = k - 1 \quad \text{with} \quad k = 1, \ldots, 2^M; i = 1, \ldots, K$$

6.4 DETECTION PERFORMANCE

In this section we present the numerical results of the detection performance for each operating option. We have examined the numerical results for four different values of the system parameters $\{p, p_0, p_1\}$, namely, $\{0.5, 0.2, 0.7\}$, $\{0.1, 0.2, 0.7\}$, $\{0.1, 0.1, 0.5\}$ and $\{0.1, 0.5, 0.9\}$. Similar results have been obtained. Thus in this section and the following sections, we present numerical results only for $p = 0.5, p_0 = 0.2$, and $p_1 = 0.7$.

6.4.1 Comparison for Three Options

We evaluate the detection performance of the three operating options in terms of P_f, P_d, and P_e. Here we adopt the optimal quantization algorithm for the quantized option. We fix $K = 4, M = 2, p = 0.5, p_0 = 0.2$, and $p_1 = 0.7$ and vary T from 3 to 10. Figures 6.3–6.5 show P_f, P_d, and P_e versus T for three options.

As we see in general, the centralized option has the best detection performance in the sense that it achieves the highest P_d and lowest P_f and P_e, while the distributed option has the worst performance. This is consistent with our expectation since the centralized option has a complete information of the observation data at the control center, while the distributed option has the least information at the control center.

Generally the detection performance is improved with the increase of T except for some variations from the monotonicity. The variations are due to the integer fluctuations of the parameters.

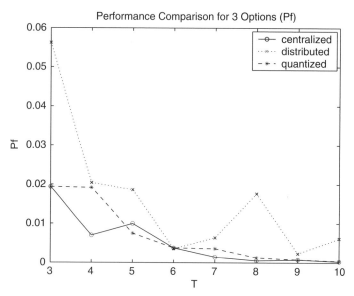

Figure 6.3 Comparison for P_f versus T.

6.4.2 Optimal versus Suboptimal

Figure 6.6 compares the detection performance in terms of P_e between the optimal quantization algorithm and the suboptimal quantization algorithm for the quantized option, where we fix $K = 4, M = 2, p = 0.5, p_0 = 0.2$, and $p_1 = 0.7$; and vary T from 3 to 10.

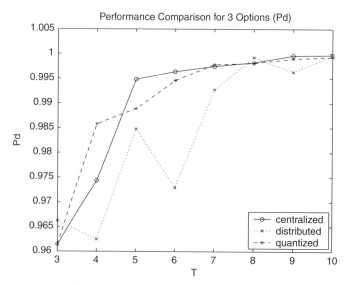

Figure 6.4 Comparison for P_d versus T.

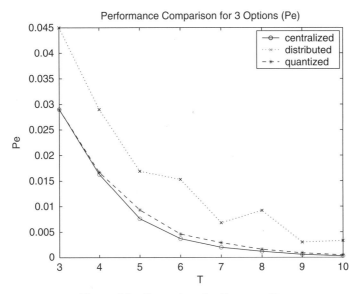

Figure 6.5 Comparison for P_e versus T.

As we can see, the suboptimal quantization algorithm performs almost indistinguishably as well as the optimal one for small values of K and T. Therefore we apply the suboptimal quantization algorithm to compute the probabilities of error and energy consumption for large K and T for the quantized option.

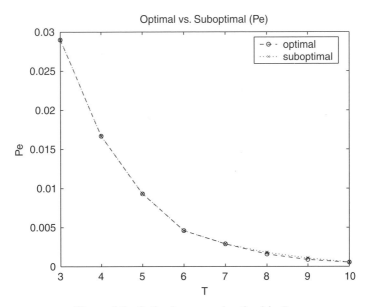

Figure 6.6 Optimal versus suboptimal in P_e.

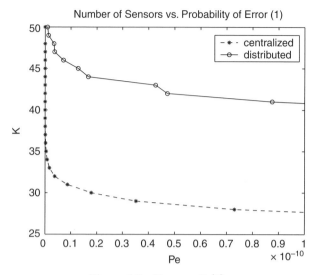

Figure 6.7 K versus $P_e(1)$.

6.4.3   Centralized versus Distributed

Shi et al. [10] observed that to achieve a probability of error of 10^{-5}, the number of binary sensors needed for every SNR is fewer than twice the number of infinite-precision sensors. Note that the binary sensor and infinite-precision sensor correspond to the distributed option and the centralized option in our model, respectively; we compare the detection performance of the two options to verify the result for our model.

We fix $T = 5, p = 0.5, p_0 = 0.2$, and $p_1 = 0.7$ and vary K from 15 to 50. Figures 6.7 and 6.8 show the results as the number of sensors (K) versus probability of error (P_e) for the two options.

The results show that the distributed option needs 10–15 more sensors, which represents roughly 40–50% of the sensors for the centralized option, to achieve the same probability of error. This is consistent with the results shown on Shi's paper [10], namely, that the number of sensors needed for the distributed option is always fewer than twice the number of sensors for the centralized option.

6.5   ENERGY-EFFICIENCY ANALYSIS

6.5.1   Energy Consumption Model

In our analysis we consider only the energy consumed at sensor nodes, and we do not take into account the energy consumption for the control center, which is assumed to have fewer stringent energy constraints. At each sensor node energy is consumed for data processing and data transmission.

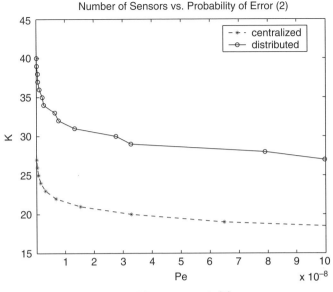

Figure 6.8 K versus $P_e(2)$.

- *Energy for Data Processing*. Energy consumed for data processing depends on the quantity of processed data and the complexity of the processing operations. Here, by assuming that T, p, p_0, and p_1 are known prior to the availability of observation data, all the thresholds needed can be computed in advance. Therefore we simply adopt "comparison" and "counting" as the basic operations for data processing, and we assume that the energy consumption for one "comparison" is the same as that for one "counting." Also we adopt the suboptimal quantization algorithm for the quantized option, so that the thresholds at sensor nodes are determined before the detection. Hence we have a simple model to represent the energy consumed for data processing at sensor nodes as

$$E_P = E_c * c \qquad (6.25)$$

where E_c represents the energy consumed for one comparison or one counting, and c is the total number of comparisons and counts involved.

- *Energy for Data Transmission*. For the transmission of data from the sensor nodes to the control center, we assume that all the sensor nodes adopt the same communication system and there exists an error-free communication channel over which sensor nodes send data to the control center. Therefore the energy consumed for successfully transmitting one bit of data over a fixed distance is a fixed value for each sensor node. Thus, for our simplified wireless sensor network model, the energy consumption for data transmission is determined by the distances from sensor nodes to the control center and the number of bits transmitted, given other parameters as fixed:

$$E_T = E_t * d^\alpha * t \tag{6.26}$$

Here E_t represents the energy consumed for transmitting one bit of data over a unit distance for some fixed communication system, d represents the distance from the sensor nodes to the control center (here we assume that all the sensor nodes have the same distance to the control center), t is the total number of bits transmitted, and α is the path loss exponent. We also assume $\alpha = 2$.

From Equations (6.25) and (6.26), the total energy consumption is given by

$$E = E_P + E_T = E_c * c + E_t * d^2 * t \tag{6.27}$$

For each option, we calculate the energy consumption as follows:

- *Centralized Option.* Since the number of 1s out of the T observations $\{n_1, n_2, \ldots, n_K\}$ are a sufficient statistic, we have two suboptions that have the same detection performances.

  Suboption 1 — sensor nodes transmit all observations to the control center, which means that there is no local data processing and T bits of data is transmitted from each sensor node to the control center. The energy consumed per node therefore is

  $$E = E_t * d^2 * T \tag{6.28}$$

  Suboption 2 — sensor nodes transmit the numbers of 1s (i.e., $\{n_1, n_2, \ldots, n_K\}$) to the control center, which means that each node performs counting T times to obtain the number of 1s, then transmits this $\log_2(T + 1)$ bits quantity to the control center, since $0 \le n_i \le T$. The energy consumed per node therefore is

  $$E = E_c * T + E_t * d^2 * \log_2(T + 1) \tag{6.29}$$

- *Distributed Option.* Each sensor node counts all the observations to obtain the number of 1s; then a single comparison with the threshold γ_d is performed to make a local decision, and exactly one bit of data is sent to the control center. The energy consumed per node therefore is

  $$E = E_c * (T + 1) + E_t * d^2 \tag{6.30}$$

- *Quantized Option.* Each sensor node first counts T times to obtain the number of 1s; then the mapping is performed for the suboptimal quantization algorithm. Let x represent the expected number of comparisons needed for the mapping. Obviously x is a function of T, M, p, p_0, and p_1, which is given by

$$x = \sum_{j=0}^{T} x(j)P[n_i = j] = \sum_{j=0}^{T} x(j) \binom{T}{j} [p_0^j(1-p_0)^{T-j}(1-p) + p_1^j(1-p_1)^{T-j}p]$$

$$(6.31)$$

where n_i is the number of 1s at S_i and $x(j)$ is the number of comparisons needed for the mapping when $n_i = j$. Here we suppose that the comparisons start from $I(2^{M-1} + 1)$ and continue to the adjacent threshold one by one. Specifically, j is first compared with $I(2^{M-1} + 1)$; if $j \geq I(2^{M-1} + 1)$, j is next compared with $I(2^{M-1} + 2)$, otherwise j is compared with $I(2^{M-1})$, and so on until $I(k) \leq j < I(k+1)$ is found, then $q_i = k - 1$ is determined. For example, when $T = 20, M = 3, p = 0.5, p_0 = 0.2$, and $p_1 = 0.7$, the set of thresholds for the suboptimal quantization algorithm is calculated to be $\{0, 6, 7, 8, 9, 10, 11, 12, 21\}$; then x can be computed as

$$x(j) = \begin{cases} 4 & \text{if} \quad 0 \leq j \leq 6 \text{ or } 11 \leq j \leq 20 \\ 3 & \text{if} \quad j = 7, 10 \\ 2 & \text{if} \quad j = 8, 9 \end{cases}$$

The total energy consumed per node is given by

$$E = E_c * [T + x(T, M, p, p_0, p_1)] + E_t * d^2 * M \qquad (6.32)$$

6.5.2 Numerical Results

In the following numerical examples, we adopt the suboptimal quantization algorithm for the quantized option to evaluate the energy consumption and detection performance.

- *Energy Consumption Comparison for Fixed E_c, E_t, d as a Function of T.* We fix $E_c = 5$ nJ/bit, $E_t = 0.2$ nJ/(bit * m^2), $d = 10$ m, and $p = 0.5$, $p_0 = 0.2$, $p_1 = 0.7$; we vary T from 5 to 100 and M from 2 to 5. Figure 6.9 shows the energy consumption per node versus T for all schemes of three options; Figure 6.10 focuses on these schemes except for suboption 1 of the centralized option. This shows that the distributed option has the least energy consumption. On the other hand, the centralized option is affected most by the increase of T in the sense that for both suboptions, the energy consumptions increase more rapidly than in all the schemes of other options.
- *Energy Consumption Comparison for Fixed T as a Function of E_c, E_t, d.* First we fix $T = 10, M = 2, p = 0.5, p_0 = 0.2$, and $p_1 = 0.7$; then we vary d from 5 to 50 m. Figure 6.11 shows the energy consumption per node versus d for the three options with different values of E_c and E_t. Then we change T to 50

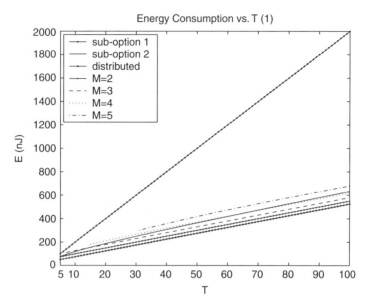

Figure 6.9 E versus T for all schemes.

and M to 4, and all the other parameters remain unchanged. Figure 6.12 shows the new curves. Here we examine the following two examples for different values of E_c and E_t and adopt suboption 2 for the centralized option:

Example 1: $E_c = 5$ nJ/bit, $E_t = 0.2$ nJ/(bit * m^2)

Example 2: $E_c = 20$ nJ/bit, $E_t = 0.05$ nJ/(bit * m^2)

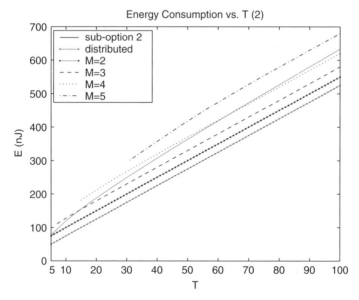

Figure 6.10 E versus T for all schemes except suboption 1.

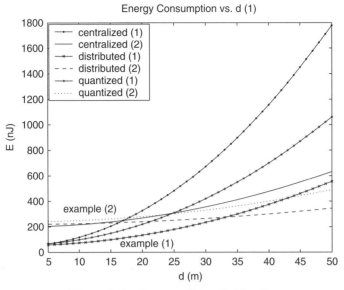

Figure 6.11 E versus $d(T = 10, M = 2)$.

As we expected, for the low E_c–high E_t case (e.g., example 1), the distributed option performs best in the sense that it has the least energy consumption, while the centralized option has the worst performance. On the other hand, for the high E_c–low E_t case (e.g., example 2), the centralized option performs best for small d; however, its energy consumption rises rapidly with the increase in

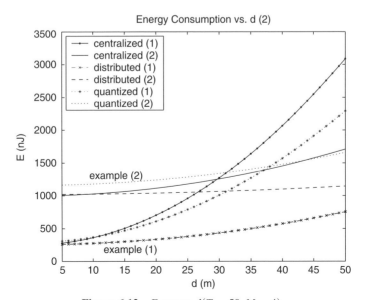

Figure 6.12 E versus $d(T = 50, M = 4)$.

d and eventually exceeds those of the other two options. The distributed option has the best performance for large d in this case.

- *Energy Consumption versus Detection Accuracy.* First we fix $K = 4$ and vary T from 3 to 10 to obtain different pairs of E and P_e for each option. Then we fix $T = 10$ and vary K from 1 to 7 to follow the same procedure. Other parameters are fixed as $M = 2$, $p = 0.5$, $p_0 = 0.2$, $p_1 = 0.7$, and $d = 10$ m. Thus we compare the three options in terms of energy consumption versus detection accuracy, as shown on Figures 6.13–6.16. Here we examine two different combinations of values for E_c and E_t for each case and adopt suboption 2 for the centralized option.

 Figures 6.13 and 6.15 show the results for $E_c = 5$ nJ/bit, $E_t = 0.2$ nJ/ (bit * m^2)

 Figures 6.14 and 6.16 show the results for $E_c = 20$ nJ/bit, $E_t = 0.05$ nJ/ (bit * m^2)

This shows, in general, for the low E_c–high E_t case, that the distributed option consumes less energy than do the other two options to achieve the same detection performance; while for the high E_c–low E_t case, it is the centralized option that needs the least energy. In other words, the distributed option performs best for the low E_c–high E_t case regarding the tradeoff between energy and accuracy, while the centralized option is best for the high E_c–low E_t case when the distance from sensor nodes to the control center is relatively low, for example, $d = 10$ m.

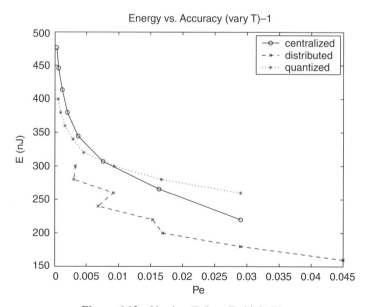

Figure 6.13 Varying T (low E_c–high E_t).

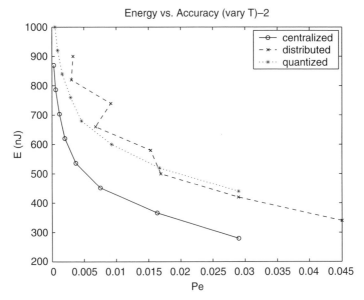

Figure 6.14 Varying T (high E_c–low E_t).

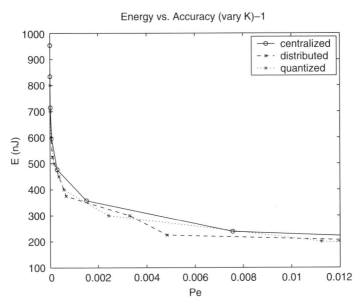

Figure 6.15 Varying K (low E_c–high E_t).

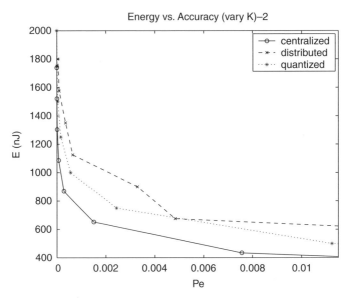

Figure 6.16 Varying K (high E_c–low E_t).

6.6 ROBUSTNESS

In this section we compare the resistance abilities of the three operating options against two forms of attack: node destruction and observation deletion.

6.6.1 Attack 1: Node Destruction

Suppose that the wireless sensor network is under attack in that sensor nodes are partially destroyed. Destroyed sensor nodes are not able to perform the detection function. Thus we investigate the detection performance of the three options for different values and numbers of sensor nodes. Figures 6.17 and 6.18 show P_e versus K for three options. Here other parameters are fixed as $T = 10, M = 2, p = 0.5$, $p_0 = 0.2$, and $p_1 = 0.7$; K is varied from 1 to 8. The suboptimal quantization algorithm is adopted for the quantized option.

Table 6.1 shows the ratio of increase in P_e when K decreases from 8 to 1; thus, for $K = i$, the ratio of increase is computed as

$$\frac{P_e(K = i) - P_e(K = 8)}{P_e(K = 8)}, \qquad i = 1, 2, \ldots, 7$$

Obviously the distributed option has the highest probability of error among three options. However, regarding the robustness to the attack, the loss of performance in terms of ratio is least for the distributed option and highest for the centralized option, which means that the distributed option is the most robust option against the attack and the centralized option is the weakest one.

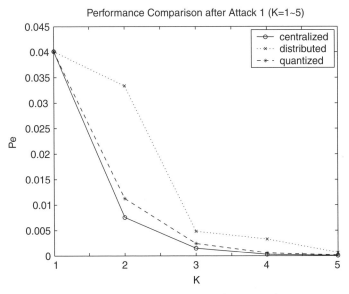

Figure 6.17 Comparison of $P_e(K = 1 \sim 5)$.

6.6.2 Attack 2: Observation Deletion

Suppose that the wireless sensor network is under attack in that observations are partially deleted. Thus the number of observations at each sensor node is not necessarily identical as before. We assume after attack $T = [T(1), T(2), \ldots, T(K)]$,

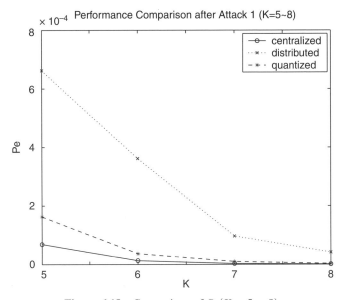

Figure 6.18 Comparison of $P_e(K = 5 \sim 8)$.

TABLE 6.1   Ratio of Increase in P_e after Attack 1

K	7	6	5	4	3	2	1
Centralized	2	12	68	289	1512	7559	40070
Distributed	1.3	7.6	14.8	77.4	114.1	793.7	953.1
Quantized	2.3	11.3	53.3	189	810.3	3744.7	13356

where $T(i)$ represents the number of observations to S_i. The decision rules for three options are slightly modified, as follows:

- *Centralized Option.* Letting $T_s = \sum_{i=1}^{K} T(i)$ denote the total number of observations at the control center, the optimal decision rule is given by

$$\hat{H} = \begin{cases} H_1 & \text{if} \quad n \geq \gamma_s \\ H_0 & \text{if} \quad n < \gamma_s \end{cases} \tag{6.33}$$

where n is the number of 1s out of T_s observations and

$$\gamma_s = \frac{\ln\dfrac{1-p}{p} + T_s \ln\dfrac{1-p_0}{1-p_1}}{\ln\dfrac{p_1(1-p_0)}{p_0(1-p_1)}}$$

is the new threshold.

- *Distributed Option.* The local optimal decision rule at node S_i is given by

$$\hat{H}_i = \begin{cases} H_1 & \text{if} \quad n_i \geq \gamma_d(i) \\ H_0 & \text{if} \quad n_i < \gamma_d(i) \end{cases} \tag{6.34}$$

where the unique threshold for S_i is

$$\gamma_d(i) = \frac{\ln\dfrac{1-p}{p} + T(i) \ln\dfrac{1-p_0}{1-p_1}}{\ln\dfrac{p_1(1-p_0)}{p_0(1-p_1)}}$$

The final optimal decision rule at the control center is given by

$$\hat{H} = \begin{cases} H_1 & \text{if} \quad \dfrac{P[b_1 b_2 \cdots b_K | H_1]}{P[b_1 b_2 \cdots b_K | H_0]} \geq \dfrac{1-p}{p} \\[4mm] H_0 & \text{if} \quad \dfrac{P[b_1 b_2 \cdots b_K | H_1]}{P[b_1 b_2 \cdots b_K | H_0]} < \dfrac{1-p}{p} \end{cases} \tag{6.35}$$

- *Quantized Option.* We assume that sensor nodes simply apply to the quantization algorithm of the T fixed case where $T = T(i)$ for S_i. While for the control center, the optimal decision rule can be expressed in the same form of the identical T case, shown as Equation (6.20), where the computation of $P[q_1 q_2 \cdots q_K | H_\alpha]$ ($\alpha = 0, 1$) is slightly changed to

$$P[q_1 \cdots q_K | H_\alpha] = \sum_{n_1 \in N_1} \cdots \sum_{n_K \in N_K} \binom{T(1)}{n_1} \cdots \binom{T(K)}{n_K} p_\alpha^{n_1 + \cdots + n_K} (1 - p_\alpha)^{T_s - n_1 - \cdots - n_K}$$

(6.36)

For the numerical example, we fix $K = 4, p = 0.5, p_0 = 0.2, p_1 = 0.7$, and assume that the optimal quantization algorithm is adopted for the quantized option. We assume that before the attack $T = [10, 10, 10, 10], M = 2$; and after the attack we have

- Example 1: $T = [5,7,9,10], M = 2$;
- Example 2: $T = [2,3,2,1], M = 1$;
- Example 3: $T = [1,1,1,1], M = 1$.

Figures 6.19 and 6.20 show the detection performance comparison for the three options. Table 6.2 shows the ratio of increase in P_e to the deletion of observations; thus, for example i, the ratio of increase is computed as

$$\frac{P_e(i) - P_e(T = 10)}{P_e(T = 10)}, \qquad i = 1, 2, 3$$

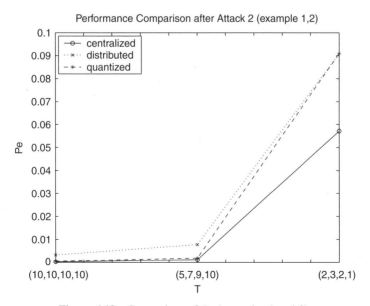

Figure 6.19 Comparison of P_e (examples 1 and 2).

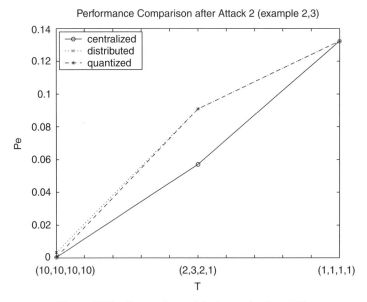

Figure 6.20 Comparison of P_e (examples 2 and 3).

Similar to the case of attack 1, the distributed option is the most robust since it achieves the least loss of performance in terms of ratio among the three options, while the centralized option is the weakest.

6.7 CONCLUSIONS

We have constructed a simplified wireless sensor network model that performs an event detection mission. We have implemented three operating options on the model, developed the optimal decision rules and evaluated the corresponding detection performance of each option. As we expected, the centralized option performs best while the distributed option is the worst regarding the accuracy of the detection. However, it is shown that the distributed option needs fewer than twice the sensor nodes for the centralized option to achieve the same detection performance.

TABLE 6.2 Ratio of Increase in P_e after Attack 2

T	(5,7,9,10)	(2,3,2,1)	(1,1,1,1)
Centralized	3	189	440
Distributed	1.4	26.5	39.1
Quantized	2.8	180.8	263.6

We have modeled the energy consumption at the sensor nodes. The energy efficiency as a function of system parameters has been compared for the three options. More importantly, we have shown a tradeoff between energy efficiency and detection accuracy. This follows from an examination of Figures 6.9–6.16. The distributed option has the best performance for low values of E_c and high values of E_t. For high E_c and low E_t, the centralized option is the best for relatively short distances from sensor nodes to the control center, while the distributed option is the best for long distances.

Furthermore, we have examined the robustness of the wireless sensor network model by implementing two attacks. For both of them, the distributed option shows the least loss of performance in terms of ratio while the centralized option has the highest loss.

The results we have presented in this chapter are based on the simplified wireless sensor network model. A number of subsequent questions arise naturally. Specifically, we need to study a less restrictive model (e.g., nonbinary data, spatial and temporal correlation among measurements), and we need to consider multihop routing to the control center. In that case we need link metrics that capture the detection performance and energy consumption measures.

REFERENCES

1. J.-F. Chamberland and V. V. Veeravalli, Decentralized detection in sensor networks, *IEEE Trans. Signal Process.* **51**(2):407–416 (Feb. 2003).

2. J. N. Tsitsiklis, Decentralized detection by a large number of sensors, *Math. Control Signals Syst.* **1**(2):167–182 (1988).

3. P. Chen and A. Papamarcou, New asymptotic results in parallel distributed detection, *IEEE Trans. Inform. Theory* **39**:1847–1863 (Nov. 1993).

4. Y. Zhu, R. S. Blum, Z.-Q. Luo, and K. M. Wong, Unexpected properties and optimum-distributed sensor detectors for dependent observation cases, *IEEE Trans. Autom. Control* **45**(1) (Jan. 2000).

5. Y. Zhu and X. R. Li, Optimal decision fusion given sensor rules, *Proc. 1999 Int. Conf. Information Fusion*, Sunnyvale, CA, July 1999.

6. I. Y. Hoballah and P. K. Varshney, Distributed Bayesian signal detection, *IEEE Trans. Inform. Theory* **IT-35**(5):995–1000 (Sept. 1989).

7. R. Niu, P. Varshney, M. H. Moore, and D. Klamer, Decision fusion in a wireless sensor network with a large number of sensors, *Proc. 7th Int. Conf. Information Fusion*, Stockholm, Sweden, June 2004.

8. P. Willett and D. Warren, The suboptimality of randomized tests in distributed and quantized detection systems, *IEEE Trans. Inform. Theory* **38**(2) (March 1992).

9. W. W. Irving and J. N. Tsitsiklis, Some properties of optimal thresholds in decentralized detection, *IEEE Trans. Automatic Control* **39**:835–838 (April 1994).

10. W. Shi, T. W. Sun, and R. D. Wesel, Quasiconvexity and optimal binary fusion for distributed detection with identical sensors in generalized Gaussian noise, *IEEE Trans. Inform. Theory* **47**:446–450 (Jan. 2001).

11. Q. Zhang, P. K. Varshney, and R. D. Wesel, Optimal bi-level quantization of i.i.d. sensor observations for binary hypothesis testing, *IEEE Trans. Inform. Theory* (July 2002).

12. V. Raghunathan, C. Schurgers, S. Park, and M. Srivastava, Energy-aware wireless sensor networks, *IEEE Signal Process.* **19**(2):40–50 (March 2002).

13. E. J. Duarte-Melo and M. Liu, Analysis of energy consumption and lifetime of heterogeneous wireless sensor networks, *Proc. IEEE GlobeCom Conf.*, Taipei, Taiwan, Nov. 2002.

14. W. Rabiner Heinzelman, A. Chandrakasan, and H. Balakrishnan, Energy-efficient communication protocol for wireless microsensor networks, *Proc. HICSS '00*, Jan. 2000.

15. C. Schurgers, V. Tsiatsis, S. Ganeriwal, and M. Srivastava, Optimizing sensor networks in the energy-latency-density design space, *IEEE Trans. Mobile Comput.* **1**(1) (Jan.–March 2002).

16. B. Krishnamachari, D. Estrin and S. Wicker, The impact of data aggregation in wireless sensor networks, *Proc. ICDCSW'02*, Vienna, Austria, July 2002.

17. D. Maniezzo, K. Yao, and G. Mazzini, Energetic trade-off between computing and communication resource in multimedia surveillance sensor network, *Proc. IEEE MWCN2002*, Stockholm, Sweden, Sept. 2002.

18. H. V. Poor, *An Introduction to Signal Detection and Estimation*, 2nd ed., Springer-Verlag, 1994.

CHAPTER 7

Mobile Target Tracking Using Sensor Networks

ASHIMA GUPTA, CHAO GUI, and PRASANT MOHAPATRA

Department of Computer Science, University of California at Davis

7.1 INTRODUCTION

Target tracking has been a classical problem since the early years of electrical systems. Sittler, in 1964, gave a formal description of the multiple-target tracking (MTT) problem [17]. In a given field of surveillance interest, there are varying number of targets. They arise in the field at random locations and at random times. The movement of each target follows an arbitrary but continuous path, and it persists for a random amount of time before disappearing in the field. The target locations are sampled at random intervals. The goal of the MTT problem is to find the moving path for each target in the field. Traditional target tracking systems are based on powerful sensor nodes, capable of detecting and locating targets in a large range. Tracking methods using distributed multisensor systems have also been investigated [2,5,14,15].

Target tracking using a sensor network was initially investigated 2002 [1,3,6,11–13,18,21,23,24]. With the advances in the fabrication technologies that integrate the sensing and the wireless communication technologies, tiny sensor motes can be densely deployed in the desired field to form a large-scale wireless sensor network. The numbers of sensor nodes are two to three magnitudes greater than those in traditional multisensor systems. On the other hand, each sensor mote can have only limited sensing and processing abilities. A target tracking system in this model can have several advantages: (1) the sensing unit can be closer to the target, and thus the sensed data will be of a qualitatively better geometric fidelity; (2) the advances in wireless sensor network techniques will guarantee quick deployment

Mobile, Wireless, and Sensor Networks: Technology, Applications, and Future Directions
Edited by Rajeev Shorey, Akkihebbal L. Ananda, Mun Choon Chan, and Wei Tsang Ooi
Copyright © 2006 John Wiley & Sons, Inc.

of such a system — the sensed data can be processed and delivered within the network, so that the final report about the target is accurate and timely; and (3) with a dense deployment of sensor nodes, the information about the target is simultaneously generated by multiple sensors and thus contains redundancy, which can be used to increase the system's robustness and increase the accuracy of tracking.

Challenges and difficulties, however, also exist in a target tracking sensor network:

1. Tracking needs collaborative communication and computation among multiple sensors. The information generated by a single node is usually incomplete or inaccurate.

2. Each sensor node has very limited processing power. Traditional target tracking methods based on complex signal processing algorithms may not be applicable to the nodes.

3. Each node also has tight budget on energy source. Every node cannot be always active in sensing and data forwarding. Thus, all the network protocols for data processing and tracking should consider the impact of power saving mode in each node.

For a target tracking sensor network, the tracking scheme should be composed of two components. The first component is the method that determines the current location of the target. It involves localization as well as the tracing of the path that the moving target takes. The second component involves algorithms and network protocols that enable collaborative information processing among multiple sensor nodes. The goal of this component is to devise techniques for efficient and distributed schemes for collaborations between nodes of a sensor network. Distributed algorithms for detection and tracking of mobile targets are designed within the constraints of various resources (especially power constraints).

This chapter provides a comprehensive study of the approaches for tracking mobile targets using sensor networks, and is organized as the follows. In Section 7.2, we discuss different target localization methods focusing on information-driven dynamic sensing, tracking using binary sensors, and smart sensor tracking. In Section 7.3, we study the supporting protocols for sensor collaboration discussing distributed group management and tracking tree management schemes. Finally, in Section 7.4, we focus on the necessary architectural support for the target tracking task. The issues considered are optimal placement of sensors and power conservation for the nodes. The concluding remarks are presented in Section 7.5.

7.2 TARGET LOCALIZATION METHODS

To track mobile targets, it is first essential to develop algorithms to locate the target and track their paths of mobility. In this section, we discuss the methodologies proposed for tracking mobile targets.

7.2.1 Traditional Tracking Methods

Traditional tracking methods make use of a centralized database or computing facility. As the number of sensors increase in the network, the central facility becomes a bottleneck both as a resource and in terms of the network traffic directed toward it. This approach therefore lacks scalability and is not fault-tolerant. Another distinguishing feature of traditional tracking approach is that usually the sensing task is performed by any one node in the network at a time. These techniques are therefore computationally heavy on that one node.

7.2.1.1 Information-Driven Dynamic Sensor Collaboration for Tracking Applications

An information-driven dynamic sensor collaboration technique has been proposed [24] for tracking applications. This approach is a type of traditional scheme, where the participants for collaboration in a sensor network were determined by dynamically optimizing the information utility of data for a given cost of computation and communication. The detection, classification, and tracking of objects and events require aggregation of data among the sensor nodes. However, not all sensors may have useful information; hence an informed selection of sensors that have the best data for collaboration will save both power and bandwidth cost. Thus flooding can be avoided, and tracking reports can be more accurate. The metrics used to determine the participant nodes (who should sense and whom the information must be passed to) are (1) *detection quality*, which includes detection resolution, sensitivity and dynamic range, misses, false alarms, and response latency; (2) *track quality*, which includes tracking errors, track length, and robustness against sensing gaps; (3) *scalability* in terms of network size, number of events, and number of active queries; (4) *survivability*, meaning fault tolerance; and (5) *resource usage* in terms of power/bandwidth consumption.

The focus is on sensor collaboration during the tracking phase, rather than on detection phase. There is one leader node that is active at any moment, and it selects and routes tracking information to the next leader. If the current state of the target is x, each new sensor measurement z_j will be combined with the current estimate $p(x|z_1, \ldots, z_{j-1})$ to form a new belief state about the tracked target (belief of the state of the target), $p(x|z_1, \ldots, z_{j-1}, z_j)$. The problem of selecting a sensor j, such that j provides the greatest improvement in the estimate at the lowest cost, is an optimization problem defined in terms of information gain and cost:

$$M(p(x|z_1, \ldots, z_j)) = \alpha \cdot \phi_{\text{utility}}(p(x|z_1, \ldots, z_j)) - (1 - \alpha) \cdot \phi_{\text{cost}}(z_j) \quad (7.1)$$

where

ϕ_{utility} = Information utility measure (e.g., if sensor provides a range constraint, usefulness can be measured by how close the sensor is to the mean state)

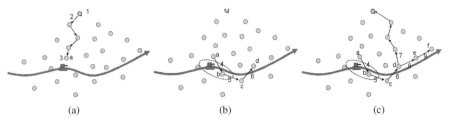

Figure 7.1 Example steps for information-driven sensor network tracking: (a) steps 1–3; (b) steps 4–6; (c) steps 7–9.

ϕ_{cost} = cost of communication and other resources characterized by link bandwidth, transmission latency, and residual battery power

α = relative weightage of utility and cost

The tracking protocol functions as follows. A user sends a query that enters the sensor network. Metaknowledge then guides this query toward the region of potential events. The leader node generates an estimate of the object state and determines the next best sensor based on sensor characteristics such as sensor position, sensing modality, and its predicted contribution. It then hands off the state information to this newly selected leader. The new leader combines its estimate with the previous estimate to derive a new state, and selects the next leader. This process of tracking the object continues and periodically the current leader nodes send back state information to the querying node using a shortest-path routing algorithm.

The tracking process described above can be best illustrated by the following example, shown in Figure 7.1:

1. User sends a query that enters the sensor network at node Q.
2. The query is directed toward a region of potential events.
3. Node a, the current leader, computes an initial estimate of the object state x_a, determines the next best sensor b among all its neighbors on the basis of their sensor characteristics, and hands off the state information to b.
4. Node b computes a new estimate by combining its measurement z_b with the previous estimate x_a using, say, a Bayesian filter: $x_b = x_a + z_b$. It also computes the next leader, node c, to pass on this state information.
5. Node c computes the new estimate $x_c = x_b + z_c$ and the next leader, node d.
6. Node d computes state information $x_d = x_c + z_d$, and next leader is node e; node d also sends current estimate back to the querying node Q.
7. Node e computes $x_e = x_d + z_e$; next leader is node f.
8. Node f computes $x_f = x_e + z_f$; next leader is node e, which sends current estimate back to querying node. And so the process continues.

The main assumption made by this approach is that each node in the network can locally estimate the cost of sensing, processing, and communicating data to another node and can monitor its power usage.

Although the algorithm described by Zhao et al. [24] is power-efficient in terms of bandwidth used since only a few nodes are up at any given time, the selection of sensors is a local decision. Thus, if the first leader is incorrectly elected, it could have a cascading effect and overall accuracy could suffer. It is also computationally heavy on leader nodes. This approach is applied to tracking a single object only, although an extension for tracking multiple objects using group management schemes is discussed in Section 7.3.

7.2.2 Tracking Using Binary Sensors

Binary sensors are so called because they typically detect one bit of information. This one bit could be used to represent indicate whether the target is (1) within the sensor range or (2) moving away from or toward the sensor.

Two approaches to the problem of target tracking using binary sensors are discussed below. The first technique uses a centralized method, while the second one is a distributed protocol.

7.2.2.1 *Tracking a Moving Object with a Binary Sensor Network*
In this binary sensor model [1] each sensor node detects one bit of information, namely, whether an object is approaching or moving away from it. This bit is forwarded to the basestation along with the node id.

The detection is performed as follows. Each sensor performs a detection and compares its measurement with a precomputed threshold (e.g., likelihood radio test). If the probability of presence is greater than the probability of absence, also called the *likelihood ratio*, the detection result is positive. The model assumes that sensors can identify whether a target is moving away from or towards it and that the sense bits are made available to a centralized processor. It also assumes that the basestation knows the location of each sensor and that a secondary binary sensor can be used in conjunction with this sensor to discover the precise location of the target.

The protocol makes use of the particle filtering approach. Multiple copies of the same object are kept. Each copy has an associated weight based on a discrete probability vector. With every sensor reading a new set of particles are created as follows:

1. A previous position is chosen according to the old weights.
2. A possible successor position is chosen.
3. If the successor position meets the acceptance criteria, it is added to the new set of particles and its weight is computed.
4. The weights are normalized so that they sum up to one.

The direction of motion of the target can be determined by calculating the normal plane to the velocity of the object tracked as described further.

The normal plane to the velocity separates the positive and negative convex hulls. A positive convex hull is the polygon enclosing all sensors that record a

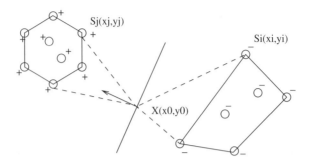

Figure 7.2   Illustration on determining the moving direction.

positive detection; in other words, the target is approaching them. Similarly, the negative convex hull is the convex hull of all sensors that detect that the target is moving away from them. As shown in Figure 7.2, the normal plane to the velocity of the object separates the two convex hulls.

In Figure 7.2, $X(x_0, y_0)$ denotes the current position of the object tracked. $S_j(x_j, y_j)$ represents the sensors toward which the target is approaching, and $S_i(x_i, y_i)$ represents all sensors from which the target is moving away. The slope of the line that is perpendicular to the velocity is the line separating the two convex hulls. This normal, denoted as m_0, can be computed given the positive and negative convex hulls that can be determined from the sensor readings. Once m_0 is known, the velocity vector can be computed and the direction of the vector will be toward the positive convex hull.

This approach conserves both communication and sensing energy. It has been shown [1] that only a small number of bits are being processed; however, precise location of the object cannot be determined, only its direction of motion can be known. Another drawback of using this algorithm is that trajectories that have parallel velocities and are a constant distance apart cannot be separated.

To ascertain the precise location of the object, this model can be extended by adding another binary sensor to each node to detect proximity information using a proximity bit. Assuming that the detection range is calibrated so that at most one sensor detects one object at a time, the protocol described above can be extended as follows:

```
if sensor S sees the object then
    forall accepted particles P not within the range of S do
        Let Pʳ ( newly generated particle by the previous protocol) =
            intersection between the range of S and semi-line (PS];
        Let P₁,.., Pₖ = ancestors of P since the last time the object was
            detected;
        for i = 1 to k do
            Pᵢ = Pᵢ - (P - Pʳ)/(k + 1);
        end
    end
end
```

This protocol states that when an object is detected by a sensor node, the ancestors of every particle that is not inside the range of the node are shifted as far as the last time the object was spotted by proportional amounts.

As mentioned earlier, this is a power-efficient scheme in terms of bits that need to be transmitted and processed; however, it uses a centralized database and hence is prone to all the shortcomings associated with centralizations, primarily loss of scalability and lack of fault tolerance.

7.2.2.2 Cooperative Tracking with Binary Detection Sensor Networks

A distributed approach to target tracking using binary sensors has been proposed [13]. Sensors determine whether the object is within their detection range and then collaborate with neighbor node data and perform statistical approximation techniques to predict the trajectory of the object. This cooperative tracking scheme improves accuracy by combining information from several nodes rather than relying on one node only. Assuming that sensors are uniformly distributed in the environment, a sensor with range R will (1) always detect an object at a distance of less than or equal to $(R - e)$ from it, (2) sometimes detect objects that lie at a distance ranging between $(R - e)$ and $(R + e)$, and (3) never detect any object outside the range of $(R + e)$ where $e = 0.1R$ but could be user-defined. Objects can move with arbitrary speed and direction. Hence the trajectory of the object is linearly approximated to a sequence of line segments along which the object moves with constant speed. The degree of divergence of this approximation from the actual path will vary depending on the speed and change in direction of the object. Each node records the duration for which the object is in its range. Neighboring nodes then exchange the timestamps and their locations. For each point in time, the object's estimated position is computed as a weighted average of the detecting node locations. The weights assigned are proportional to a function of the duration for which the target is within range of a sensor. The target will remain within range of sensors closer to the target path for a longer period. A line fitting algorithm (least-squares regression) is executed on the resulting set of points. The object path is predicted by extrapolating the target trajectory to enable asynchronous wakeup of nodes along that path. In this technique it is assumed that nodes know their locations and that their clocks are synchronized. Note that the density of sensor nodes should be high enough for sensing ranges of several sensors to overlap for this algorithm to work, and also sensors should be capable of differentiating the target from the environment.

A number of different weighting schemes could be used, three of which are outlined below:

1. Assigning equal weights to all readings. This yields the most imprecise results, namely, a higher rate of error between actual target path and its sensed path

2. Heuristic: $w_i = \ln(1 + t_i)$, where t_i is the duration for which the sensor heard the object. This error rate is lower and this method gives a better approximation of the object trajectory.

3. Assigning weights inversely proportional to the estimated perpendicular distance to the path of the object. Using the formula $w_i = 1/\sqrt{r^2 - 0.25(v * (t_i - 1/f))^2}$, where r is the sensor radius, v is the estimated speed of the object, t_i is the duration of object detection, and f is the sensor sampling frequency. This is the most precise method but requires estimation of the velocity of the object, which is too costly in terms of the communication costs required to make the estimate.

Hence the second approach is the most appropriate. The line fitting computation requires collection of all position estimates at a centralized location for processing. To minimize latency and bandwidth usage, some nodes are designated as gateways to outside networks with more computing resources. The sensor network is logically organized into trees rooted at each gateway, and each node collects data from its children and sends it up to the nearest or least busy gateway. This algorithm works very well for time-critical applications. It can continuously refine the path estimated using old data in conjunction with new data. Also, as it is a collaborative approach, it provides more accurate results than do those based on a single sensor detection of objects and is less computationally expensive on one particular node.

The protocol indicates that the higher the node density, the better the estimate on the trajectory; however, this means that we keep all nodes near the object awake. There exists a tradeoff between power usage in terms of active sensor nodes detecting the object and the preciseness of the estimation. Hence a denser network should not necessarily mean turning on all nodes that are near the object, but only a certain number required to make an acceptable estimate. Also there is no way of detecting multiple targets. This protocol also requires that the sensors be able to distinguish clearly between environment and sensed phenomena; that is, it ignores any interference, which does not seem to be a realistic assumption.

7.2.3 Other Sensor-Specific Methods

A number of tracking methodologies use specific sensor capabilities. Some are tailored for a specific application, while others depend on target attributes. One such approach is the smart sensor network approach [8], which uses sensors that have high processing abilities. Sensor nodes with onboard processing capability and radio interface are set up in a self-configuring and fault-tolerant network to perform collaborative sensing in a power-efficient manner. This algorithm detects and tracks a moving object, and alerts sensor nodes along the predicted trajectory of the object. The algorithm is power-efficient since communication between the sensor nodes is limited in the vicinity of the object and its predicted course.

Sensor nodes are randomly scattered in the geographic region. The algorithm assumes that each node is aware of its absolute location via a GPS or a relative location. The sensors must be capable of estimating the distance of the target from the sensor readings. The presence of target is detected on the basis of the sensor reading. If the sensor reading is greater than a particular threshold, then the detection is positive. Once a positive detection take place, a `TargetDetect` message containing the node location and its distance from the target is broadcast to the sensor network. This detection process repeats every second.

The location of the object is determined using the triangulation method. Either a node that has already detected the target waits for two other `TargetDetect` messages and performs triangulation, or a node that receives three `TargetDetect` messages can perform triangulation to locate the position of the object. The next step is trajectory estimation. To estimate the target trajectory, object location is estimated at a minimum of two instants in time and a straight line is drawn through it. However, a higher number of readings and curve fitting algorithms can provide a better trajectory estimation and hence a better prediction and more accurate sensor awakening.

Once the trajectory is determined, nodes within a specified perpendicular distance d, from the trajectory of the object are sent a *warning* message containing the location of sender and parameters describing the straight-line trajectory. Nodes within the specified distance rebroadcast this warning message. To optimize the propagation of the warning messages and prevent flooding of the network, the direction of propagation of warning messages is limited to the direction of motion of the target. This is done using the following technique. The node receiving a warning message computes a line perpendicular to the trajectory passing through itself as shown in the Figure 7.3. This line divides the area into two regions, R_1 and R_2. R_2 is the direction of motion. Therefore a node forwards the warning message if it lies within the specified distance and the message was received from a node in region R_1. $E_1 - E_3$ define the object trajectory. To avoid origination of multiple warning messages for the same object in order to conserve bandwidth and power, a node refrains from sending a warning message for some time after it has forwarded one.

This algorithm assumes that a sensor knows or learns the sensor reading to distance mapping. It also assumes that the network density is such that the subset of the sensor network that lie in the region where warning messages must be

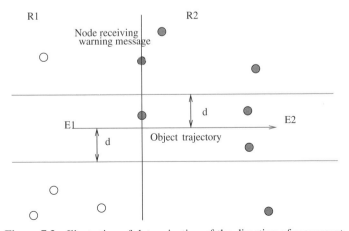

Figure 7.3 Illustration of determination of the direction of movement.

propagated must form a connected graph among themselves. No warning message can be propagated in a sparse network.

The protocol uses a distributed approach to target tracking and hence is resilient as well as scalable. Another advantage of this approach is that the sensor node wakeup process is localized in the vicinity of the object and its predicted course, thereby saving on communication energy.

Note that trajectory estimation could be inaccurate as two originating nodes could derive different trajectories for the same object. The protocol also does not handle tracking of multiple targets. Multiple target tracking issues are discussed in the following section.

7.3 PROTOCOL SUPPORT FOR DISTRIBUTED TRACKING

As opposed to centralized sensor array processing, in which all processing occurs on a central processor, in a distributed model, sensor networks distribute the computation among sensor units. Each sensor unit acquires local, partial, and relatively coarse information from its environment. The network then collaboratively determines a fairly precise estimate based on its coverage and multiplicity of sensing modalities

7.3.1 Distributed Group Management for Track Initiation and Maintenance in Target Localization Applications

A distributed approach to object tracking is discussed in the paper by liu et al. [12], which is an extension of the approach discussed in Section 7.2 (above) with the added ability to track multiple targets.

This is a cluster-based distributed tracking scheme. The sensor network is logically partitioned into local collaborative groups. Each group is responsible for providing information on a target and tracking it. Sensors that can jointly provide the most accurate information on a target, in this case, those that are nearest to the target form a group. As the target moves, the local region must move with it; hence groups are dynamic with nodes dropping out and others joining in.

A leader-based tracking algorithm similar to the one described by Zhao et al. [24] is used. All sensor nodes that record a detection greater than the threshold form a collaborative group and elect a leader. At any time t, each group has a unique leader who knows the geographic region of the collaboration yet does not need to know the exact members of the group. The leader measures and updates its estimate of the target location, called the "belief state." On the basis of the new information, the leader selects the most informative sensor and sends it the updated information. This sensor then becomes the leader at time $t + x$, where x is the communication delay.

The leader suppresses other nodes from further detection, thereby limiting power dissipation and also preventing creation of multiple tracks for the same target. The leader node initializes the belief state and kicks off the tracking algorithm.

7.3.1.1 Group Formation and Leader Election

Group formation is based on geographic nearness to the target and is done as follows. Initially all nodes are in sensing mode. Once a detection is made at a node, it sends the likelihood ratio to all other nodes within twice its detection range: $2R$. The only thing known about the target is that it is within R distance of the node. Hence other detector nodes could be within a distance of $2R$ from this node (R from the target).

Each node that detects the target checks and compares all detection messages received within a time period of t_{comm}, where t_{comm} is set to a value that is greater than the time taken for the detection messages to reach their respective destination yet less than the time required for the target to move. The group leader is elected according to the timestamp of the detection message. A sensor node declares itself the leader if its message is timestamped earlier than all other messages or if an identically timestamped message has a lower likelihood ratio.

7.3.1.2 Group Management

As the target moves, groups have to be broken and reformed with new members dynamically. Group management is therefore a key aspect of the algorithm. The selected leader initializes the belief state $p(x_0|z_0)$ as a uniform disk of radius R centered at its own location. The belief state gets refined with each successive measurement. This area will contain the target with high probability. Different algorithms are used to compute the suppression region, defined as the region in which all sensor nodes will form part of the group. In this case a regression region containing all the sensor nodes that detect the target with a probability greater than a specified threshold is identified. A margin R is added to this region to compute the suppression region. Hence in the initial case the suppression region will be a concentric circle of radius $2R$ centered at the leader.

As the target moves, nodes that were previously not detecting may begin detecting the target. This can cause multiple tracks for a single target to exist, in other words, contention. Hence, these nodes have to be suppressed because the algorithm is designed to be optimal for single-node tracking only. SUPPRESSION messages are used to minimize this scenario and to claim group membership. The leader sends a SUPPRESS message to all other nodes in the collaborative region to tell them to stop sensing and join the group. The leader will perform the sensing for the entire group. To avoid the overhead of sending SUPPRESSION messages to all the nodes in the new group, part of which overlaps with the previous group, the leader sends SUPPRESS messages only to new nodes. The leader also sends UNSUPRESS messages only to those nodes that now are no longer part of the region. Each node can be in any one of the following four states :

1. *Detecting* — this node is not a part of any group and periodically checks for possible targets.
2. *Leader* — this node performs the sensing and updates the track and the group.
3. *Idle* — this node belongs to a collaborative group but is not performing any sensing. It is waiting passively for a possible hand off from the leader.

4. *Waiting for timeout* — intermediate states waiting for potential detections to arrive from other nodes.

Another message type used by the algorithm is the HANDOFF message. HANDOFF messages are used to hand off the leadership to another node. Each HANDOFF message comprises of (1) the belief state, (2) the sender ID, (3) the receiver ID, (4) a flag indicating successful or lost track, and (5) a timestamp. This scheme assumes that all sensor nodes are time-synchronized and are aware of their one-hop neighborhood. In this chapter we also assume that the routing protocol used will limit the propagation of detection messages to the specified region (in order to avoid flooding the network)

It is clear that time synchronization is a major prerequisite for this approach to work. Consider the case where a few nodes miss some detection messages because they did not arrive within the t_{comm} window; then multiple groups will be formed for tracking the same object. Since these tracks correspond to the same target, they may collide; hence a merge mechanism for redundant paths is required, which is discussed below.

7.3.1.3 Distributed Track Maintenance

The algorithm can handle multiple target tracking since each target is tracked by a single group at any point of time, and the sensor network consists of many such groups. Multiple tracking is easy if tracks are far apart and the collaborative regions are nonoverlapping. However, multiple tracks, whether for the same object or different objects, could collide. We therefore need a mechanism to handle just such a scenario and perform track maintenance accordingly.

Each track is assigned a unique ID, for example, in terms of the timestamp of the track initiation. All messages originating from that group are tagged with the ID. Each node can now keep track of its multiple membership. A node that belongs to more than one group and is not a leader in any group would require as many UNSUPPRESSION messages as SUPPRESSION messages in order to free it.

If a leader node receives a SUPPRESSION message with an ID different from its own, this implies that a group collision has taken place. In such a case the algorithm supports group merging, and one track should be dropped on the basis of the timestamp of the SUPPRESSION messages. Each leader compares the timestamp in the newly received SUPPRESSION $(t_{suppression})$ message with its own (t_{leader}), and the older one is retained on the assumption that the belief state of the older track would be more refined and hence more reliable. Therefore, if $t_{suppression} < t_{leader}$, the leader drops its own track and relays the new SUPPRESSION message to its group and then relinquishes leadership. Hence the two groups merge into one, and the new group leader is now the collective group's leader. If $t_{suppression} \geq t_{leader}$, then the leader's track survives.

This algorithm works well for merging multiple tracks corresponding to the same target. If two targets come very close to each other, the two groups merge into one group and track the two targets as a single virtual target. Once the targets separate, one of the targets will be redetected as a new target and another group will be formed to track it.

Note that if two targets come very close to each other, then the mechanism described will be unable to distinguish between them.

7.3.2 Tracking Tree Management

A dynamic convoy tree-based collaboration (DCTC) framework has been proposed [21,22]. The convoy tree includes sensor nodes around the detected target, and the tree progressively adapts itself to add more nodes and prune some nodes as the target moves (see Fig. 7.4). When the target first enters the surveillance zone, active (not in sleeping mode) sensor nodes that are close to the target will detect the target. These nodes will collaborate with each other to select a root and construct an initial convoy tree. Relying on the convoy tree, the information about the target generated from all the on-tree nodes will be gathered to the root node, which will process the gathered information and generate a more accurate report on the location and direction of movement of the target. As the target moves, some nodes lying upstream of the moving path will drift farther away from the target and will be pruned from the convoy tree. On the other hand, some free nodes lying on the projected moving path will soon need to join the collaborative tracking. Since they normally are under power saving mode, it is necessary to wake them up before the target actually arrives. These issues are discussed in detail in Section 7.4.2. As the tree further adapts itself according to the movement of the target, the root will be too far away from the target, which introduces the need to relocate a new root and reconfigure the convoy tree accordingly.

If the moving target's trail is known a priori and each node has knowledge about the global network topology, it is possible for the tracking nodes to agree on an optimal convoy tree structure. An algorithm that optimizes the energy consumption for data gathering along the convoy tree is discussed by Zhang and Cao [21]. However, in the real scenario, this global information may not be available, and their paper has given practical solutions.

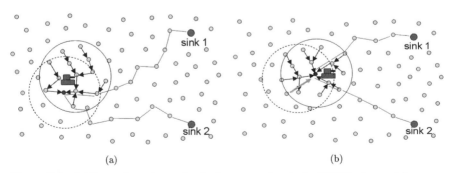

Figure 7.4 Adding and pruning nodes for the convoy tree in the DCTC scheme: (a) convoy tree at current time; (b) convoy tree at next time.

7.3.2.1 Construction of Initial Tree

When a target first enters the surveillance zone, the sensor nodes that can detect the target can collaborate to construct the initial convoy tree. First, a root node should be elected among the initial nodes, which is the election phase of the initialization process. The root election is based on the heuristic that the root is the closest to the target, namely, the geometric center of the nodes in the tree. Each node i will need to broadcast to its neighbors an $election(d_i, id_i)$ message with its distance to the target (d_i) and its own id. If a node does not receive any election message with (d_j, id_j) that is smaller than (d_i, id_i), it becomes a root candidate. Otherwise, it gives up and selects the neighbor with the smallest (d_j, id_j) to be its parent. It is possible that multiple root candidates will come up. Thus, the second phase is needed by letting the candidate i flooding a $winner(d_i, id_i)$ message to other nodes in the initial convoy tree. When a root candidate i receives a $winner(d_j, id_j)$ message with smaller (d_i, id_i) values, it gives up the candidacy. It will further attach itself into the tree rooted at the candidate with the smallest (d_i, id_i).

7.3.2.2 Tree Expansion and Pruning

For each time interval, the root of the convoy tree adds some nodes and removes some nodes according to the target's movement. To identify which nodes are to be added and removed, a prediction-based method has been discussed [21]. It is assumed that the location of the target in the next time interval can be predicted given the estimated moving speed of the target. If the target moving direction does not change frequently, the chance of correctly predicting the target's future position is high. Figure 7.5 shows the set of added nodes and the set of removed nodes.

7.3.2.3 Tree Reconfiguration

With the movement of the target, the nodes that participate in the tracking change continuously. When the target moves farther away, more and more nodes drift farther from the root node. Thus, the root should be replaced by a node closer to the target, and the convoy tree needs to be reconfigured accordingly. This can be

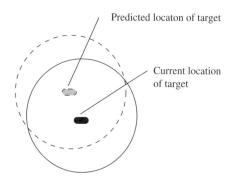

Figure 7.5 Prediction-based expansion–pruning scheme.

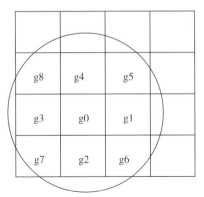

Figure 7.6   The sequential tree reconfiguration.

triggered by a simple heuristic, that if the distance between the current target location and the root becomes larger than a threshold, the tree needs to be reconfigured. The reconfiguration threshold can be set as $d_m + \alpha \cdot v_t$, where d_m is a parameter specifying the minimum distance that triggers the reconfiguration, v_t is the velocity of the target at time t, and α is a parameter specifying the impact of the velocity.

After the reconfiguration is triggered, a sequential scheme can be used. It is based on the grid structure, which is normally used for the power saving mode for the free nodes [19]. In order to save power for the free nodes, the network is divided into grids. At each time, only one node is selected to be the grid head and remains active continuously. Other nodes will be in power saving mode. The grid size is smaller than the transmission range so that the grid heads can form a connected topology.

The main idea of this sequential reconfiguration scheme, proposed by Zhang and Cao [21], can be explained using Figure 7.6. Suppose that the newly selected root is at the grid g_0. The reconfiguration procedure starts by the new root broadcasting the `reconf` message to the nodes in its grid. The new root also needs to send the `reconf` message to the heads of the neighboring grids (namely, g_1, g_2, g_3, g_4). Thereafter, the nodes in grid g_0 can all set their parent as the new root. Also the nodes in the neighboring four grids will be informed by their grid head, and can adjust their parent by the information provided by the new root. The heads of grids g_5, g_6, g_7, g_8 may not be able to receive the `reconf` message directly from the new root at g_0. They should be informed by the heads in grids g_1, g_2, g_3, g_4. A repetitive procedure can further adjust the tree structure in those four corner grids.

7.4 NETWORK ARCHITECTURE DESIGN FOR DISTRIBUTED TRACKING

In this section we discuss optimal sensor deployment strategies to ensure maximum sensing coverage with minimal number of sensors, as well as power conservation in sensor networks, a key aspect in the design of any sensor network architecture.

7.4.1   Deployment Optimization for Target Tracking

An important issue in designing a target tracking sensor network is the placement of sensors within the surveillance zone. First, the sensors should fully cover the surveillance zone. In most cases, when a target is detected, a single sensor node is enough for pinpointing the location of the target; thus, it should be ensured that every point in the zone is covered by at least k sensor nodes. Huang and Tseng [9], show how each node can check if the local area in its sensing range satisfies the k-coverage condition. Further, the placement of the sensor nodes can also affect the way how target localization is conducted. This issue is discussed elsewhere in the literature [4,25].

Chakrabarty et al. [4] have studied the sensor placement issues for target tracking analytically. The paper provides a modified problem model for target localization, based on a grid manner discretization of the space. In some applications or systems, it is sensible to find the gridpoint closest to the target's estimated location, instead of pinpointing the exact coordinates of the target. In such a problem model, an optimized placement of sensors will satisfy the requirement that every gridpoint in the sensor field be covered by a unique subset of sensors. In this way, the set of sensors reporting a target at time t uniquely identifies the grid location for the target at time t. Thus, the sensor placement problem can be modeled as a special case of the alarm placement problem described by Rao [16]. The problem is described as follows. Given a graph G, which models a system, one must determine how to place "alarms" on the nodes of G so that any single node fault can be diagnosed. It has been shown [16] that the minimal placement of alarms for arbitrary graphs is an NP-complete problem. However, it was also shown [4], that for special topologies such as a set of gridpoints, minimal placements can be found with efficient algorithms.

To achieve the optimal placement, the concept of *covering coding* [10] is used. For a node v in the graph G, the coverage of v with radius r is defined as the subset of nodes in G that are within r hops away from v. A covering coding of G is a covering of nodes in G such that any node can be uniquely identified by examining the nodes that cover it. For a regular graph, such as a set of gridpoints, the results in two other studies [4,10] have shown schemes of optimal covering codes. Let a $(3, p)$ grid denote a set of three-dimensional gridpoints, with p nodes on each dimension. If $p > 4$ and p is even, the minimum number sensors needed is $p^3/4$. If p is odd, the lower bound on the minimum sensors can be derived using the results from covering coding problem as well. Figure 7.7 shows an optimal placement scheme of sensors in a 13×13 two-dimensional grid. The minimum number of needed sensors is 65 to cover a total of 169 gridpoints (sensor density $= 0.38$).

7.4.2   Power Conservation for Target Tracking

For a sensor network to be viable in terms of cost-effectiveness, the network must work for a certain amount of time, and the longer this period, the better it is. Wireless sensor nodes typically do not have access to a continuous power supply and rely on their battery for power. It is thus important to minimize power usage in

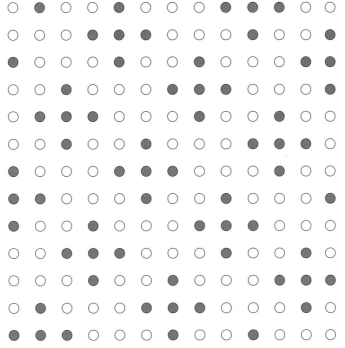

Figure 7.7 An optimal placement of sensors in a 13×13 grid.

order to prolong the lifetime of the network without any significant loss in the accuracy of the information provided by the network. In this section we describe architectural and protocol support used to achieve maximal power efficiency.

7.4.2.1 Power Conservation and Quality of Surveillance in Target Tracking Sensor Networks

In this section, we discuss the sleep–awake pattern of each node during the tracking stage as described by Gui and Mohapatra [7], to obtain power efficiency. Consider a distributed sensor network monitoring a large operational area. The network operations have two stages: the *surveillance stage* during the absence of any event of interest, and the *tracking stage*, which is in response to any moving targets. From a sensor node's perspective, it should initially work in the low-power mode when there are no targets in its proximity. However, it should exit the low-power mode and be active continuously for a certain amount of time when a target enters its sensing range, or more optimally, when a target is about to enter within a short period of time. Finally, when the target passes by and moves farther away, the node should decide to switch back to the low-power mode.

Intuitively, a sensor node should enter the tracking mode and remain active when it senses a target during a wakeup period. However, it is possible that a node's sensing range is passed by a target during its sleep period, so that the target can

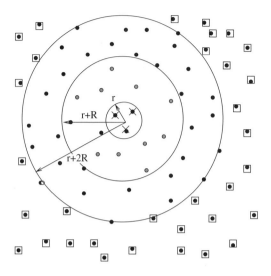

▣	Node in Waiting (low power) mode
•	Node in Prepare mode
◦	Node in SubTrack mode
✖	Node in Tracking mode

Figure 7.8   Layered onion-like node state distribution around the target.

pass across a sensor node without being detected by the node. Thus, it is necessary that each node be proactively informed when a target is moving toward it.

In this section, we discuss the proactive wakeup (PW) algorithm. Each sensor node has four working modes: waiting, prepare, subtrack, and tracking mode. The *waiting* mode represents the low power mode in surveillance stage. *Prepare* and *subtrack* modes both belong to the preparing and anticipating mode, and a node should remain active in both modes. Figure 7.8 shows the layered onion-like node state distribution around the target.

At any given time, if we draw a circle centered at the current location of the target where radius r is the average sensing range, any node that lies within this circle should be in tracking mode. It actively participates a collaborative tracking operation along with other nodes in the circle. Regardless of the tracking protocol, the tracking nodes form a spatiotemporal local group, and tracking protocol packets are exchanged among the group members. Let us mark these tracking packets so that any node that is awake within the transmission range can overhear and identify these packets. Thus, if any node receives tracking packets but cannot sense any target, it should be aware that a target may be coming in the near future. From the overheard packets, it may also get an estimation of the current location and moving speed vector of the target. The node thus transits into the subtrack mode from either waiting mode or prepare mode. At the boundary, a subtrack node can be $r + R$ away from the target, where R is the transmission range. To carry the wakeup wave

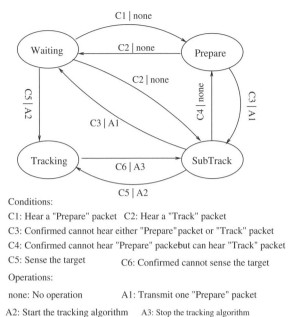

Figure 7.9   State transition diagram for tracking group management algorithm.

farther away, a node should transmit a prepare packet. Any node that receives a pre-pare packet should transit into prepare mode from waiting mode. A prepare node can be as far as $r + 2R$ away from the target.

Figure 7.9 shows the state transition diagram of the proactive wakeup (PW) algorithm. If a tracking node confirms that it can no longer sense the target, it tran-sits into the subtrack mode. Further, if it later confirms that it can no longer receive any tracking packet, it transits into the prepare mode. Finally, if it confirms that it can receive neither tracking nor prepare packet, it transits back into the waiting mode. Thus, a tracking node gradually turns back into low-power surveillance stage when the target moves farther away from it. In essence, the PW algorithm makes sure that the tracking group is moving along with the target.

7.4.3   Target Tracking Using Hierarchical and/or Broadband Sensor Networks

Target tracking is a generic problem that can be applied to different kinds of targets in various types of environments (hostile, benign, environmentally friendly or intense), for example, tracking vehicles in a *hostile* environment such as battlefield surveillance. Another example could be tracking a moving fire. This variety in application demands a corresponding variety in sensor nodes in terms of size, processing power, radio interface capabilities, and sensing abilities to enable multimodal sensing. For example, in battlefield surveillance it would be more helpful to know the vehicle type, onboard armament, and personnel. This information could

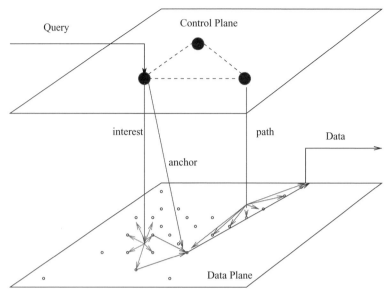

Figure 7.10   Hierarchical sensor network architecture.

be gained using image sensors. So the network can consist of a few sensors that are image sensors and a large number of low-level sensors for the other functions. It would be desirable if this information could be sent in a secure manner using encryption algorithms and authentication techniques. These are usually outside the scope of normal sensors because of the high demand on memory and power resources that are at a premium with sensor nodes. However, if we use heterogeneous sensor nodes, where some have high processing power and others are low-level sensors, we could achieve a fair degree of security.

A novel approach explored by Yuan et al. [20] describes the use of hierarchical architecture for a heterogeneous broadband sensor network to facilitate interaction between sensor nodes and improve energy efficiency. The possibility of combining a few high-powered sensor nodes, H nodes having full-scale wireless cards with a large number of low power sensor motes, and L nodes in one networked system is considered. Sensors can be organized into a higher layer consisting of a few H nodes forming a control plane and a lower layer consisting of L nodes that may be randomly deployed, as illustrated in Figure 7.10. The H nodes form a broadband backbone for data delivery. They can set up dedicated collision-free transmission paths while minimizing overhearing and idle listening. H nodes are assumed to have a tunable radio transmission range of R, large enough for H nodes to communicate with each other. H nodes are deployed in a grid to ensure complete coverage, and ideally every L node is associated with at least one H node; however, an H node need not know the L nodes that are associated with it. The entire sensor network is divided into many small regions. In case of grid deployment with four H nodes, as shown in the Figure 7.11, the unit square sensor field is divided into 13 pieces and an ID is created for each region on the basis of H node information.

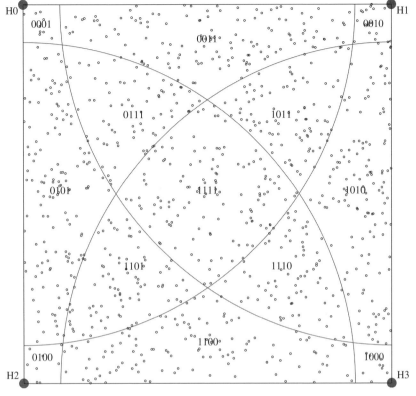

Figure 7.11 Single-grid cell of HSN architecture.

During the operational phase, L nodes wake up periodically to listen to instructions from associated (parent) H nodes and act accordingly. If an L node loses connectivity with all H nodes, it can continue to function using traditional sensor network protocols. This chapter assumes that all H nodes are clock-synchronized. The sleep–awake–active pattern of L nodes can be described as follows. At the beginning of each cycle, all L nodes wake up to listen to WAKEUP messages. If an L node receives a WAKEUP message that does not match its area ID, it will keep listening for further instructions. However, if a node determines that it cannot play a role in the specified instruction, it will sleep until the next cycle.

A WAKEUP message is broadcasted by an H node, so all L nodes in its coverage area receive it. This accounts for $(n\pi R/4)L$ nodes statistically. After receiving a WAKEUP message with area ID 1101, only L nodes in that area will remain awake. This approach considerably reduces the number of L nodes hearing the INSTRUCTION message. Hence this sleep–wakeup cycle will reduce the power consumed by the network, thereby increasing the lifetime of the network.

The power consumed by a sensor consists of sensing energy and transmission energy. To minimize power consumed while sensing, ideally sensor nodes should be turned on only for the duration of an event of interest, and only those sensor nodes near the event should be turned on. This is possible only if the sensors have some external guiding information that wakes them up, in this case the control plane in the HBSN as described earlier. The H sensors can identify the region of interesting events using their sensing capabilities and wake up only those L sensors within that area. If the H sensors are binary sensors with range R and the entire area covered by the sensor network is divided into n regions, then, assuming a uniform distribution of events, the probability of events occurring in a region i is proportional to the area of that region denoted by A_i. The number of sensors in a region is also proportional to the areas of the region. Hence the average proportion of sensors turned on for an event is $P \propto \sum_i^n A_i^2 n$. If the interesting event occurs for a constant portion of the time, given by β, then the energy consumed by the sensing circuits can be given by expression $E_{HBSN}^S \propto \beta \sum_i^n A_i^2 \times E_{ESAT}$.

The next factor to be considered is the transmission energy. This chapter assumes that an AODV-type algorithm will be used to transmit information to the anchor area from the region of data origination. Typically energy inefficiency during transmission is caused by (1) idle listening, (2) overhearing, (3) collision, and (4) control packet overhead. However, in this approach a PATH message is used to reserve a path from the data origination region to the anchor region; hence collisions are avoided. Also sensor nodes that don't lie on the reserved path are turned to sleep mode, hence avoiding the idle listening and overhearing inefficiencies.

7.5 CONCLUSION

Various approaches for target tracking using wireless sensor networks are described in this chapter. The key advantages of using sensor networks for this application are increased accuracy in target localization, better fault tolerance, and ease of deployment. However, each sensor mote has limited capabilities in terms of power, sensing, and processing abilities. Therefore, comprehensive and accurate data can be obtained only through the collaboration of sensor nodes in the network as a single node does not have the capability to provide this information. In this chapter we discuss a number of approaches used to perform target collaboration and data aggregation, ranging from the traditional centralized approaches to the more robust distributed schemes. A sensor network approach to target tracking is viable in terms of cost-effectiveness, only if the network has a reasonably long lifetime. Ideally wireless sensors do not have access to a continuous power supply and rely on their battery. Hence, all communication and processing protocols for sensor collaboration and data processing have to focus on maximizing power efficiency. Therefore, optimal sensor deployment and power-efficient algorithms to support sensor communication and data aggregation are key aspects in the design of any sensor network. In this chapter, we discussed a number of target localization and tracking methods, each with different tradeoffs. We focused on the underlying architectural support required for the problem of target tracking specifically with respect to deployment and power efficiency. Power utilization combined with data communication are the main metrics for evaluating most of the schemes discussed in the chapter. Finally, we introduced the concept and motivation for using multi-modal heterogeneous sensor networks. Such networks are power-efficient and can be used for a variety of target tracking applications.

REFERENCES

1. J. Aslam, Z. Butler, V. Crespi, G. Cybenko, and D. Rus, Tracking a moving object with a binary sensor network, *Proc. ACM Int. Conf. Embedded Networked Sensor Systems (SenSys)*, 2003.

2. Y. Bar-Shalom and X.-R. Li, *Multitarget-Multisensor Tarcking: Principles and Techniques*, Artech House, 1995.

3. R. R. Brooks, P. Ramanathan, and A. M. Sayeed, Distributed target classification and tracking in sensor network, *Proc. IEEE*, **91**(8) (2003).

4. K. Chakrabarty, S. S. Iyengar, H. Qi, and E. Cho, Grid coverage for surveillance and target location in distributed sensor networks, *IEEE Trans. Comput.* **51**(12) (2002).

5. C. Y. Chong, K. C. Chang, and S. Mori, Distributed tracking in distributed sensor networks, *Proc. American Control Conf.*, 1986.

6. M. Chu, H. Haussecker, and F. Zhao, Scalable information-driven sensor querying and routing for ad hoc heterogeneous sensor networks, *Int. J. High Perform. Comput. Appl.* **16**(3) (2002).

7. C. Gui and P. Mohapatra, Power conservation and quality of surveillance in target tracking sensor networks, *Proc. ACM MobiCom Conf.*, 2004.

8. R. Gupta and S. R. Das, Tracking moving targets in a smart sensor network, *Proc VTC Symp.*, 2003.

9. C. F. Huang and Y. C. Tseng, The coverage problem in a wireless sensor network, *Proc. ACM Workshop on Wireless Sensor Networks and Applications* (*WSNA*), 2003.

10. M. G. Karpovsky, K. Chakrabaty, and L. B. Levitin, A new class of codes for covering vertices in graphs, *IEEE Trans. Inform. Theory* **44** (March 1998).

11. J. Liu, M. Chu, J. Liu, J. Reich, and F. Zhao, Distributed state representation for tracking problems in sensor networks, *Proc. 3rd Int. Symp. Information Processing in Sensor Networks* (*IPSN*), 2004.

12. J. Liu, J. Liu, J. Reich, P. Cheung, and F. Zhao, Distributed group management for track initiation and maintenance in target localization applications, *Proc. Int. Workshop on Information Processing in Sensor Networks* (*IPSN*), 2003.

13. K. Mechitov, S. Sundresh, Y. Kwon, and G. Agha, *Cooperative Tracing with Binary-Detection Sensor Networks*, Technical report UIUCDCS-R-2003-2379, Computer Science Dept., Univ. Illinois at Urbaba — Champaign, 2003.

14. L. Y. Pao, Measurement reconstruction approach for distributed multisensor fusion, *J. Guid. Control Dynam.* (1996).

15. L. Y. Pao and M. K. Kalandros, Algorithms for a class of distributed architecture tracking, *Proc. American Control Conf.*, 1997.

16. N. S. V. Rao, Computational complexity issues in operative diagnosis of graph based systems, *IEEE Trans. Comput.* **42**(4) (April 1993).

17. R. W. Sittler, An optimal data association problem in surveillance theory, *IEEE Trans. Military Electron.* (April 1964).

18. Q. X. Wang, W. P. Chen, R. Zheng, K. Lee, and L. Sha, Acoustic target tracking using tiny wireless sensor devices, *Proc. Int. Workshop on Information Processing in Sensor Networks* (*IPSN*), 2003.

19. Y. Xu, J. Heidemann, and D. Estrin, Geography informed energy conservation for ad hoc routing, *Proc. ACM MobiCom Conf.*, 2001.

20. L. Yuan, C. Gui, C. Chuah, and P. Mohapatra, Applications and design of hierarchical and/or broadband sensor networks, *Proc. BASENETS Conf.*, 2004.

21. W. Zhang and G. Cao, Dctc: Dynamic convoy tree-based collaboration for target tracking in sensor networks, *IEEE Trans. Wireless Commun.* **11**(5) (Sept. 2004).

22. W. Zhang and G. Cao, Optimizing tree reconfiguration for mobile target tracking in sensor networks, *Proc. IEEE InfoCom.*, 2004.

23. W. Zhang, J. Hou, and L. Sha, Dynamic clustering for acoustic target tracking in wireless sensor networks, *Proc. 11th IEEE Int. Conf. Network Protocols* (*ICNP*), 2003.

24. F. Zhao, J. Shin, and J. Reich, Information-driven dynamic sensor collaboration for tracking applications, *IEEE Signal Proces. Mag.* (March 2002).

25. Y. Zhou and K. Chakrabarty, Sensor deployment and target localization in distributed sensor networks, *ACM Trans. Embedded Comput. Syst.* **3** (Feb. 2004).

CHAPTER 8

Field Gathering Wireless Sensor Networks

ENRIQUE J. DUARTE-MELO and MINGYAN LIU

Department of Electrical and Computer Engineering, University of Michigan, Ann Arbor

8.1 INTRODUCTION

The technologies that enable the creation of wireless networks have enjoyed rapid development since the mid-1990s. Antennas, radio transceivers, and processors have been greatly improved in terms of form, size, power efficiency, and so on. This progress, together with marked advances in the area of microsensors, has allowed small, inexpensive, energy-efficient, and reliable sensors with wireless networking capabilities to quickly become a reality.

The development of these wireless sensors has given rise to the increasingly popular concept of *wireless sensor networks*, which has been the subject of extensive studies that have enabled a broad range of applications. These applications range from scientific data gathering, environmental hazard monitoring, aiding in managing inventories in large warehouses, intrusion detection and surveillance, and battlefield monitoring. Wireless sensor networks are ideally suited for these applications because of their rapid and inexpensive deployment (e.g., compared to wired solutions). They can be easily deployed (e.g., airborne) to areas otherwise inaccessible by the land. The low-cost low-energy nature of these sensors (if made biodegradable) also makes them easily disposable.

Many of these applications have been investigated in the literature. For example, one can find environmental monitoring applications in several studies [1–5], where wireless sensor networks are used for purposes such as flood detection and pollution study. Cases of using wireless sensors as a part of a bigger "smart" environment at home or in the laboratory, for example, can be found in the literature [6–9]. Other

Mobile, Wireless, and Sensor Networks: Technology, Applications, and Future Directions
Edited by Rajeev Shorey, Akkihebbal L. Ananda, Mun Choon Chan, and Wei Tsang Ooi
Copyright © 2006 John Wiley & Sons, Inc.

application scenarios include habitat monitoring [10], health care [11], and the detection and monitoring of car theft [12].

This chapter focuses on a class of applications, referred to as *field gathering*, and considers a number of design and performance issues that arise in such applications. Specifically, a *field gathering wireless sensor network* is a network of sensor nodes deployed over a field (1D, 2D, or 3D) with the purpose of taking spatial and temporal measurements of a given set of parameters about the field (e.g., temperature). Nodes within the network may act as *sources* (sensor nodes that take measurements of the specified parameters) and/or *relays* (sensors nodes that do not take measurement themselves but receive data from sources or other relays and pass on the data to other nodes). The placement of these nodes may be deterministic or random. There is typically one or more *destination* or *sink* node for whom the measured data are destined. These are also often called *collectors*. They may be located within or outside the sensing field. In a field gathering sensor network the communication pattern is of the many-to-one (or many-to-few) type, in that sensed data are eventually gathered at the collector(s). At a given instant sensors take measurements at their respective locations. Then, through a series of transmissions and retransmissions, the data are relayed to the collector(s), where they are processed and put together to form a *snapshot* of the field for that particular time instance. This process is then repeated. As sensors take measurements periodically over time, a sequence of snapshots are formed at the collector. Without loss of generality, for the rest of our discussion it will be assumed that there is a single collector/sink node that serves as the destination for all gathered data.

Many research issues and challenges related to design and performance arise in this type of network. They include distributed data compression, distributed data dissemination, collaborative signal processing, and energy-efficient networking. This chapter examines two performance aspects associated with such networks, and provides a survey on more recent studies and results. There is particular interest in the limits achievable in both aspects. The first concerns the *lifetime* of the network, and is a direct indication of energy efficiency. This is a measure of how long the network lasts and is limited by individual sensor's energy constraint. The second concerns the *throughput* of the network, a direct indication of the effectiveness of the communication and networking strategies used. This is a measure of how fast the network can deliver data to the collector and is limited by techniques employed at the physical, MAC, and network layers. Both performance limits are affected by a range of factors, including the network architecture as well as data compression schemes used.

Subsequent sections formally define these measures, introduce a number of models taking into account some of the aforementioned factors, and explore corresponding results. The intention of this chapter is to present a fairly comprehensive picture of existing models and results. However, it has to be mentioned that this is a rapidly developing research area with a large volume of literature and new ideas and results emerging continually. As a result, this survey is a collection of selected representative work in this area and is not meant to be exhaustive.

The rest of the chapter is organized as follows. Section 8.2 presents in more detail the network lifetime measure, and Section 8.3 presents results on network throughput. Both sections first examine data dissemination alone and then take into account data compression. Section 8.4 discusses open problems in this area, and Section 8.5 concludes the chapter.

8.2 LIFETIME LIMIT IMPOSED BY ENERGY CONSTRAINTS

The *lifetime* of a network is typically defined as the time until a certain number (or percentage) of the sensor nodes run out of energy. Thus lifetime depends on but is not limited to the following factors within the context of a field gathering sensor network: initial energy, size of the network, sensor deployment, location of the collector, sampling/measurement interval, data rate, transmission power and range, and routing strategy used,

It's worth mentioning that simply measuring the longevity of the network in units of time in many cases may not accurately capture the energy efficiency with which the network is operated. For instance, if one network takes measurements half as often as another (and therefore produces half as many snapshots within the same amount of time), one may find the former to last longer in terms of time. This, however, does not imply that the former is designed or operated in a more energy-efficient way than the latter, which is simply used more often. In such cases, one should compare the lifetime measure under exactly the same assumptions and models on the data sources. Alternatively, rather than asking how long the network can last, it may be more pertinent to ask what is the total amount of data (or snapshots) that the network can deliver (to the collector) before a predetermined number of nodes run out of energy. This alternative "total data" measure essentially evaluates energy efficiency by number of bit (or snapshots) delivered per unit of energy. The discussion that follows uses both definitions to investigate the energy efficiency of the network.

To begin a more detailed discussion of lifetime, it is first necessary to outline the network model employed. The model described below is a basic one used in many of the studies on lifetime. Variations and extensions will be specified later on as needed.

8.2.1 Model and Assumptions

The following are a set of assumptions used in the network model:

- The network is deployed in a two-dimensional field of finite area. This assumption is mostly for the convenience of discussion and does not prevent the analysis from being applied to a higher dimension.
- There is a single collector situated either in or outside the field. The collector is assumed to have sufficient power and energy.
- Every node in the field may be both a data source and a relay. When taking measurements, each sample is quantized and encoded into bits. The measurements may be taken at regular intervals, and data compression may also be

used. Ultimately, all these measurements are abstracted into the form of a data source that generates data at a certain rate.

- A fixed amount of power is needed to reach a receiver at a fixed distance away. All transmissions are assumed to be successful within this distance. This assumption thus does not take into account randomness in the channel.
- Nodes have an arbitrarily adjustable transmission power and range, in the sense that they can use the exact minimum necessary power to reach a receiver node. They are also assumed to have sufficient power to reach every other node in the network as well as the collector in a direct transmission. These assumptions maximize the set of routes that can be considered. They can be relaxed in many cases, as will be pointed out.
- Nodes consume energy when transmitting, receiving, and sensing, but not while idling.

These assumptions collectively form a set of idealized conditions in that lifetime estimation based on these assumptions is in general an upper bound on the actual lifetime achievable in a real network.

8.2.2 Basic Mathematical Framework

Under the aforementioned assumptions, one can create a mathematical framework that takes into account the initial energy available to each node; the network layout; and the energy cost of transmitting, receiving, and sensing; and that tries to maximize the lifetime of the network (or the amount of data that the network can collect), by considering all possible routing strategies in transmitting data to the collector.

A typical linear program that can be established to achieve this goal is as follows:

$$\max_{f} \quad t \tag{8.1}$$

$$\text{S.t.} \quad \sum_{j \in M} f_{i,j} + f_{i,C} \quad = \sum_{j \in M} f_{j,i} + r_i t, \quad \forall i \in M \tag{8.2}$$

$$\sum_{j \in M} f_{i,j} e_{tx}^{i,j} + f_{i,C} e_{tx}^{i,C} = \sum_{j \in M} f_{j,i} e_{rx} + r_i t e_s < E_i, \quad \forall i \in M \tag{8.3}$$

$$f_{i,j} \geq 0, \quad \forall i,j \in M \tag{8.4}$$

$$f_{i,i} = 0, \quad \forall i \in M \tag{8.5}$$

$$f_{C,i} = 0, \quad \forall i \in M \tag{8.6}$$

This linear program will be referred to as **P1**. Here M denotes the set of all the nodes in the network; C denotes the collector; $f_{i,j}$, also known as *flow*, denotes the amount of data i delegates to be routed via j, of all the data i has; r_i denotes the data generation rate at node i in bits per second; $e_{tx}^{i,j}$, e_{rx}, and e_s denote the amount of energy used (in joules per bit) in transmitting from node i to node j, in receiving, and in sensing and processing, respectively; $e_{tx}^{i,C}$ is the energy spent per bit transmitted from node i directly to the collector; and E_i is the initial amount of energy available to node i. This formulation is also often considered a *fluid flow model*.

The objective function (8.1) is the lifetime of the network, defined as the time until the first node's death. This objective is maximized over all flows; that is, it is maximized over each node's routing choices in terms of the amount to be routed by other nodes. Equation (8.2) is the flow conservation constraint. It forces the amount of data sensed at node i plus the amount of data received by node i to be equal to the amount sent from node i to other nodes and the collector. Expression (8.3) is the energy constraint. It implies that the network only operates until the death of the first node. Finally, constraints (8.4–8.6) ensure that there is no negative data flow, that a node does not transmit to itself, and that data flowing in the opposite direction are not considered, that is, from the collector to the sensors (or equivalently, once the data reach the collector, that they stay there and do not go back into the field).

Looking at this formulation, one can observe that the notion of lifetime expressed here is related only to the total amount of data generated (i.e., $r_i t$ for all i). It does not imply anything on the rate at which the data are transmitted or the amount of time it takes to accomplish all the transmissions. Therefore this notion of lifetime implies that nodes have infinite transmission capacity such that there is no constraint on how rapidly a node may transmit, or that multiple transmissions and receptions can occur simultaneously without causing interference. This is certainly an ideal scenario.

An alternative way of interpreting this lifetime concept is to view it as the *functional* lifetime of the network, in that it is simply an indication of how much data the network generates (and delivers), rather than how long it takes the network to deliver, as long as one assumes that whenever a node is not engaged in transmission, reception or sensing it consumes zero energy. Following this interpretation, interference can be taken into account so that a certain schedule (timeshare) may be followed for these transmissions to be accomplished. The optimization itself does not provide such a schedule, only the total amount to be transmitted to each node. Additional work is needed to derive a schedule that is causal, realizes such flow allocation, and satisfies interference requirement (this point is discussed further). In essence, by adopting the notion of functional lifetime, we allow transmissions to take place sequentially in time.

This basic formulation can be modified to model a variety of scenarios. For example, the flow conservation (8.2) in its current form implies zero buffering and no data aggregation at the nodes. It can be easily modified by subtracting from the right-hand side (RHS) an appropriate amount that represents buffered data or data eliminated as a result of aggregation. Similarly, (8.3) can also be modified to model the scenario where nodes transmit at a fixed power or have an upper bound on their transmission range.

One might wonder how the linear program result would differ if the network lifetime were defined to end when all nodes, rather than only the first node, run out of energy. Note that in order to maximize the lifetime via **P1**, the energy consumption of different nodes has to be balanced to the maximum degree possible. Whenever sufficient energy balancing is feasible, running **P1** would result in a solution where all nodes die at the same time; thus in effect the two definitions are indistinguishable. This can be shown to be valid for example, where all nodes start with the same amount of initial energy, nodes are uniformly distributed within a field, and the distance between sensor nodes are small compared to the distance between them and the collector, which is the case of interest in many applications. However, energy balancing is not always feasible. An example is where all nodes except one are clustered together close to the collector, while a single node is isolated and far away from the collector. In this case the single node will run out of energy before the rest.

8.2.3 Variations of the Framework

Some of the earlier studies on using a linear program (LP) to determine the lifetime of the network or to design a more energy-efficient routing algorithm include the papers by Chang and Tassiulas [13] and Bhardwaj and Colleagues [14,15]. We discuss the results from these studies in more detail in the following text.

Chang and Tassiulas [13] employ a linear program very similar to **P1**, where the network lifetime is maximized subject to a flow conservation constraint and an energy constraint. One difference is that the energy model they use [13] does not include energy spent in receiving data; thus, the energy constraint is as follows:

$$\sum_{j \in M} e_{i,j} f_{i,j} \le E_i, \quad \forall i \in M \tag{8.7}$$

where $e_{i,j}$ is the energy spent per bit transmitted from node i to j. They also develop distributed algorithms [13] to determine routing patterns that approximate the solution produced by the linear program. These are divided into the *flow redirection* algorithms and the *flow augmentation* algorithms. The former simply redirect a portion from each flow at every node in such a way that the minimum lifetime of every node will not decrease.

On the other hand, the flow augmentation algorithms start by calculating the shortest-cost path to the destination node according to a certain cost definition. Then the flow on this path is increased. After this increment the shortest-cost path is recalculated and the procedure is repeated. The performance of this algorithm depends on the cost function, which includes energy spent transmitting, nodes' residual energy and initial energy as inputs as suggested by Chang and Tassiulas [13]. The basic cost function proposed by those authors [13] is $c_{i,j} = e_{i,j}^{x_1} E_{\mathrm{res}_i}^{-x_2} E_i^{x_3}$, where E_{res_i} is the residual energy of node i. By adjusting the values of $\{x_1, x_2, x_3\}$, different cost functions and augmentation algorithms of the same family may be obtained. Two special examples are $\{x_1, x_2, x_3\} = \{0, 0, 0\}$, which reduces to the minimum-hop path, and $\{x_1, x_2, x_3\} = \{1, 0, 0\}$, which is the minimum-transmitted-energy path.

The same linear program is modified to consider the cases when only a single power level is available and when multiple power levels are present. Chang and Tassiulas [13] provide local algorithms that converge to the optimal solution when a single power level is considered, and show that the algorithms are close to optimal in the latter case.

Bhardwaj and Chandrakasan [15] use a similar linear programming approach, motivated by the idea of *role assignment*. The key of this idea is a set of roles assigned to the nodes that result in data being relayed to the collector. Possible roles for sensors are sensing, relaying, or aggregating. For example, the possible role assignments for a network consisting of two nodes and one collector, where only the first node senses, are

$$f_1 : 1 \rightarrow C$$
$$f_2 : 1 \rightarrow 2 \rightarrow C$$

In f_1 node 1 has the role of a source, while C is the collector. In f_2 node 1 has the role of a source and 2 has the role of a relay.

Then a collaborative strategy can be defined where different role assignments are used for a certain fraction of the time. For example, f_1 and f_2 can each be used half of the time. To maximize lifetime, one would simply need to find the set of fractions that maximizes lifetime subject to constraints, including nonnegative fractions and total energy. Bhardwaj and Chandrakasan [15] show that this role assignment formulation is equivalent to a network flow problem[1]. Transforming it into a network flow formulation results in a linear program very similar to **P1**. The main difference with respect to **P1** is that Bhardwaj and Chandrakasan [15] consider the possibility of data aggregation, as will be discussed in Section 8.2.5.

In another related paper, Bhardwaj et al. [14] determine an upper bound on the lifetime of sensor networks via simple nonconstructive proofs. Specifically, they first assume that there is a single source and a single sink/collector at fixed locations while an arbitrary number of relays may be placed at arbitrary locations to assist the data transmission from the source to the sink. Naturally these relays should be placed along a straight line connecting the source and the sink, and the optimal number of relays or hops and the optimal distances between them can be determined that minimizes the total amount of energy used in transmitting a fixed amount of the data from the source to the sink. The network lifetime resulting from this construction immediately provides an upper bound for the lifetime of a sensor network with a single source at a fixed location. This result is then generalized to the case where the source is located with uniform probability somewhere along a line, and the case where the source is located with uniform probability somewhere inside a square region. In all these cases the results are in the form of an upper bound.

As mentioned earlier, a more general definition of lifetime aims at the time of death of a certain percentage of nodes rather than the first node. This motivated the

---

[1]The role assignment approach itself has the limitation that the number of feasible role assignments grows exponentially as the number of nodes grows, which makes it less practical for large networks as pointed out by those authors [15].

work by Shi et al. [16], who seek a formulation that maximizes both the time until the death of the first node and the time until the death of all nodes. To achieve this goal, they [16] attempt to maximize the time until a set of nodes runs out of energy while minimizing the size of this set. Nodes in the network are thus classified into different sets, and this process is repeated until no nodes remain. This problem is referred to as the *lexicographic max–min node lifetime problem* [16]. This problem is solved through a serial linear program with parametric analysis. The result of this program is a sorted network lifetime vector $[t_1, t_2, \cdots, t_n]$, and a corresponding sensor set $\{S_1, S_2, \cdots, S_n\}$, where t_i is the time where the ith set of nodes S_i runs out of energy. Shi et al. [16] also provide an algorithm for determining a flow routing schedule that will achieve the lexicographic max–min optimal node lifetime vector, based on the vector computed from the serial linear program.

8.2.4 An Approach that Does Not Depend on Specific Network Layout

The nature of **P1** is such that the results obtained are specific to a particular network layout (i.e., precise locations of each node in the network). However, in many applications the placement of nodes is random rather than perfectly controlled. In such cases what is known may be the deployment probability distribution instead of precise locations of nodes. A question thus arises as to whether a formulation similar to **P1** can be established to obtain the lifetime estimate on a network of unknown layout but known deployment distribution.

The present authors [17] establish such a formulation by assuming a very dense sensor network. We argue that as node density increases, the node distribution can be described with increasing accuracy by a continuous density function $\rho(\sigma)$, where $\sigma = (x, y)$ denotes any point within the sensing field. Similarly, an energy density $e(\sigma)$ models the amount of energy in the field (in joules per unit area), and the rate at which the source generates data can be modeled by an information density function $i(\sigma)$ (in bits per unit time per unit area). These functions are in general related.

Using these definitions, the following continuous linear program is obtained:

$$\max_f \ t \cdot \int_{\sigma \in A} i(\sigma) d\sigma \sim \max_f t \tag{8.8}$$

$$\text{S.t.} \int_{\sigma' \in A} f(\sigma, \sigma') d\sigma' + \int_{\sigma' \in C} f(\sigma, \sigma') d\sigma'$$

$$= \int_{\sigma' \in A} f(\sigma', \sigma) d\sigma' + i(\sigma) \cdot t, \quad \forall \sigma \in A \tag{8.9}$$

$$\int_{\sigma' \in A} f(\sigma, \sigma') p_{tx}(\sigma, \sigma') d\sigma' + \int_{\sigma' \in C} f(\sigma, \sigma') p_{tx}(\sigma, \sigma') d\sigma'$$

$$+ \int_{\sigma' \in A} f(\sigma', \sigma) p_{rx} d\sigma' + t \cdot \epsilon_s(\sigma, i(\sigma)) \leq e(\sigma), \quad \forall \sigma \in A \tag{8.10}$$

$$f(\sigma, \sigma') \geq 0, \quad \forall \sigma, \sigma' \in A \cup C \tag{8.11}$$

$$f(\sigma, \sigma') = 0, \quad \forall \sigma = \sigma' \tag{8.12}$$

$$f(\sigma, \sigma') = 0, \quad \forall \sigma \in C, \forall \sigma' \in A \tag{8.13}$$

This linear program will be referred to as **P2**. Here A is the area where the nodes are deployed and C is the area where the sink(s) is (are) located (this could be a single point). Equations (8.8)–(8.13) are the continuous equivalent of (8.1)–(8.13). Note that here the flow $f(\sigma, \sigma')$ denotes the amount transmitted from location σ to σ' and has a unit of bits per unit area squared. Assuming $i(\sigma)$ to be known, maximizing lifetime t is equivalent to maximizing the total amount of data delivered up to t, as indicated in the objective function (8.8).

A major difference between the applicability of **P1** and **P2** is that **P2** does not depend on a specific deployment layout, but only the distribution of the deployment. The resulting objective function value approximates the *expected* lifetime of a randomly deployed network (averaged over all possible deployment realizations) under idealized conditions.

Clearly the continuous linear program cannot be solved as is. We [17] suggest discretizing the density functions and solving the resulting discrete linear program. This ultimately leads to a grid of points where the nodes, information, and energy densities are concentrated and a linear program similar to **P1** (using these grid-points) is solved. The result is therefore the lifetime estimate of a grid network created by properly discretizing the density functions. It's worth pointing out that although the end formulation is the same, this approach takes as input the density functions rather than specific node locations.

We have shown [17] via numerical experiments that the result of this approach closely approximates that obtained by averaging a large number of results from performing **P1** over randomly generated topologies, even for cases where node density is not very high. Figure 8.1 is an example of this approximation, where we have calculated the total number of bits sent to the collector before the first sensor death as shown in the objective function (8.8). The results are obtained by varying the number of points in the grid for a network deployed over a field of 500×500 m. As can be seen, the finer the grid, the closer the approximation. We [17] also discussed the stability and robustness of the linear program and how the result may differ when the grid does not accurately reflect the deployment.

8.2.5   The Use of Data Compression

Throughout the preceding discussion, sensor nodes in the network are assumed to generate data at a preestablished rate[2]. At the same time, one of the major properties that sets apart wireless sensor networks and generic wireless ad hoc networks is the potential for in-network processing and data compression. This is especially true in a field gathering sensor network as measurements taken by neighboring sensors and measurements taken over time by the same sensor are correlated, particularly when the node density is high. Exploiting this correlation is not only desirable but also necessary for highly energy-constrained sensors as it can reduce the amount of data that need to be transmitted. Here we take a look at

---

[2]The information desities in **P1** and **P2** can potentially also be functions of time, but they still need to be predetermined functions.

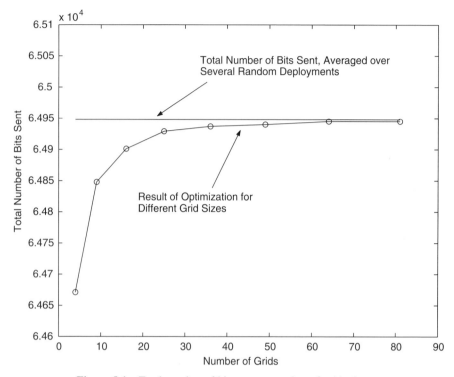

Figure 8.1   Total number of bits versus number of gridpoints.

whether data compression can be properly modeled in the framework outlined earlier, and whether the network lifetime can be derived when data compression is used.

As mentioned earlier, Bhardwaj and Chandrakasan [15] propose modifications to the linear program (of the type of **P1**) of the network lifetime to allow the possibility of data aggregation. This is done via a multicommodity flow program where the unaggregated flow and each aggregated flow are represented by one commodity. With this modification the rules of data aggregation (e.g., the identity of nodes that can aggregate as well as the amount of compression) are assumed known a priori so they can be added to the set of constraints. Therefore under this approach data aggregation is predetermined rather than optimized.

In another work the present authors [18] incorporate a method of data compression into the framework presented by **P1** and propose using a Slepian–Wolf [19] type of distributed data compression. The objective is to maximize the number of snapshots that can be delivered by the network, over all possible routing paths and over all possible rate combinations of the nodes. Assuming a single collector, this approach results in the following linear formulation.

$$\max_{n} \quad n$$

$$\text{S.t.} \quad \sum_{j \in M} f_{i,j} + f_{i,C} = \sum_{j \in M} f_{j,i} + n \cdot R_i, \quad \forall i \in M \tag{8.14}$$

$$\sum_{j \in M} f_{i,j} \cdot e_{tx}^{i,j} + f_{i,C} \cdot e_{tx}^{i,C} + \sum_{j \in M} f_{j,i} \cdot e_{rx} + n \cdot R_i \cdot e_s \leq E_i, \quad \forall i \in M \tag{8.15}$$

$$\sum_{i \in S} R_i \geq H_d(S|S^c), \quad \forall S \subseteq S_o \tag{8.16}$$

$$\sum_{i=1}^{M} f_{i,C} = n \cdot \sum_{i=1}^{M} R_i \tag{8.17}$$

$$f_{i,j} \geq 0, \quad \forall i,j \in M \cup \{C\} \tag{8.18}$$

$$f_{i,i} = 0, \quad \forall i \in M \tag{8.19}$$

$$f_{C,i} = 0, \quad \forall i \in M \tag{8.20}$$

This formulation is referred to as **P3**, where n is the number of snapshots delivered to the collector. Most of the constraints remain the same as in previous formulations. The main difference is the addition of constraint (8.16), where R_i is the rate (in bits per snapshot) of the ith node, $H_d(\cdot)$ denotes the differential entropy, $S_o = \{X_1, \ldots, X_M\}$, and $X_i, i = 1, .., M$, is the sample taken by node i. This constraint specifies the region of all feasible combinations of rates. Unfortunately this formulation is no longer a linear program and its solution requires considerably more computing time. As an example, Figure 8.2 shows the optimal rate allocation

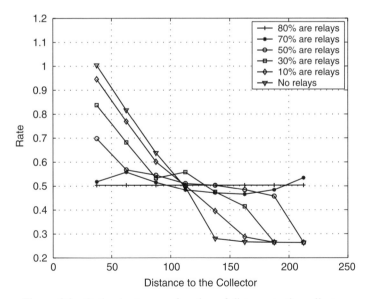

Figure 8.2 Optimal rates as a function of distance to the collector.

for a small linear network of eight sensing nodes. We vary the number of additional relays and the results are shown for cases with different percentage of relays.

Data compression and dissemination are also jointly considered in the paper by Critescu et al. [20], where routing and rate allocation are jointly determined with the objective of minimizing the cost and/or energy consumed in *transmitting a single snapshot* rather than to maximize the lifetime of the network (Thus energy balancing was not an issue in that study [20]). A single sink and a tree communication structure are assumed, and two compression strategies are compared: Slepian–Wolf and joint entropy coding.

In the case of the Slepian–Wolf model, it is shown that the shortest-path tree is always optimal. It is also shown that when using the shortest-path tree, exact determination of the optimal rates to be used by the nodes can be based on the cost of transmission from each node to the collector. Specifically, assume that nodes are labeled 1 through N starting with the one with the lowest cost and ending with the one with the highest cost and denoting by $H(X_i)$ the entropy of the quantized sample taken at node i. Then the rate for node 1 should be $H(X_1)$, $H(X_2|X_1)$ for node 2, $H(X_3|X_1, X_2)$ for node 3, and so on. As a result, this approach has a simple communication structure but complex encoding procedure as using Slepian–Wolf requires all nodes to have knowledge of the correlation structure among measurements taken by nodes in the network.[3] To overcome this limitation, Critescu et al. [20] further note that if one assumes that correlation decreases with distance, then the rate of a given node is affected mostly by its closest neighbors. According to this observation, a distributed algorithm is proposed that first determines the shortest-path tree using the Bellman–Ford algorithm and then determines the rate used by each node by exchanging information on correlation among nodes in a close neighborhood. As the neighborhood becomes larger, the solution obtained by this algorithm approaches the optimal.

By contrast, using the joint entropy coding model results in simpler encoding but a more complex communication structure. This is because a node acting as relay for another node encodes its data given the data it receives. Therefore the amount of data compression that can be achieved depends on the path taken by the data. On the other hand, the shortest path does not necessarily maximize the amount of compression that can be obtained. Unfortunately, for this case there is no polynomial time algorithm that can determine the optimal routing paths.

8.3 THROUGHPUT LIMIT IMPOSED BY NETWORK AND COMMUNICATION ARCHITECTURE

A second performance measure we are interested in is how fast the data can be delivered. This is limited by many physical and network layer schemes as well as the overall architecture and organization of the network (e.g., flat vs. hierarchical).

---

[3]This observation applies equally to the approach used in the paper by the present authors [18], where the compression model is also assumed to be Slepian–Wolf.

It is also affected by the channel models used to describe attenuation, interference, and fading.

From the perspective of a single node, there are two common ways to measure data delivery. It may be measured by the number of bits per second per node that the network can deliver, referred to as the *per node throughput*. It may also be measured by the number of bits-meter per second per node, referred to as the *per node transport capacity* (first defined by Gupta and Kumar [21]). Similarly, we can define the *network throughput* and the *network transport capacity*, which are the sums of corresponding per-node quantities. In the remaining discussion the simpler term *throughput* (transport capacity) will often be used to mean either per node or network throughput (transport capacity). Later it will be seen that when data compression is used, it may be more appropriate to use snapshots per second rather than bits per second within the context of field gathering.

It's worth noting the relationship between these two measures. In a case where the distance from the source to the destination cannot be controlled (or is determined by the model or assumptions), throughput and transport capacity are equivalent in measuring the efficiency and usefulness of the network in that they are off by a constant multiplier (e.g., the average pathlength). On the other hand, if one can control the distance from the source to the destination, then maximizing transport capacity is different from maximizing throughput and the former is more meaningful. This is because the notion of throughput is insufficient in measuring the usefulness of the network as two nodes placed far apart may not achieve a very high throughput between them (compared to two placed close to each other), for reasons such as interference, but the data delivered traverse a longer distance, representing more work done by the network. This is not captured by the throughput measure.

The seminal work by Gupta and Kumar [21] first revealed that the per node transport capacity in a random network is $\Theta(1/\sqrt{n\log n})$ bit-m/s.[4] A more detailed discussion of this work is given in Section 8.3.2. This work has inspired numerous other studies, including similar studies for the case of directional antenna [22], and the case where mobility is present. For the latter, Grossglauser and Tse [23] showed that the per node throughput of $\Theta(1)$ is possible when node mobility is considered, via a special type of two-stage packet relaying. This indicates that the throughput can be independent of n when mobility is exploited, although this comes with unbounded delay. More recently it was shown [24] that there is a fundamental relationship between achievable throughput and delay in the presence of mobility, where one could be traded off for the other. As our focus is on static field gathering networks with a many-to-one communication pattern, these studies will not be discussed further.

In what follows, as in the previous section, the basic network model and assumptions that apply to most of the subsequent discussion are presented. The discussion will be limited to the case of the omnidirectional antenna. Section 8.3.2 presents the main results from three studies [21,25,26], where different physical

---

[4]$f(n) = \Theta(g(n))$ means $f(n) = O(g(n))$ and $g(n) = O(f(n))$ by Knuth's notation.

layer features are assumed. These studies all apply to the scenario where sources and destinations are randomly chosen, forming multiple one-to-one communication flows. This is different from the basic field gathering model. However, as these studies are fundamental to the notion of throughput capacity, it helps to present a more general picture before introducing Section 8.3.3, which examines the more specific case of many-to-one communication, so that this notion may be put into perspective. Section 8.3.4 discusses how data compression affects the throughput (in terms not of raw bits but of snapshots) for a given quality (or distortion) measure of the snapshots. Most of the throughput studies presented here are existence studies, in that they show certain throughput as being achievable by some channel sharing or scheduling algorithm that exists. Section 8.3.5 presents some research effort in deriving practical algorithms that may approach the achievable throughput.

8.3.1 Basic Model and Assumptions

An initial common set of assumptions that apply to most of the subsequent discussions is stated first (additional assumptions adopted by different studies will be introduced during the discussion):

- The network consists of n randomly deployed nodes. Once deployed, the nodes remain static. The sensing field is two-dimensional with unit area.
- All n nodes use omnidirectional antennas and transmit with a fixed power P. It should be noted that in general P is a function of n. In some of the work surveyed below, instead of assuming a fixed per node transmission power, the total transmission power of all nodes is assumed fixed.
- A node cannot simultaneously transmit and receive, and it can receive only one transmission at a time. All interferences are treated as noise.
- The power received at node j from a transmission from node i is determined by one of two propagation models. Under propagation model **A1**, the power received is $P/x_{i,j}^{\alpha}$, where $x_{i,j}$ is the distance between nodes i and j and α is a positive constant that represents the path loss. Under propagation model **A2**, the power received is $P/(1 + x_{i,j})^{\alpha}$. **A1** is typically considered a farfield propagation model, in that it fairly accurately describes the power received over a long distance. When distance becomes very short, this model exhibits a power amplification of received signal (i.e., when $x_{i,j}$ falls below 1), which obviously does not reflect reality. **A2** voids this amplification while remaining very similar to **A1** over large distances.
- When **A1** is used, a transmission from node i to node j will be considered successful if

$$\frac{P/x_{i,j}^{\alpha}}{N + (1/G)\sum_{k=1,\,k\neq i}^{k=n} P/x_{k,j}} \geq \beta \tag{8.21}$$

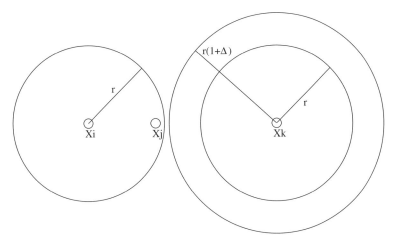

Figure 8.3   Transmission range r and interference range Δ.

where N is the ambient noise, G is the processing gain, and β is the SNR threshold. The same transmission will be considered successful when **A2** is used if

$$\frac{P/(1+x_{i,j})^{\alpha}}{N+(1/G)\sum_{k=1,\,k\neq i}^{k=n}P/(1+x_{k,j})^{\alpha}} \geq \beta \qquad (8.22)$$

These are commonly known as the *physical model* in the literature [21]. A *protocol model* has also been introduced [21] as follows. Assume a fixed transmission range r and a fixed interference range Δ. Then a transmission from node i to node j will be successful if $x_{i,j} \leq r$ and for any other transmitting node $k, x_{j,k} > r(1+\Delta)$ (see Fig. 8.3). This model was shown [21] to be equivalent to **A1**, in that one can select the proper values of these two parameters such that all simultaneous transmissions allowed by the protocol model satisfy (8.21).

8.3.2   Some Results on One-to-One Communication

Under the model given in the previous section, this section first considers the throughput and transport capacity in a scenario consisting of multiple one-to-one communication connections, each with its source and destination randomly chosen. The results are asymptotic in the form of throughput or transport capacity achievable with high probability as the network size n goes to infinity.

This problem was first considered by Gupta and Kumar [21] under propagation model **A1** and the additional assumption that each node has a maximum transmission rate of W bits per second. It was found that under these assumptions the per node transport capacity is $\Theta(1/\sqrt{n\log n})$ bit-m/s. This transport capacity is achieved when the transmission power P is the minimum needed to maintain connectivity. Gupta and Kumar [21] also shown that subdividing the channel or using clustering cannot improve the achievable per node transport capacity. The same results hold under the protocol model.

The key to obtaining the upper bound is to determine the level of spatial reuse. Note that under both propagation models, there is a limit on how many simultaneous transmissions can take place and be successful. The fact that the power is reduced to the minimum needed to maintain connectivity and that the received power is $P/x_{i,j}^{\alpha}$ allow the number of simultaneous transmissions to grow with n.

What if one uses **A2** instead of **A1**? Arpacioglu and Haas [26] consider the same problem under **A2**, with the additional assumption that nodes may receive more than one transmission at a time (this relaxation should only increase the transport capacity). Note that this assumption is more realistic over small distances, in that power should not be amplified at the receiver. It is shown that under this model, the number of simultaneous transmissions is bounded above by a constant independent of n, rather than growing with n as was the case under **A1**. An intuitive explanation of the difference is that under **A1** as nodes get arbitrarily close, P can be decreased rapidly enough such that the spatial reuse (simultaneous transmissions) increases with n. One the other hand, under **A2** P cannot be decreased beyond a certain point even as nodes get arbitrarily close.

This difference has a huge effect on the resulting throughput, which scales as $\Theta(1/n)$ instead of $\Theta(1/\sqrt{n \log n})$. This immediate follows from the result that the number of simultaneous transmission is upper-bounded by a constant, which is shared by n nodes. Thus it can be seen that **A1** tends to produce more optimistic results as it allows the number of simultaneous transmissions to grow with n.

Xie and Kumar [25] further consider the same problem with more sophisticated communication schemes where collaborative transmissions are used. They also consider the case where the total transmission power is fixed while the transmission rate is determined by the transmission power. They adopt a slightly different propagation model where the received power is $P(e^{\gamma x_{i,j}}/x_{i,j}^{\delta})$, where $\gamma \geq 0$ is the absorption constant and $\delta > 0$ is the path loss exponent. This propagation model will be known as **A3**. They [25] also investigate a *large attenuation* scenario defined by $\gamma > 0$ and $\delta > 3$, and a *low attenuation* scenario defined by $\gamma = 0$ and $\frac{1}{2} < \delta < 1$. It is shown that for networks whose channel has a large attenuation, multihop operation is optimal; and that coherent multistage relaying with interference cancellation is optimal for a single source–destination pair when attenuation is low. The fact that multihop is optimal under large attenuation confirms Gupta and Kumar's [21] results.

8.3.3 Some Results on Many-to-One Communication

As mentioned earlier, one prominent feature of a field gathering wireless sensor network is that the traffic pattern is of the many-to-one type, where data converge to a collector from different sensors.

When following the model laid out earlier and assuming that the collector cannot receive data from more than one node at a time, the collector immediately becomes a single bottleneck. The present authors [27] have studied this problem under the protocol model (or equivalently propagation model **A1**), and the assumption that each node has a maximum transmission rate of W bits per second. This article shows that the per node throughput of this network scales as $\Theta(1/n)$. The upper bound is a trivial one, as the collector cannot receive simultaneous transmissions. However, other than the case where each sensor transmits directly to the collector, the exact W/n throughput is not in general achievable in a multihop scenario. It is also shown [27] that using load-aware routing the achievable throughput is nearly twice that of one that is not load aware. One important observation from this study [27] is that a larger transmission range (which is the result of a higher transmission power P under model **A1**) results in higher throughput in the many-to-one case. In particular, the achievable per node throughput is $\Theta(1/nf(r,\Delta))$, where r is the transmission range and $f(r,\Delta)$ is a function that decreases with r. This is in direct contrast with the multiple one-to-one case where higher throughput is obtained with the lowest possible transmission range. Although we [27] considered only propagation model **A1**, it's worth mentioning that the same results also hold under propagation model **A2**. This is because the limit on throughput in this case is determined entirely by the bottleneck created at the collector.

An interesting question is whether the throughput under these assumptions can be improved by the use of clusters. Earlier it was mentioned that in the multiple one-to-one case there was no advantage in using clusters. It turns out that by using clusters the upper bound of W/n is achievable, by assuming that the cluster heads can reach the collector in one hop and that there is no interference between the two layers [27].

El Gamal [28] studies a similar many-to-one problem, while exploring whether collaborative transmission schemes could improve the throughput result obtained by the present authors [27]. In addition to the added assumption of collaboration, El Gamal [28] no longer assumes a fixed rate of transmission W, but instead adopts a total transmission power constraint. The main idea is to have a two-stage transmission, where a source first uses some of its power to transmit to certain number of nearby nodes. Then together they transmit the data to the collector in a collaborative way. This operation is followed by each source. Under model **A1**, it is found that the per node throughput could be improved to $\Theta(\log n/n)$. Again this result is highly dependent on the assumption that the received power is $P/x_{i,j}^{\alpha}$. As neighbors become arbitrarily close, the source can increase link capacity and its number of neighbors with the same transmission power. Hence the improved throughput.

It turns out that by simply using the total power constraint instead of the fixed range of transmission assumption, one does not actually need to use collaboration to obtain a throughput of $\Theta(\log n/n)$ [29]. Simply by noting that with fixed power the link capacity grows with n, once the distance between nodes is small enough, one can obtain the same result with a multihop operation of the network. This result is also limited by the fact that when dealing with short distances, assuming that the received power is $P/x_{i,j}^{\alpha}$ is not very realistic.

8.3.4   Effect of Data Compression

The discussion so far has been on measuring how fast (and how far) the network can deliver its data in bits. As argued earlier, when there is sufficient correlation between measurements taken by different sensors, data compression can effectively reduce the amount of raw data that need to be sent. This, of course, does not affect the speed with which the network can deliver data in bits, but it does change the overall efficiency of the network. Similar to the lifetime study, for field gathering purposes, throughput in the presence of data compression would be better measured in snapshots per second rather than bits per second. This will also require the introduction of a quality measure such as distortion, and the question would be to see how quickly the network can deliver snapshots of prespecified quality.

From the results presented previously it is known that the per node throughput (measured in bits per second) diminishes as the network grows. However, one may expect the amount of bits that need to be sent to also diminish as n grows because of the increased correlation. The key issue then becomes whether the amount of compression is sufficient to offset the diminishing throughput so that the number of snapshots per second is kept constant as n grows.

The paper by Scaglione and Servetto [30] is one of the first studies that considered joint data compression and routing in a sensor network where the communication model is many-to-many in that the goal is to allow all sensor to be able to reconstruct a snapshot of the field. Marco et al. [31] study the network throughput in terms of snapshots per second for the many-to-one field gathering scenario with data compression. By assuming that sensors use identical scalar quantizers and entropy coding, and by adopting the basic model presented earlier with model **A1**, they show that the savings achieved by compression are insufficient to counter the diminishing throughput. Thus the time needed between the delivery of successive snapshots of the field grows to infinity as the network size grows to infinity.

This conclusion is due to a surprising result shown by Marco et al. [31] that although the number of bits that each node needs to transmit goes to zero as n increases, the total number of bits that the network needs to deliver still goes to infinity. This, along with the fact that the total throughput of the network is a constant regardless of n, gives the main result [31].

8.3.5   Practical Algorithms

The previous discussion centers around the issue of deriving the scaling laws of the network throughput or transport capacity as the network becomes very dense. These are asymptotic results that cannot be directly applied to a specific network of a finite size. Here we survey some of the research effort in trying to derive the maximum throughput obtainable in a given network. Such maximization typically aims at constructing the feasible transmission schedules and rates for each node such that the overall network throughput is maximized.

As we have seen, a major limiting factor in achieving higher throughput is interference, which requires transmission schedules that avoid simultaneous transmissions that interfere with each other. Jain et al. [32] model interference via conflict graphs. To create a conflict graph, a connectivity graph is first established, in which vertices represent the nodes and edges represent links between nodes (whether there is a link between two nodes may be determined by assumptions on transmission range and power). Then the connectivity graph is mapped into a conflict graph; first, links in the connectivity graph are mapped into vertices in the conflict graph, and then in the conflict graph an edge is added to connect two vertices if the links that they represent in the connectivity graph interfere with each other.

The connectivity graph is used to create a linear program that maximizes the flow of information toward the destination. The resulting flows will provide the routing information. The conflict graph is used to determine the schedule that should be used. An important limitation of this methodology as noted by Jain et al. [32] is that given a network and all the source–destination pairs, obtaining the optimal throughput is NP-hard. In fact it is NP-hard to approximate the optimal throughput. Experiments using an effort parameter are performed [32] to determine how much effort is put into getting closer to the optimal solution.

Coleri and Varaiya [33] use a similar approach to determine the transmission schedule. The main difference is that in this approach a simplification is adopted, and the schedule from the conflict graph is determined by mapping the conflict graph from the original tree network to a linear network. This simpler linear case is solved, and the solution is used as an approximation to the original problem.

The protocol proposed by Coleri and Varaiya [33] uses this schedule to create a time-division multiple access (TDMA) scheme. The authors argue that a scheduled approach is better suited for wireless sensor networks than would be a contention-based approach. The contention-based approach wastes a lot of energy listening to the channel and thus decreases the lifetime of the network. Moreover, lifetime is not the only advantage. The authors show that other performance metrics, such as delay, are improved when using a scheduled approach.

8.4 OPEN PROBLEMS

The previous sections have surveyed a number of studies and results on the lifetime and throughput performance of a field gathering sensor network. There remain numerous open problems. Some of these are related to developing models with more general assumptions or with assumptions that more accurately reflect practical scenarios. Some concern practical algorithms that can approximate optimal results. A subset of these problems that are most relevant to this chapter is discussed below.

Propagation Model Assumption   Results have been presented under two different propagation models: **A1** and **A2**. As has been mentioned, assuming that power received is $P/x_{i,j}^\alpha$ does not accurately model the increasing network density scenario, as diminishing distance between nodes results in power amplification under this model. Modifying this to $P/(1 + x_{i,j})^\alpha$ has a dramatic effect in the results. Developing alternate propagation models and studying the scenarios discussed in previous sections under more accurate propagation models might yield more insight into advantages and disadvantages not previously known.

Lifetime Throughput Tradeoff This survey has presented the lifetime and throughput studies as separate problems, while in fact they are closely related. Some qualitative nature of the tradeoff between the two can be observed in certain cases. For example, the choice of routes that maximize lifetime might create bottlenecks in the network that reduce throughput. However, a thorough understanding of how throughput and lifetime are related remains open. It is highly desirable to be able to properly characterize and quantify the tradeoff. More broadly, addressing the tradeoffs between different performance measures (not limited to lifetime and throughput) remains a key challenge.

Decentralized Algorithms Section 8.2 presented different approaches to obtaining upper bounds on the lifetime of a network. However, results on developing practical algorithms that may achieve or come close to achieving these bounds are relatively limited. The same applies to the throughput study. It would be very interesting and highly desirable to be able to take some of these theoretical studies and use them to guide the design of efficient practical MAC and network layer algorithms.

Hierarchical Architectures The concept of clustering has been extensively studied. It is intuitively seen as a way of dealing with scaling issues in large-scale sensor networks. However, while many clustering protocols have been proposed [34,35], most of the studies that characterize network performance focus on a flat architecture. More thorough studies of the performance of a hierarchical architecture are needed, so that more efficient protocols can be created.

Modeling of In-network Processing and Collaboration As sensors are highly energy- and processing-power-constrained devices, in-networking processing and collaboration, including distributed signal processing and data compression, are quickly becoming part of the sensor functionality. This enhances the performance of the sensor network but also makes the modeling of such performance more difficult, as we have seen in some of our earlier discussions. Good abstraction and models is another research challenge.

8.5 CONCLUSION

This chapter surveys recent studies and results concerning the performance of the class of field gathering wireless sensor networks. Of particular interest are the lifetime and throughput limits of such networks. In either case different models and approaches are presented along with a fairly detailed discussion of corresponding results, based on a number of relevant papers. Some open research problems are also briefly discussed.

ACKNOWLEDGMENT

This work is partially supported by NSF grants ANI-0112801, ANI-0238035, and CCR-0329715.

REFERENCES

1. J. Agre and L. Clare, An integral architecture for cooperative sensing networks, *IEEE Comput. Mag.*, (May 2000).

2. T. Imielinski and S. Goel, Dataspace: Querying and monitoring deeply networked collections in physical space, *Proc. ACM Int. Workshop on Data Engineering for Wireless and Mobile Access (MobiDE)*, Seattle, WA, 1999.

3. C. Jaikaeo, C. Srisathapornphat, and C. Shen, Diagnosis of sensor networks, *Proc. IEEE Int. Conf. Communications (ICC)*, Helsinki, Finland, June 2001.

4. P. Bonnet, J. Gehrke, and P. Seshadri, Querying the physical world, *IEEE Pers. Commun.* (Oct. 2000).

5. N. Bulusu, D. Estrin, L. Girod, and J. Heidemann, Scalable coordination for wireless sensor networks: Self-configuring localization systems, *Proc. Int. Symp. Communication Theory and Applications (ISCTA)*, July 2001.

6. G. D. Abowd and J. P. G. Sterbenz, Final report on the interagency workshop issues for smart environments, *IEEE Pers. Commun.* (Oct. 2000).

7. I. A. Essa, Ubiquitous sensing for smart and aware environment, *IEEE Pers. Commun.* (Oct. 2000).

8. C. Herring and S. Kaplan, Component-based software systems for smart environments, *IEEE Pers. Commun.* (Oct. 2000).

9. E. M. Petriu, N. D. Georganas, D. C. Petriu, D. Makrakis, and V. Z. Groza, Sensor-based information appliances, *IEEE Instrum. Meas. Mag.* (Dec. 2000).

10. A. Cerpa, J. Elson, M. Hamilton, and J. Zhao, Habitat monitoring: Application driver for wireless communications technology, *Proc. ACM SigComm Conf.*, Costa Rica, April 2001.

11. J. M. Kahn, R. H. Katz, and K. S. J. Pister, Next century challenges: Mobile networking for smart dust, *Proc. Int. Conf. Mobile Computing and Networking (MobiCom)*, Seattle, WA, 1999.

12. G. J. Pottie and W. J. Kaiser, Wireless integrated network sensors, *Commun. ACM* **43**(5) (2000).

13. J. Chang and L. Tassiulas, Energy conserving routing in wireless ad-hoc networks, *Proc. Annual Joint Conf. IEEE Computer and Communication Societies (InfoCom)*, Tel-Aviv, Israel, March 2000.

14. M. Bhardwaj, T. Garnett, and A. P. Chandrakasan, Upper bounds on the lifetime of sensor networks, *Proc. IEEE Int. Conf. Communications (ICC)*, Helsinki, Finland, June 2001.

15. M. Bhardwaj and A. P. Chandrakasan, Bounding the lifetime of sensor networks via optimal role assignments, *Proc. Annual Joint Conf. IEEE Computer and Communication Societies (InfoCom)*, New York, June 2002.

16. Y. Shi, Y. T. Hou, and H. D. Sherali, *On Lexicographic Max-Min Node Lifetime Problem for Energy-Constrained Wireless Sensor Networks*, Technical Report, Bradley Dept. ECE, Virginia Tech, Sept. 2003.

17. E. J. Duarte-Melo, M. Liu, and A. Misra, An efficient and robust computational framework for studying lifetime and information capacity in sensor networks, *ACM Kluwer MONET* (special issue on energy constraints and lifetime performance in wireless sensor networks) (2004).

18. E. J. Duarte-Melo, M. Liu, and A. Misra, A computational approach to the joint design of distributed data compression and data dissemination in a field-gathering wireless sensor network, *Proc. 41st Annual Allerton Conf. Communication, Control, and Computing*, Allerton, IL, Oct. 2003.

19. D. Slepian and J. Wolf, Noiseless coding of correlated information sources, *IEEE Trans. Inform. Theory*, **IT-19**: 471–480 (July 1973).

20. R. Cristescu, B. Beferull-Lozano, and M. Vetterli, On network correlated data gathering, *Proc. Annual Joint Conf. IEEE Computer and Communication Societies (InfoCom)*, Hong Kong, March 2004.

21. P. Gupta and P. R. Kumar, The capacity of wireless networks, *IEEE Trans. Inform. Theory*, (March 2000).

22. C. Peraki and S. D. Servetto, On the maximum stable throughput problem in random networks with directional antennas, *Proc. 4th ACM Int. Symp. Mobile Ad Hoc Networking and Computing (MobiHoc)*, Annapolis, MD, June 2003.

23. M. Grossglauser and D. Tse, Mobility increases the capacity of ad-hoc wireless networks, *Annual Joint Conf. IEEE Computer and Communication Societies (InfoCom)*, Anchorage, AK, April 2001.

24. A. El Gammal, J. Mammen, B. Prabhakar, and D. Shah, Throughput-delay trade-off in wireless networks, *Proc. Annual Joint Conf. IEEE Computer and Communication Societies (InfoCom)*, Hong Kong, March 2004.

25. L. Xie and P. R. Kumar, A network information theory for wireless communication: Scaling laws and optimal operation, *IEEE Trans. Inform. Theory* **50** (May 2004).

26. O. Arpacioglu and Z. Haas, On the scalability and capacity of wireless networks with omnidirectional antennas, *Proc. Int. Workshop on Information Processing in Sensor Networks (IPSN)*, Berkeley, CA, April 2004.

27. E. J. Duarte-Melo and M. Liu, Data-gathering wireless sensor networks: Organization and capacity, *Wireless Sensor Networks* (special issue on computer networks) **43** (2003).

28. H. El Gamal, On the scaling laws of dense wireless sensor networks, *IEEE Trans. Inform. Theory* (in press).

29. A. Chakrabarti, A. Sabharwal, and B. Aazhang, Multi-hop communication is order-optimal for homogeneous sensor networks, *Proc. Int. Workshop on Information Processing in Sensor Networks (IPSN)*, Berkeley, CA, April 2004.

30. A. Scaglione and S. D. Servetto, On the interdependence of routing and data compression in multi-hop sensor networks, *Proc. Int. Conf. Mobile Computing and Networking (MobiCom)*, Atlanta, GA, 2002.

31. D. Marco, E. J. Duarte-Melo, M. Liu, and D. Neuhoff, On the many-to-one transport capacity of a dense wireless sensor network and the compressibility of its data, *Proc. Int. Workshop on Information Processing in Sensor Networks (IPSN)*, Palo Alto, CA, April 2003.

32. K. Jain, J. Padhye, V. N. Padmanabhan, and L. Qiu, Impact of interference on multi-hop wireless network performance, *Proc. 9th Annual Int. Conf. Mobile Computing and Networking (MobiCom)*, San Diego, CA, Sept. 2003.

33. S. Coleri and P. Varaiya, Pedamacs: Power efficient and delay aware medium access protocol for sensor networks, *Proc. IEEE Global Communications Conf.* (*GlobeCom*), Dallas, Texas, July 2004.

34. W. R. Heinzelman, A. Chandrakasan, and H. Balakrishnan, Energy-efficient communication protocol for wireless microsensor networks, *Proc. Hawaii Int. Conf. System Sciences* (*HICCS*), Maui, HL, Jan. 2000.

35. A. Manjeshwar and D. P. Agrawal, Teen: A routing protocol for enhanced efficiency in wireless sensor networks, *Proc. Int. Workshop on Parallel and Distributed Computing Issues in Wireless Networks and Mobile Computing in Conjunction with the International Parallel and Distributed Processing Symposium* (*IPDPS*), San Francisco, CA, April 2001.

CHAPTER 9

Coverage and Connectivity Issues in Wireless Sensor Networks

AMITABHA GHOSH* and SAJAL K. DAS

Department of Computer Science and Engineering, University of Texas at Arlington

9.1 INTRODUCTION

Wireless sensor networks [33,34] have inspired tremendous research interest in since the mid-1990s. Advancement in wireless communication and microelectro-mechanical systems (MEMSs) have enabled the development of low-cost, low-power, multifunctional, tiny sensor nodes that can sense the environment, perform data processing, and communicate with each other untethered over short distances. A typical wireless sensor network consists of thousands of sensor nodes, deployed either randomly or according to some predefined statistical distribution, over a geographic region of interest. A sensor node by itself has severe resource constraints, such as low battery power, limited signal processing, limited computation and communication capabilities, and a small amount of memory; hence it can sense only a limited portion of the environment. However, when a group of sensor nodes collaborate with each other, they can accomplish a much bigger task efficiently. One of the primary advantages of deploying a wireless sensor network is its low deployment cost and freedom from requiring a messy wired communication backbone, which is often infeasible or economically inconvenient.

Wireless sensor networks ensure a wide range of applications [2], starting from security surveillance in military and battlefields, monitoring previously unobserved environmental phenomena, smart homes and offices, improved healthcare, industrial diagnosis, and many more. For instance, a sensor network can be deployed

---

*Current address: Department of Electrical Engineering, University of Southern California, Los Angeles (amitabhg@usc.edu).

Mobile, Wireless, and Sensor Networks: Technology, Applications, and Future Directions
Edited by Rajeev Shorey, Akkihebbal L. Ananda, Mun Choon Chan, and Wei Tsang Ooi
Copyright © 2006 John Wiley & Sons, Inc.

in a remote island for monitoring wildlife habitat and animal behavior [25], or near the crater of a volcano to measure temperature, pressure, and seismic activities. In many of these applications the environment can be hostile where human intervention is not possible and hence, the sensor nodes will be deployed randomly or sprinkled from air and will remain unattended for months or years without any battery replacement. Therefore, energy consumption or, in general, resource management is of critical importance to these networks.

Sensor nodes are scattered in a sensing field with varying node densities. Typical node densities might vary from nodes 3 m apart to as high as 20 nodes/m^3. Each node has a sensing radius within which it can sense data, and a communication radius within which it can communicate with another node. (We will discuss the models [52] for sensing and communication later.) Each of these nodes will collect raw data from the environment, do local processing, possibly communicate with each other in an optimal fashion to perform neighborhood data or decision fusion (aggregation) [23], and then route back those aggregated data in a multihop fashion to data sinks, usually called the *basestations*, which link to the outside world via the Internet or satellites. Since an individual node measurement is often erroneous because of several factors, the need for collaborative signal and information processing (CSIP) [49] is critical. Here the assumption is that the more a sensor network has access to the information scattered across different nodes, the greater the likelihood that it would be able to provide more reliable and correct information about the underlying stochastic process.

One important criterion for being able to deploy an efficient sensor network is to find optimal node placement strategies. Deploying nodes in large sensing fields requires efficient topology control [35]. Nodes can either be placed manually at predetermined locations or be dropped from an aircraft. However, since the sensors are randomly scattered in most practical situations, it is difficult to find a random deployment strategy that minimizes cost, reduces computation and communication, is resilient to node failures, and provides a high degree of area coverage [20]. The notion of area coverage can be considered as a measure of the quality of service (QoS) in a sensor network, for it means how well each point in the sensing field is covered by the sensing ranges. Once the nodes are deployed in the sensing field, they form a communication network, which can dynamically change over time, depending on the topology of the geographic region, internode separations, residual battery power, static and moving obstacles, presence of noise, and other factors. The network can be viewed as a communication graph, where sensor nodes act as the vertices and a communication path between any two nodes signifies an edge.

In a multihop sensor network, communication nodes are linked by a wireless medium, which is often unreliable and insecure. These links can be formed by radio, infrared, or optical media. Although infrared communication is license-free, cheap, and robust against interference from electrical devices, it requires line of sight between the sender and the receiver. "Smart dust" [21], which is an autonomous sensing, computing, and communication system based on optical media for transmission, also needs line of sight. Most of the current hardware for internode communication is based on radiofrequency (RF) circuit design in which securing the wireless communication links is of great concern because of the potential malicious users and eavesdroppers who can modify and corrupt data packets, insert rogue packets in the network, or launch denial-of-service (DoS) attacks. Therefore, designing proper authentication protocols and encryption algorithms for sensor networks is very important and a challenging task as well, especially because of severe resource constraints as mentioned earlier.

Routing protocols and node scheduling are two other important aspects of wireless sensor networks because they significantly impact the overall energy dissipation. Routing protocols involve primarily discovery of the best routing paths from source to destination, considering latency, energy consumption, robustness, and cost of communication. Conventional approaches such as flooding and gossiping waste valuable communication and energy resources, sending redundant information throughout the network. In addition, these protocols are neither resource-aware nor resource-adaptive. Challenges lie in designing cost-efficient routing protocols [39,37], which can efficiently disseminate information in a wireless sensor network using resource-adaptive algorithms. On the other hand, node scheduling for optimal power consumption requires identification of redundant nodes [40] in the network, which can be switched off at times of inactivity.

In this chapter, we discuss primarily the node deployment issues that are related to area coverage and network connectivity in wireless sensor networks. In Section 9.2, we introduce the notion of coverage and connectivity and state their importance with respect to different application scenarios. Section 9.3 describes the different models for sensing, communication, coverage, and other functions. We also introduce some mathematical notations and describe a few appropriate mobility models that will be applicable to mobile sensor networks. In Section 9.4 we describe the coverage algorithms based on exposure paths. In Section 9.5 we describe various deployment strategies and compare these strategies with respect to their goals, assumptions, complexities, and usefulness in practical scenarios. Section 9.6 discusses miscellaneous techniques based on node redundancy, which are used to optimize coverage and ensure connectivity. We provide a summary of our work and discuss open research problems and challenges in Section 9.7.

9.2 COVERAGE AND CONNECTIVITY

Optimal resource management and assuring reliable QoS are two of the most fundamental requirements in ad hoc wireless sensor networks. Sensor deployment strategies play a very important role in providing better QoS, which relates to the issue of how well each point in the sensing field is covered. However, due to severe resource constraints and hostile environmental conditions, it is nontrivial to design an efficient deployment strategy that would minimize cost, reduce computation, minimize node-to-node communication, and provide a high degree of area coverage, while at the same time maintaining a globally connected network is nontrivial. Challenges also arise because topological information about a sensing field is rarely available and such information may change over time in the presence of obstacles.

Many wireless sensor network applications require one to perform certain functions that can be measured in terms of area coverage. In these applications, it is necessary to define precise measures of efficient coverage that will impact overall system performance.

Historically, three types of coverage have been defined by Gage [12]:

1. *Blanket coverage* — to achieve a static arrangement of sensor nodes that maxi mizes the detection rate of targets appearing in the sensing field

2. *Barrier coverage* — to achieve a static arrangement of sensor nodes that minimizes the probability of undetected penetration through the barrier

3. *Sweep coverage* — to move a number of sensor nodes across a sensing field, such that it addresses a specified balance between maximizing the detection rate and minimizing the number of missed detections per unit area

In this chapter, we will focus mainly on the blanket coverage, where the objective is to deploy sensor nodes in strategic ways such that an optimal area coverage is achieved according to the needs of the underlying applications. Here, it is worth mentioning that the problem of area coverage is related to the traditional *art gallery problem* (AGP) [30] in computational geometry. The AGP seeks to determine the minimum number of cameras that can be placed in a polygonal environment, such that every point in the environment is monitored. Similarly, the coverage problem basically deals with placing a minimum number of nodes, such that every point in the sensing field is optimally covered under the aforementioned resource constraints, presence of obstacles, noise and varying topography.

Before proceeding further, let us introduce the notion of the *degree* of coverage. In the simplest term, the *degree* of coverage at a particular point in the sensing field can be related to the number of sensors whose sensing range cover that point. It has been observed and postulated that different applications would require different degrees of coverage in the sensing field. For instance, a military surveillance application would need a high degree of coverage, because it would want a region to be monitored by multiple nodes simultaneously, such that even if some nodes cease to function, the security of the region will not be compromised, as other nodes will still continue to function, whereas some of the environmental monitoring applications, such as animal habitat monitoring or temperature monitoring inside a building, might require a low degree of coverage. On the other hand, some specific applications might need a framework, where the degree of coverage in a network can be dynamically configured. An example of this kind of application is intruder detection, where restricted regions are usually monitored with a moderate degree of coverage until the threat or act of intrusion is realized or takes place. At this point, the network will need to self-configure and increase the degree of coverage at possible threat locations. A network that has a high degree of coverage will clearly be more resilient to node failures. Thus, the coverage requirements vary across applications and should be kept in mind while developing new deployment strategies.

Along with coverage, the notion of connectivity is equally important in wireless sensor networks. If a sensor network is modeled as a graph with sensor nodes as vertices and the communication link, if it exists, between any two nodes as an edge, then, by a *connected network* we mean that the underlying graph is connected, that is, between any two nodes there exists a single-hop or multihop communication path consisting of consecutive edges in the graph. Similar to the notion of degree of coverage, we shall also introduce the notion of degree of network connectivity. A sensor network is said to have k connectivity or be k-node-connected if removal of any $(k-1)$ nodes does not render the underlying communication graph disconnected. In latter sections, we shall provide formal definitions of k connectivity and k coverage from graph theory perspectives. Like single degree of coverage, single-node connectivity is not sufficient for many sensor network applications because the failure of a single node would render the network disconnected. It should be noted that robustness and throughput of a sensor network are directly related to connectivity.

Area coverage and connectivity in wireless sensor networks are not unrelated problems. Therefore, the goal of an optimal sensor deployment strategy is to have a globally connected network while optimizing coverage at the same time. By optimizing coverage, the deployment strategy would guarantee that optimum area in the sensing field is covered by sensors, as required by the underlying application. By ensuring that the network is connected, it is also ensured that the sensed information is transmitted to other nodes and possibly to a centralized basestation that can make valuable decisions for the application.

9.3 MATHEMATICAL FRAMEWORK

In this section, we introduce the basic mathematical framework for sensing models, communication models, coverage models, mobility models, and graph-theory-based network connectivity models applicable to wireless sensor networks. These will be used in subsequent sections for describing and analyzing the existing algorithms on coverage and connectivity and to provide future research directions.

9.3.1 Sensing Model

Each node has a sensing gradient, whose radius, although ideally extending to infinity, attenuates gradually as the distance increases. The *sensitivity* S of a sensor s_i at point P is usually modeled as follows [26]

$$S(s_i, P) = \frac{\lambda}{[d(s_i, P)]^\gamma} \tag{9.1}$$

where λ and K are positive sensor-dependent parameters and $d(s_i, P)$ is the Euclidean distance between the sensor and the point. Typically the value of γ is dependent on environmental parameters and varies between 2 and 5. Since the sensitivity rapidly decreases as the distance increases, we define a maximum sensing range for each sensor. It is customary to assume a binary sensing model, according

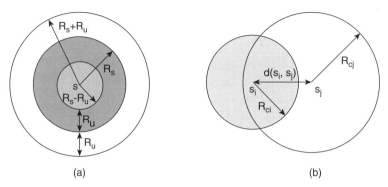

Figure 9.1   (a) Probabilistic sensing model; (b) communication model.

to which a sensor is able to sense from all the points that lie within its sensing range and any point lying beyond it is outside its sensing range. Thus, according to this model the sensing range for each sensor is confined within a circular disk of radius R_s. In a heterogeneous sensor network, the sensing radii of different types of sensors might vary, but in this chapter, to simplify the analysis of coverage algorithms, we assume that all the nodes are homogeneous and the maximum sensing radius for all of them is the same, R_s.

This binary sensing model can be extended to a more realistic one and expressed in probabilistic terms [52]. This is illustrated in Figure 9.1a. Let us define a quantity $R_u < R_s$, such that the probability that a sensor would detect an object at a distance less than or equal to $(R_s - R_u)$ is 1, and at a distance greater than or equal to $(R_s + R_u)$ is 0. In the interval $((R_s - R_u), (R_s + R_u))$, there is a certain probability p, that an object will be detected by the sensor. The quantity R_u is a measure of uncertainty in sensor detection. This probabilistic sensing model reflects the sensing behavior of devices such as infrared and ultrasound sensors.

9.3.2   Communication Model

Similar to the sensing radius, we define a communication radius R_{c_i} (see Fig. 9.1b) for each sensor s_i. Two sensors, s_i and s_j, are able to communicate with each other if the Euclidean distance between them is less than or equal to the minimum of their communication radii, that is, when $d(s_i, s_j) \leq \min\{R_{c_i}, R_{c_j}\}$. This basically means that the sensor with smaller communication radius falls within the communication radius of the other sensor. Two such nodes that are able to communicate with each other are called *one-hop neighbors*. The communication radii might vary depending on the residual battery power (energy) of an individual sensor. In this chapter, we assume that the communication radii for all the nodes are the same, denoted by R_c.

9.3.3   Coverage Model

Depending on the sensing range, an individual node will be able to sense a part of the sensing field. From the probabilistic sensing model, we define the notion of probabilistic coverage [52] of a point $P(x_i, y_i)$ by a sensor s_i by the following equations:

$$c_{x_i y_i}(s_i) = \begin{cases} 0, & R_s + R_u \leq d(s_i, P) \\ e^{-\gamma a^\beta}, & R_s - R_u < d(s_i P) < R_s + R_u \\ 1, & R_s - R_u \geq d(s_i P) \end{cases} \tag{9.2}$$

Here, $a = d(s_i, P) - (R_s - R_u)$ and γ and β are parameters that measure the detection probabilities when an object is within a certain distance from the sensor. All points that lie within a distance of $(R_s - R_u)$ from the sensor are said to be 1-covered and all points lying within the interval $((R_s - R_u), (R_s + R_u))$ have a coverage value that exponentially decreases as the distance increases and is less than 1, as observed in Equation (9.2). Beyond the distance $(R_s + R_u)$, all the points have 0 coverage by this sensor. However, a point might be covered by multiple sensors at the same time, each contributing a certain value of coverage. In the following, we define the concept of *total coverage* [52] of a point.

Definition 9.1 (Total Coverage of a Point)   Let $S = \{s_i, i = 1, 2, \ldots, k\}$ be the set of nodes whose sensing ranges cover the point $P(x_i, y_i)$. We define the total coverage of the point P as follows:

$$C_{x_i y_i}(S) = 1 - \prod_{i=1}^{k}(1 - c_{x_i y_i}(s_i)) \tag{9.3}$$

Since $c_{x_i y_i}(s_i)$ is the probabilistic coverage of a point as defined in Equation (9.2), the term $(1 - c_{x_i y_i}(s_i))$ is the probability that the point is not covered by sensor s_i. Now, since the probabilistic coverage of a point by one node is independent of another node, the product $\prod_{i=1}^{k}(1 - c_{x_i y_i}(s_i))$ of all such k terms will denote the joint probability that the point is not covered by any of the nodes. Hence, one minus this product would give the probability that point P is covered jointly by its neighboring sensors, and is defined as its total coverage. Clearly, the total coverage of a point lies in the interval [0,1].

9.3.4 Graph-Theoretic Perspective of Wireless Sensor Networks

9.3.4.1 Geometric Random Graph

Over the years, several natural phenomena have been modeled using different graph-theoretic abstractions, more specifically, using random graphs. Understanding the structural properties of such graphs provides valuable insights into the underlying physical phenomena. In this section, we provide some concepts related to graph theory that concern the notion of coverage and connectivity.

Under the particular sensing, communication and coverage models described in the previous sections, the structure of geometric random graphs (GRGs) provides the closest resemblance to wireless sensor networks. A number of probabilistic aspects related to setup of an ad hoc sensor network, such as sprinkling nodes randomly in a sensing field and simultaneous routing of information through different paths, motivates the study of GRGs in the networking community. Furthermore, it has been observed in practice that a sensor network cannot be too dense because of spatial reuse; specifically, when a particular node is transmitting, all other nodes within its transmission radius must remain silent to avoid collision and corruption of data. In this chapter, we consider the generic GRG model $G(n, r, l)$, where, instead of limiting the locations of the graph vertices within a unit square, we assume that the vertices are distributed according to a probability distribution function (pdf) in a d-dimensional space, having a length l for each dimension; and that an edge exists between any two vertices if the Euclidean distance between them is less than the communication radius. In this generic GRG model, the node density n/l^2 can converge to zero, to a constant $c > 0$, or diverge as $l \to \infty$, depending on the relative values of n, r, and l. Therefore, this model is applicable to both sparse and dense communication networks. Next, we provide a formal definition of GRGs.

Definition 9.2 (Geometric Random Graph)   We define a generic *geometric random graph* as $G(n, r, l) = (V, E)$, where a total number n of vertices are distributed according to a pdf f, in a d-dimensional space $[0, l]^d$ to form the nodes in V and an edge $(u, v) \in E$ exists between any two nodes u and v if the distance between them is less than r, namely, $d(u, v) < r$, for some $0 < r \leq l$.

Some of the results from GRG can be applied to study the connectivity in ad hoc wireless networks. For instance, if we assume that a communication graph is induced on a wireless network, then the minimum common transmission range required for all the sensors, such that the communication graph that is connected is equal to the longest Euclidean edge of the minimum spanning tree built on the GRG [31].

These kinds of results from GRGs can be analyzed using the continuum percolation theory [3]. In the theory of continuum percolation, nodes are distributed according to a Poisson density λ. The main result of the theory states that there exists a finite, positive value of λ, say, λ_c, which is called the *critical density*, such that a phase transition occurs in the graph. This means that when the node density crosses a particular threshold λ_c, the detectability of an ad hoc network becomes 1; that is, an object moving within the sensor network can be detected with probability almost equal to 1.

9.3.4.2 Graph Connectivity

In the previous sections, we introduced the concept of degree of coverage and connectivity; here we provide formal definitions for those concepts in terms of node degree and connectivity in a graph [45].

Figure 9.2   A 3-connected graph and a disconnected graph.

Definition 9.3 (Node Degree)   Let $G(V, E)$ be an undirected graph. The degree $\deg(u)$, of a vertex $u \in V$ is defined as the number of neighbors of u. The minimum node degree of G is defined as $\delta(G) = \min_{\forall u \in G} \{\deg(u)\}$.

Definition 9.4 (k-Node Connectivity)   A graph is said to be connected if for every pair of nodes, there exists a single-hop or a multihop path connecting them; otherwise the graph is called *disconnected*. A graph is said to be k-connected if for any pair of nodes there are at least k mutually independent (node-disjoint) paths connecting them. In other words, there is no set of $(k - 1)$ nodes, whose removal would render the graph disconnected or result in a trivial graph (single vertex).

Definition 9.5 (k-Edge Connectivity)   In a similar fashion, the notion of k-edge connectivity is defined when there are at least k edge-disjoint paths between every pair of nodes. In other words, there is no set of $(k - 1)$ edges whose removal will result in a disconnected graph or a trivial graph.

It can be proved [45] that if a graph is k-node-connected, then it is also k-edge-connected, but the reverse is not necessarily true. In this chapter, we shall use the term *connectivity* to mean node connectivity. In Figure 9.2, a 3-connected and a disconnected graph are shown.

Mapping these graph connectivity definitions to the wireless sensor networks scenario, we say that the communication graph formed by the sensor nodes is connected, if between every pair of nodes there exists a single-hop or multihop communication path. A sensor network would be k-connected if at least k other nodes fall within the transmission range R_c of each node. The connectivity problem in sensor networks has been approached from different angles in the literature. One such way is to assign different transmission ranges to the sensors such that the network is connected. This problem has been defined as the *critical transmission range* (CTR) assignment problem [36], which can be formulated for the case of homogeneous sensor network as follows. Given a total number *(N)* of nodes to be deployed in an area A, what is the minimum value of the transmission range to be assigned to *all* the sensors, such that the network ensures global connectivity?

We are now ready to describe various techniques that are used to ensure optimal network coverage and connectivity. In the following sections, we classify these approaches into three main categories and analyze them in terms of their goals, assumptions, algorithm complexities, and practical applicability:

1. Coverage based on exposure paths
2. Coverage based on sensor deployment strategies
3. Miscelleneous strategies

9.4 COVERAGE BASED ON EXPOSURE PATHS

Approaches to solve the coverage problem in wireless sensor networks using exposure paths is basically a combinatorial optimization problem. Two kinds of optimization viewpoints exist in formulating the coverage problem: worst-case and best-case coverage.

In the worst-case coverage, usually the problem is tackled by trying to find a path through the sensing region, such that an object moving along that path will have the least observability by the nodes. Hence, the probability of detecting the moving object would be minimum. Finding such a worst-case path is important because if such a path exists in the sensing field, a user can change the locations of the nodes or add new nodes to increase the coverage and hence observability. Two well-known methods of approaching the worst-case coverage problem are *minimal exposure path* [26] and *maximal breach path* [24,27].

On the other hand, in the best-case coverage, the goal is to find a path that has the highest observability, and hence an object moving along that path will be most probable to be detected by the nodes. Finding such a path can be useful for certain applications, including those that require the best coverage path in regions where security is of highest concern, or those that would like to maximize some predefined benefit function from the nodes while traversing the sensor field. An example of the latter kind is a solar-powered autonomous robot traversing in a light detecting sensor network so as to accumulate the most light within a certain timeframe. By using the best coverage path, the solar powered robot can gain the maximum amount of light within its limited time. Two approaches to solve the best-case coverage problem are *maximal exposure path* [42] and *maximal support path* [27]. In the following text, we describe several methods to calculate the worst-case and best-case coverage paths and the algorithms that use the concept of exposure to derive analytical results.

9.4.1 Minimal Exposure Path: Worst-Case Coverage

Exposure is directly related to the area coverage problem in sensor networks. It is a measure of how well a sensing field is covered with sensors. Informally stated, it can be defined as the *expected average ability* of observing a target moving in the sensing field. The *minimal exposure path* provides valuable information about the worst-case coverage in sensor networks. Let us first explain the notion of *exposure*, which is defined as an integral of a sensing function that is inversely proportional to the distance from the sensors, along a path between two specified points during a certain time interval [28,42]. We can state this formally as follows.

Definition 9.6 (Exposure)   The exposure of a moving object in a sensing field during time interval $[t_1, t_2]$ along a path $p(t)$ is defined as the integral:

$$E(p(t), t_1, t_2) = \int_{t_1}^{t_2} I(F, p(t)) \left| \frac{dp(t)}{dt} \right| dt \qquad (9.4)$$

where the sensing function $I(F, p(t))$ is a measure of sensitivity at a point on the path by the closest sensor or by all the sensors in the sensing field.

In the first case, it is called the *closest sensor field intensity*, defined as $I_C(F, P(t)) = S(s_{min}, P)$, where the sensitivity S is given by Equation (9.1) and s_{min} is the sensor closest to point P. In the latter case, it is called the *all-Sensor field intensity*, defined as $I_A(F, P(t)) = \sum_1^n S(s_i, P)$, where the n active sensors, $s_1, s_2, \ldots, s_n$, contribute a certain value of sensitivity to point P depending on their distance from it. In Equation (9.4), the quantity $|dp(t)/dt|$ is an arc element of the path. If the path is defined in parametric coordinates as $p(t) = (x(t), y(t))$, then

$$|dp(t)/dt| = \sqrt{(dx(t)/dt)^2 + (dy(t)/dt)^2} \tag{9.5}$$

This definition of exposure as given by Equation (9.4) makes it a path-dependent value. Given two endpoints A and B in the sensing field, different paths between them, as shown in Figure 9.3a, are likely to have different exposure values. The

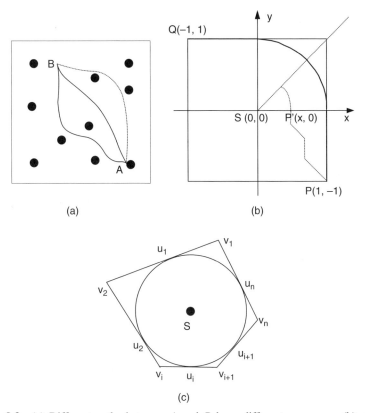

(a)   (b)

(c)

Figure 9.3 (a) Different paths between A and B have different exposures; (b) minimal exposure path for single sensor in a square sensing field; (c) minimal exposure path for single sensor in a sensing field bounded by a convex polygon.

problem of minimal exposure path is to find a path $p(t)$ in the sensing field such that the value of the integral $E(p(t), t_1, t_2)$ is minimum. In the following, we describe a few strategies to calculate the minimal exposure path.

As an example, illustrated in Figure 9.3b, it can be proved [28] that the minimal exposure path between two given points $P(1, -1)$ and $Q(-1, 1)$ in a sensing field, restricted within the region $|x| \leq 1$, $|y| \leq 1$ and having only one sensor located at $(0,0)$, consists of three segments: (1) a straight-line segment from P to $(1,0)$, (2) a quarter-circle from $(1,0)$ to $(0,1)$, and (3) another straight-line segment from $(0,1)$ to Q. The basis of the proof lies in the fact that, since any point on the dotted curve is closer to the sensor than any point lying on the straight-line segment along the edge of the square, the exposure is more in the former case. Also, since the length of the dotted curve is longer than the line segment, the dotted curve would induce more exposure when an object travels along it, given that the time duration is the same in both cases. The calculations show that the exposure along the arc of the quarter-circle in Figure 9.3b is $\pi/2$.

This method can be extended in the following way to more generic scenarios when the sensing region is a convex polygon $v_1, v_2, \ldots, v_n$ and the sensor is located at the center of the inscribed circle, as illustrated in Figure 9.3c. Let us define two curves between points v_i and v_j of the polygon as

$$\Gamma_{ij} = \overline{v_i u_i} \circ \widehat{u_i u_{i+1}} \circ \widehat{u_{i+1} u_{i+2}} \circ \cdots \circ \widehat{u_{j-2} u_{j-1}} \circ \widehat{u_{j-1} v_j}$$

$$\Gamma'_{ij} = \overline{v_i u_{i-1}} \circ \widehat{u_{i-1} u_{i-2}} \circ \widehat{u_{i-2} u_{i-3}} \circ \cdots \circ \widehat{u_{j+1} u_j} \circ \widehat{u_j v_j}$$

where $\overline{v_i u_i}$ is the straight-line segment from point u_i to v_i, $\widehat{u_i u_{i+1}}$ is the arc on the inscribed circle between two consecutive points u_i and u_{i+1}, $\circ$ denotes concatenation, and all $\pm$ operations are modulo n. It can be shown that the minimum exposure path between vertices v_i and v_j is either of the curves Γ_{ij} or Γ'_{ij}, whichever has less exposure.

Next, we extend the preceding two methods of calculating minimum exposure path under the scenario of many sensors. To simplify, the problem can be transformed from the continuous domain into a tractable discrete domain by using an $m \times n$ grid [28]. The minimal exposure path is then restricted to straight-line segments connecting any two consecutive vertices of a grid square. This approach transforms the grid into an edge–weighted graph and computes minimal exposure path using Djikstra's single-source shortest-path algorithm (SSSP) or Floyd-Warshal's all–pair shortest-path algorithm (APSP). The SSSP algorithm complexity is dominated by the grid generation process, which has a time complexity $O(n)$, where n is the total number of gridpoints. On the other hand, the APSP algorithm is dominated by the shortest-path calculation process, which has a time complexity $O(n^3)$.

A different approach based on variational calculus, due to Euler and Lagrange, has been used [42] to find a closed-form expression for minimal exposure path in case of a single sensor. In the following, we state the fundamental theorem of variational calculus and briefly describe the method from the paper by Veltri et al. [42] to derive an analytic solution for minimal exposure path. Informally stated, variational calculus is an approach to solving a class of optimization problems that seek a functional (y) to make some integral function (J) an extreme. The fundamental theorem of variational calculus states the following [11] theorem.

Theorem 9.1   Let $J[y]$ be a function of the form $J[y] = \int_b^a F(x, y, y')dx$ defined on the set of functions $y(x)$, which have continuous first-order derivatives in $[a, b]$ and satisfy the boundary condition $y(a) = A$ and $y(b) = B$. Then a necessary condition for $J[y]$ to have an extremum for a given function $y(x)$ is that $y(x)$ satisfies the Euler–Lagrange equation:

$$\frac{\partial F}{\partial y} - \frac{d}{dx}\left(\frac{\partial F}{\partial y'}\right) = 0 \tag{9.6}$$

Assuming the sensitivity of a sensor at a point P as given by $S(s_i, P) = 1/d(s_i, P)$ [$\lambda = 1$ and $\gamma = 1$ in Eq. (9.1)], the minimal exposure path between two arbitrary points A and B can be expressed in the following form using Equation (9.6) in polar coordinates $\rho(\theta) = ae^{\{[\ln(b/a)]/c\}\theta}$, where the constant a is the distance from sensor s_i to A, b is the distance from sensor s_i to B, and c is the angle formed by $\angle ASB$, as shown in Figure 9.4a. The function F in this case is given by $F = (1/\rho)\sqrt{\rho^2 + (d\rho/d\theta)^2}$, after the transformation $x = \rho \cos \theta$ and $y = \rho \sin \theta$.

For the case of multiple sensors, a grid-based approximation algorithm [42] using the Voronoi diagram can be applied. In this approach, the gridpoints are placed along the Voronoi edges and gridpoints that are part of the same Voronoi cell are connected via an edge. The weight of such an edge is determined by the single sensor minimal exposure path weight between the two points. Each node exchanges a set of messages to find topological information and uses it in the localized Voronoi-based approximation algorithm to calculate the minimal exposure path.

In addition to the methods of calculating minimum exposure path, the solution to the *unauthorized traversal* (UT) problem [8] is relevant, which is to find a path P

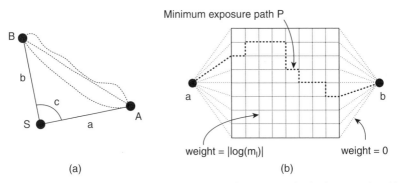

Figure 9.4   (a) Exposure path in single sensor scenario; (b) unauthorized traversal problem.

that has the least probability of detecting a moving target, given that n sensors are deployed in the sensing field. According to the coverage model described in Section 9.3.3, the probability of failure to detect a target at a point u by a sensor s is $(1 - c_u(s))$. If the decision about a target's presence is taken by a collaborative group of sensors using value fusion or decision fusion, then we can replace $c_u(s)$ by $D(u)$, where $D(u)$ is the probability of consensus target detection using value fusion or decision fusion. Thus, the net probability $G(P)$, of failure to detect a target moving in the path P, is given by

$$G(P) = \prod_{u \in P}(1 - D(u)) \Rightarrow \log G(P) = \sum_{u \in P} \log(1 - D(u)) \qquad (9.7)$$

Let us briefly describe the method of calculating a minimal exposure path in the UT algorithm. The algorithm divides the sensor field into a fine grid and assumes that the target moves only along the grid. Then finding the minimum exposure path on this grid is to find a path P that minimizes $|\log G|$.

Consider two consecutive gridpoints, v_1 and v_2. Let m_l denote the probability of failure to detect a target traveling between v_1 and v_2 along the line segment l. Then we have $\log m_l = \sum_{u \in P} \log(1 - D(u))$. Each segment l is assigned a weight $|\log m_l|$, and two fictitious points a, b and line segments with zero weights are added from them to the gridpoints as illustrated in Figure 9.4b. Thus the minimum exposure path in this configuration is to find the least-weight path from a to b, which can be identified using Dijkstra's shortest-path algorithm.

9.4.2 Maximal Exposure Path: Best-Case-Coverage

Earlier, we introduced the notion of maximal exposure path by relating it to the highest observability in a sensing field. In this section, we shall further explain the concept and state a few methods to calculate such a path. A maximal exposure path between two arbitrary points A and B in a sensing field is a path following which the total exposure, as defined by the integral in Equation (9.4), is maximum. It can be interpreted as a path having the best-case coverage. It has been proved [42] that finding the maximal exposure path is NP-hard because it is equivalent to finding the longest path in an undirected weighted graph, which is known to be NP-hard. However, there exist several heuristics to achieve near-optimal solutions under the constraints that the object's speed, pathlength, exposure value, and time required for traversal are bounded. Given these constraints, any valid path that can reach the destination before deadline is contained within an ellipse with the starting and ending points as the foci. This greatly reduces the search space for finding the optimal exposure path. In the following we describe each of the heuristics briefly [42].

1. *Random Path Heuristic.* This is the simplest heuristic to approximately calculate the maximal exposure path. In this method, a random path is created according to the rule that a node on the shortest path from source A to

destination B is selected at certain times, and a random node is selected at other times. Nodes on the shortest path are selected because of the time constraint, and random nodes are selected to collect more exposure. This approach does not depend on the network topology and is computationally inexpensive.

2. *Shortest-Path Heuristic*. In this approach, first a shortest path is calculated between the two endpoints A and B, assuming that certain topographical knowledge is available. Then, to achieve maximal exposure, an object must travel at maximum speed along this path and stop at the point with the highest exposure. However, it might not yield a good approximation because no other path, which might have more exposure, is allowed to be explored.

3. *Best-Point Heuristic*. This heuristic superimposes a grid over the ellipse and then finds the shortest path to each gridpoint from A and B. Next the total exposure of the two paths having a common gridpoint is calculated. The path that gives the maximal exposure is the optimal exposure path. The quality of the optimal path depends on the granularity of the grid; however, this approach is computationally expensive.

4. *Adjusted Best-Point Heuristic*. This method improves the best-point heuristic by considering paths that consist of multiple shortest paths. Performing one or more of the path adjustments such as moving, adding, or deleting a node on the shortest path iteratively, the optimal solution can be found.

9.4.3 Maximal Breach Path: Worst-Case Coverage

In Section 9.4.1, we discussed several methods to find a minimal exposure path in a sensing field under a single-sensor as well as multiple-sensor scenarios. We observed that finding a minimal exposure path is equivalent to finding a worst-case coverage path, which provides valuable information about node deployment density in the sensing field. A concept very similar to finding the worst-case coverage paths is the notion of maximal breach paths [27]. A maximal breach path through a sensing field starting at A and ending at B is a path such that, for any point P on the path, the distance from P to the closest sensor is maximum. The concept of the Voronoi diagram [29], a well-known construct from computational geometry, is used to find a maximal breach path in a sensing field. In two dimensions, the Voronoi diagram of a set of discrete points (also called *sites*) divides the plane into a set of convex polygons, such that all points inside a polygon are closest to only one point. In Figure 9.5a, 10 randomly placed nodes divide the bounded rectangular region into 10 convex polygons, referred to as *Voronoi polygons*. Any two nodes s_i and s_j are called *Voronoi neighbors* of each other if their polygons share a common edge. The edges of a Voronoi polygon for node s_i are the perpendicular bisectors of the lines connecting s_i and its Voronoi neighbors.

Since by construction, the line segments in a Voronoi diagram maximizes the distance from the closest sites, the maximal breach path must lie along the Voronoi edges. If it does not, then any other path that deviates from the Voronoi edges would

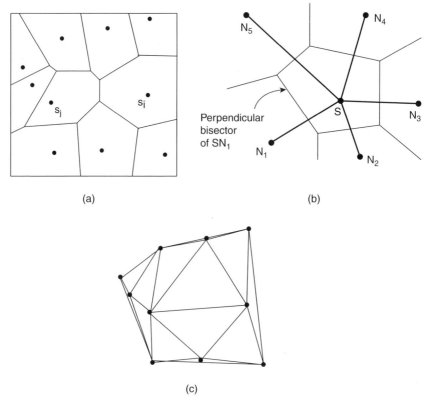

(a)

(b)

(c)

Figure 9.5 (a) Voronoi diagram of 10 randomly deployed nodes; (b) Voronoi polygon for node S, constructed by drawing perpendicular bisectors of the lines connecting S and its neighbors; (c) Delaunay triangulation for the same set of nodes.

be closer to at least one sensor, thus providing more exposure. Having said that the maximal breach path between two endpoints A and B will lie along the Voronoi edges, we now describe an algorithm that finds such a path. First a geolocation-based approach is used to determine node locations, and a Voronoi diagram based on that information is constructed. Then a weighted, undirected graph G is constructed by creating a node for each vertex and an edge corresponding to each line segment in the Voronoi diagram. Each edge is given a weight equal to the minimum distance from the closest sensor. The algorithm then checks the existence of a path from A to B using breadth-first search (BFS) and then uses binary search between the smallest and largest edge weights in G to find the maximal breach path. It should be noted that the maximal breach path is not unique. It can be proved that the worst-case time complexity of the algorithm is given by $O(n^2 \log n)$, and for sparse networks it is $O(n \log n)$.

Furthermore, the maximal breach path algorithm finds a path such that at any given time, the exposure is no more than some particular value that it tries to minimize. On the other hand, the minimal exposure path does not focus on exposure at one particular time, but rather tries to minimize the exposure acquired throughout the entire time interval in the network.

9.4.4 Maximal Support Path: Best-Case Coverage

A maximal support path through a sensing field starting at A and ending at B is a path such that for any point P on that path, the distance from P to the closest sensor is minimized. This is similar to the concept of maximal exposure path. However, the difference lies in the fact that a maximal support path algorithm finds a path at any given time instant, such that the exposure on the path is no less than some particular value that should be maximized. In contrast, the maximal exposure path does not focus on any particular time; rather, it considers all the time spent during an object's traversal.

A maximal support path in a sensing field can be found by replacing the Voronoi diagram by its dual, Delaunay triangulation as shown in Figure 9.5b, where the edges of the underlying graph are assigned weights equal to the length of the corresponding line segments in the Delaunay triangulation. (A *Delaunay triangulation* [29] is a triangulation of graph vertices such that the circumcircle of each Delaunay triangle does not contain any other vertices.) Similar to the maximal breach path approach described earlier, this algorithm also checks for the existence of a path using breadth-first search and applies binary search to find the maximal support path. The worst-case and average-case complexities for this algorithm are $O(n^2 \log n)$ and $O(n \log n)$, respectively.

So far we have described several methods to derive worst-case and best-case coverage paths exploiting the concept of exposure to detect targets in a sensing field. Now we will see that exposure paths can also be used to find the optimal number of sensors (critical node density) required for complete coverage with very high target detectability [1]. Since the sensing task is inherently probabilistic, the method for critical density calculation takes into account the nature and characteristics of both the sensor and the target. We consider the path-based exposure model as described in Equation (9.4) and that the target moves in a straight line with constant speed away from the sensor at a distance δ. Assuming the probabilistic sensing model as described in Section 9.3.1, typical values are calculated for the quantities $(R_s - R_u)$ and $(R_s + R_u)$, which are termed as *radius of complete influence* (denoted by R_{ci}) and *radius of no influence* (denoted by R_{ni}), respectively. It can be proved that for a typical threshold exposure E_{th}, the values for radius of complete influence and no influence are given by the following equations [1]

$$E_{th} = \frac{\lambda}{v R_{ci}} \left(\frac{\delta}{\delta + R_{ci}} \right) \tag{9.8}$$

$$E_{th} = \frac{2\lambda}{v R_{ni}} tan^{-1} \left(\frac{\delta}{2 R_{ni}} \right) \tag{9.9}$$

and that to cover an area A with random deployment, the number of nodes required is of the order of $O(A/R_{ni}^2)$.

9.5 COVERAGE BASED ON SENSOR DEPLOYMENT STRATEGIES

The second approach to the coverage problem is to seek sensor deployment strategies that would maximize coverage as well as maintain a globally connected network graph. Several deployment strategies have been studied for achieving an optimal sensor network architecture that would minimize cost, provide high sensing coverage, be resilient to random node failures, and so on. In certain applications, the locations of the nodes can be predetermined and hence can be hand-placed or deployed using mobile robots, while in other cases we need to resort to random deployment methods, such as sprinkling nodes from an aircraft. However, random placement does not guarantee full coverage because it is stochastic in nature, hence often resulting in accumulation of nodes at certain areas in the sensing field but leaving other areas deprived of nodes. Keeping this in mind, some of the deployment algorithms try to find new optimal sensor locations after an initial random placement and move the sensors to those locations, achieving maximum coverage. These algorithms are applicable to only mobile sensor networks. Research has also been conducted in mixed-sensor networks, where some of the nodes are mobile and some are static; and approaches are also proposed to detect coverage holes after an initial deployment and to try to heal or eliminate those holes by moving sensors. It should be noted that an optimal deployment strategy should not only result in a configuration that would provide sufficient coverage but also satisfy certain constraints such as node connectivity and network connectivity [32].

As mentioned in the introduction, the problem of sensor deployment is related to the traditional *art gallery problem* (AGP) [30] in computational geometry. The AGP seeks to determine the minimum number of cameras that can be placed in a polygonal environment, such that the entire environment is monitored. In a similar way, an optimal deployment strategy tries to deploy nodes at optimal locations, such that the area covered by the sensors is maximized. In the following, we briefly describe several sensor deployment algorithms targeted for static, mobile, and mixed-sensor networks, that aim to provide optimum sensing field architecture.

9.5.1 Imprecise Detections Algorithm (IDA)

Dhilon et al. [9] propose a grid coverage algorithm that ensures that every gridpoint is covered with a minimum confidence level. They consider a *minimalistic* view of a sensor network by deploying a minimum number of sensors on a grid that would transmit a minimum amount of data. The model assigns two probability values p_{ij} and p_{ji} for every pair of gridpoints (i,j), where p_{ij} is the probability that a target at gridpoint j is detected by a sensor at gridpoint i and p_{ji} is the probability that a target at gridpoint i is detected by a sensor at gridpoint j. In absence of obstacles, these values are symmetric: $p_{ij} = p_{ji}$. From this, a miss probability matrix M is generated where $m_{ij} = (1 - p_{ij})$. The obstacles are modeled as static objects, and the value of p_{ij} is set to zero if an obstacle appears in the line of sight between two gridpoints (i,j) (as illustrated in Fig. 9.6).

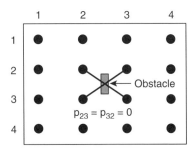

Figure 9.6 Two probability values, p_{ij} and p_{ji}, are assigned for every pair of gridpoints (i, j) under line-of-sight static obstacle modeling.

The algorithm as described by Dhilon et al. [9] takes three inputs: (1) M, M^*, $M_{\min}$, where M is the miss probability matrix as mentioned above; (2) $M^* = (M_1, M_2, \ldots, M_N)$, such that M_i is the probability that a gridpoint i is not *collectively*[1] covered by the set of sensors; and (3) $M_{\min} = 1 - T$, which is the maximum value of the miss probability that is permitted for any gridpoint. The algorithm is iterative and uses a greedy heuristic to determine the best placement of one sensor at a time. It terminates when either a preset upper limit on the number of sensors is reached or sufficient coverage of the gridpoints is achieved.

The time complexity of the algorithm is $O(n^2)$, where n is the total number of gridpoints in the sensor field. It attempts to evaluate the global impact of an additional sensor by summing up the changes in the miss probabilities for the individual gridpoints. However, the algorithm models the obstacles depending on whether they appear in the line of sight of the target and the sensor, which is applicable for infrared cameras, for example, but not for sensors that do not require line of sight, such as acoustic and temperature sensors. Also, since a complete knowledge of the terrain is assumed, the algorithm is not very applicable in cluttered environments, such as interior of buildings, because modeling obstacles becomes extremely difficult in those scenarios.

9.5.2 Potential Field Algorithm (PFA)

In contrast to static sensor networks, nodes in mobile sensor networks are capable of moving in the sensing field. Such networks are capable of self-deployment starting from an initial configuration. The nodes would spread out such that coverage in the sensing field is maximized while maintaining network connectivity. A potential field-based deployment approach using mobile autonomous robots has been proposed to maximize the area coverage. [18,32]. Poduri and Sukhatme augment the scheme such that each node has at least K neighbors. The potential field technique using mobile robots was first introduced in 1986 [22]. In the following we describe the concept of potential field and the algorithm proposed by Poduri and Sukhatme [32].

---

[1]The notion of collective or total sensor coverage of a point is expressed in Equation (9.3).

The basic concept of potential field is that each node is subjected to a force $\mathbf{F}$ (a vector)[2] that is the gradient of a scalar potential field U; that is, $\mathbf{F} = -\nabla U$. Each node is subjected to two kinds of force: (1) $\mathbf{F}_{cover}$, which causes the nodes to repel each other to increase their coverage; and (2) $\mathbf{F}_{degree}$, which constrains the degree of nodes by making them attract toward each other when they are on the verge of being disconnected. The forces are modeled as being inversely proportional to the square of the distance between a pair of nodes and obey the following two boundary conditions:

1. $\|\mathbf{F}_{cover}\|$ tends to infinity when the distance between two nodes approaches zero to avoid collision.
2. $\|\mathbf{F}_{degree}\|$ tends to infinity when the distance between critical neighbors approaches R_c, the communication radius.

In mathematical terms, if $\|X_i - X_j\| = \Delta x_{ij}$ is the Euclidean distance between two nodes, i and j, then $\mathbf{F}_{cover}(i,j)$ and $\mathbf{F}_{degree}(i,j)$ can be expressed as

$$\mathbf{F}_{cover}(i,j) = \frac{-K_{cover}}{\Delta x_{ij}^2}\left(\frac{x_i - x_j}{\Delta x_{ij}}\right) \tag{9.10}$$

$$\mathbf{F}_{cover}(i,j) = \begin{cases} \dfrac{-K_{degree}}{(\Delta x_{ij} - R_c)^2}\left(\dfrac{x_i - x_j}{\Delta x_{ij}}\right), & \text{for critical connection} \\ 0, & \text{otherwise} \end{cases} \tag{9.11}$$

In the initial configuration all the nodes are accumulated in one place, and thus each node has more than K neighbors, assuming that the total number of nodes is $\geq K$. Then, they start repelling each other using F_{cover} until there are only K neighbors left, at which point the connections reach a critical level, and none of these connections should be broken at a later point of time to ensure K connectivity. Each node continues to repel all its neighbors using $\mathbf{F}_{cover}$, but as the distance between the node and its critical neighbors increases, $\|\mathbf{F}_{cover}\|$ decreases and $\|\mathbf{F}_{degree}\|$ also increases. Finally, at some distance ηR_c, where $0 < \eta < 1$, the net force $\|\mathbf{F}_{cover} + \mathbf{F}_{degree}\|$ becomes 0, at which point each node and its neighbors reach an equilibrium and the sensing field becomes uniformly covered with nodes. At a latter point, if a new node joins the network or an existing node ceases to function, the nodes will need to reconfigure to satisfy the equilibrium criteria.

9.5.3 Virtual Force Algorithm (VFA)

Similar to the potential field approach as described in by Poduri and Sukhatme [32], a sensor deployment algorithm based on virtual forces has been proposed [50,52] to increase the coverage after an initial random deployment. Since a random placement does not guarantee effective coverage, an approach that modifies the sensor

---

[2] The bold symbol $\mathbf{X}$ represents a vector, and $\|\mathbf{X}\|$ represents the magnitude of the vector.

locations after a random placement is useful. In this section, we describe the virtual force algorithm (VFA) briefly.

A sensor is subjected to three kinds of force, which are either attractive or repulsive in nature. In the VFA model, obstacles exert repulsive forces ($\mathbf{F}_{iR}$), areas of preferential coverage (sensitive areas where a high degree of coverage is required) exert attractive forces ($\mathbf{F}_{iA}$), and other sensors exert attractive or repulsive forces ($\mathbf{F}_{ij}$), depending on the distance and orientation. A threshold distance d_{th} is defined between two sensors to control how close they can approach each other. Likewise, a threshold coverage c_{th} is defined for all gridpoints such that the probability that a target at any given gridpoint is reported as being detected is greater than this threshold value. The coverage model as described in this algorithm is given by Equations (9.2) and (9.3). The net force on a sensor s_i is the vector sum of all three forces:

$$\mathbf{F}_i = \sum_{j=1, j\neq i}^{k} \left(\mathbf{F}_{ij}\right) + \mathbf{F}_{iR} + \mathbf{F}_{iA} \tag{9.12}$$

The term $\mathbf{F}_{ij}$ can be expressed in polar coordinates with magnitude and orientation as

$$\mathbf{F}_{ij} = \begin{cases} \left(w_A\left(d_{ij} - d_{th}\right), \alpha_{ij}\right), & \text{if } d_{ij} > d_{th} \\ 0, & \text{if } d_{ij} = d_{th} \\ \left(w_R/d_{ij}, \alpha_{ij} + \pi\right), & \text{otherwise} \end{cases} \tag{9.13}$$

where d_{ij} is the distance between sensors s_i and s_j, α_{ij} is the orientation of the line segment from s_i to s_j, w_A and w_R are measures of attractive and repulsive forces, respectively. The VFA algorithm is a centralized one, and it executes in a cluster head. After the nodes are randomly placed in the sensing field, for all gridpoints, the algorithm calculates the total coverage as defined by Equation (9.3). Then it calculates the virtual forces exerted on a sensor s_i by all other sensors, obstacles, and preferential coverage area, for all i. Next, depending on the net forces, new locations are calculated by the cluster head and sent to the sensor nodes, which perform a one-time movement to the designated positions.

For an $n \times m$ grid with a total number of k sensors deployed, the computational complexity of the VFA algorithm is $O(nmk)$. The efficiency of the algorithm depends on the values of the quantities w_A and w_R. Negligible computation time and a one-time repositioning of sensors are two of its primary advantages. However, the algorithm does not provide any route plan for repositioning the sensors to avoid collision.

9.5.4 Distributed Self-Spreading Algorithm (DSSA)

Along the lines of potential field and virtual force based approaches, a distributed self-deployment algorithm (DSSA) has been proposed [16] for mobile sensor networks that maximizes coverage and maintains uniformity of node distribution. They define *coverage* as the ratio of the union of covered areas of each node to the complete area of the sensing field and *uniformity* as the average of local standard deviations of internodal distances. In uniformly distributed networks, internodal distances are almost the same and hence the energy consumption is uniform. DSSA assumes that the initial deployment is random and that each node knows its location. Similar to VFA, it uses the concept of electric force that depends on the internode separation distance and local current density (μ_{curr}). In the beginning of the algorithm, the initial density for each node is equal to the number of its neighbors. The algorithm defines the notion of expected density as the average number of nodes required to cover the entire area when the nodes are deployed uniformly. It is given by $\mu(R_c) = (N\pi R_c^2)/A$, where N is the number of sensors and R_c is the communication range. DSSA executes in steps and models the force on the ith node by the jth node at timestep n as

$$f_n^{i,j} = \frac{\mu_{curr}}{\mu^2(R_c)} \left(R_c - \left|p_n^i - p_n^j\right|\right) \left(p_n^i - p_n^j\right)/\left(\left|p_n^i - p_n^j\right|\right) \qquad (9.14)$$

where p_n^i denotes the location of ith node at timestep n. Depending on the net forces from the neighborhood, a node can decide on its next movement location. The algorithm settles down when a node moves an infinitely small distance over a period of time or when it moves back and forth between two same locations.

9.5.5   VEC, VOR and MiniMax Algorithms

Wang et al. [44], describe three distributed self-deployment algorithms (VEC, VOR, and min–max) for mobile sensors using Voronoi diagrams. After the sensors are deployed in the field, the algorithm locates coverage holes (area not covered by any sensor) and calculates new positions that would increase coverage by moving sensors from densely populated regions to sparsely ones. The Voronoi diagram, as explained in Section 9.4.3, consists of Voronoi polygons such that all the points inside a polygon are closest to the sensor that lies within the polygon, as illustrated in Figure 9.7a. Once the Voronoi polygons are constructed, each sensor within the

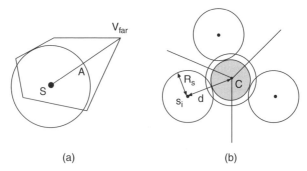

(a)   (b)

Figure 9.7   (a) The VOR algorithm moves a sensor toward the farthest Voronoi vertex, V_{far}; (b) bid estimated by sensor S_i is the area of the shaded circle with center at C.

polygon examines the existence of possible coverage holes. If such a hole is discovered, the sensors move to new positions according to certain heuristics to reduce or eliminate the coverage hole. In the following we explain the heuristics.

The vector-based algorithm (VEC) pushes sensors from densely covered areas to sparsely covered areas. Two sensors exert a repulsive force when they are too close to each other. If d_{av} is the average distance between any two sensors when they are evenly distributed in the sensing field, the virtual force between the sensors s_i and s_j will move each of them $(d_{av} - d(s_i, s_j))/2$ distance away from each other. In case, one of the sensor's sensing range completely covers its Voronoi polygon, only the other sensor should move away $(d_{av} - d(s_i, s_j))$ distance. In addition to the mutual repulsive forces between sensors, the boundaries also exert forces to push sensors too close to the boundary inside. If $d_b(s_i)$ is the distance of a sensor s_i from its closest boundary, then the repulsive force would move it a distance $d_{av}/2 - d_b s_i$ toward the inside of the region. Before actually moving to the new position, each sensor calculates whether its movement would increase the local coverage within its Voronoi polygon. If not, the sensor wouldn't move to the target location; instead it applies a *movement adjustment scheme* and will move to the midpoint position between its target location and new location.

The Voronoi-based algorithm (VOR) is a greedy algorithm that pulls sensors toward their local maximum coverage holes. If a sensor detects a coverage hole within its Voronoi polygon, it will move toward its farthest Voronoi vertex (V_{far}), such that the distance from its new location (A) to (V_{far}) is equal to the sensing range (see Figure 9.7a). However, the maximum moving distance for a sensor is limited to at most half the communication range, because the local view of the Voronoi polygon might be incorrect because of limitations in communication range. VOR also applies the *movement adjustment scheme* as in VEC and additionally applies an *oscillation control scheme* that limits a sensor's movement to opposite directions in consecutive rounds.

The min–max algorithm is very similar to VOR, but it moves a sensor inside its Voronoi polygon to a point such that the distance from its farthest Voronoi vertex is minimized. Since moving a sensor to its farthest Voronoi vertex might lead to a situation such that the vertex that was originally close now becomes a new farthest vertex, the algorithm positions each sensor such that no vertex is too far away from the sensor. The authors define the concept of min–max circle, the center of which is the new targeted position. To find the min–max circle, all circumcircles of any two and any three Voronoi vertices are found and the one with minimum radius covering all the vertices is the min–max circle. The time complexity of this algorithm is in the cubic order of the number of Voronoi vertices.

9.5.6 Bidding Protocol (BIDP)

The algorithms described in the previous sections (PFA, VFA, DSSA, VEC, VOR, min–max) deal with sensor networks where all the nodes are mobile. However, there is a high cost associated with rendering each node mobile. Instead, a balance can be achieved by using both static and mobile sensors (mixed-sensor network), while still ensuring sufficient coverage. Wang et al. [43] describe such a protocol, called the *bidding protocol*, for mixed sensor networks. They reduce the problem to the NP-hard set covering problem and provide heuristics to solve it near-optimally.

Initially, a mixture of static and mobile nodes are randomly deployed in the sensing field. Next, the static sensors calculate their Voronoi polygons and find coverage holes with their polygons and also bid to the mobile sensors to move to holes' locations. If a hole is found, a static sensor chooses the location of the farthest Voronoi vertex as the target location of the mobile sensor and calculates the bid as $\pi(d - R_s)^2$, where d is the distance between the sensor and the farthest Voronoi vertex and R_s is the sensing range (see Fig. 9.7b). A static sensor then finds a closest mobile sensor whose base price (each mobile sensor has an associated base price that is initialized to zero) is lower than its bid and sends a bidding message to this mobile sensor. The mobile sensor receives all such bids from its neighboring static sensors and chooses the highest bid and moves to heal that coverage hole. The accepted bid becomes the mobile sensor's new base price. This approach ensures that a mobile sensor does not move to heal a coverage hole when its departure generates a larger hole in its original place. The authors also incorporate a self-detection algorithm to ensure that no two mobile sensors move to heal the same coverage hole. They also apply the *movement adjustment scheme* as described in VEC, to push sensors away from each other if their movement can guarantee more coverage.

9.5.7 Incremental Self-Deployment Algorithm (ISDA)

Howard and colleagues [17,19] presented an incremental and greedy self-deployment algorithm for mobile sensor networks, in which nodes are deployed one at a time into an unknown environment. Each node makes use of the information gathered by previously deployed nodes to determine its optimal deployment location. The algorithm ensures maximum coverage but at the same time guarantees that each node remains in line of sight with at least another node. Conceptually it is similar to the frontier-based approach [46], but here, occupancy maps are built from live sensory data and are analyzed to find frontiers between free space and unknown space. In the following, we highlight the four phases of the algorithm.

1. *Initialization Phase.* In this phase, nodes are assigned one of the three states: waiting, active, or deployed with the exception of a single node that acts as an anchor and is already deployed.

2. *Goal Selection Phase.* In this next phase an optimal location is chosen for the next node to be deployed on the basis of previously deployed sensors. The concept of *occupancy grid* [10] (see Fig. 9.8b) is used as the first step to global map building. Each cell is assigned a state of either free (known to contain no obstacles), occupied (known to contain one or more obstacle) or unknown. However, not all free space represents valid deployment locations because nodes have finite size and a free cell that is close to an occupied cell may not be reachable. Hence, the occupancy grid is further processed to build a *configuration grid* (see Fig. 9.8c). In a configuration grid, a cell is free if all

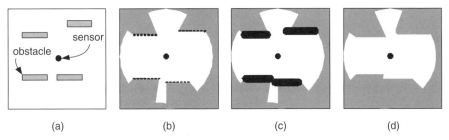

(a)   (b)   (c)   (d)

Figure 9.8   (a) Environment with obstacles and a single sensor; (b) occupancy grid — black cells are occupied, gray ones are unknown, and white ones are free; (c) configuration grid — black cells are occupied, gray ones are unknown, and white ones are free; (d) reachability grid–white cells are reachable, gray ones are unknown.

the occupancy grid cells lying within a certain distance are also free. A cell is occupied if there are one or more occupancy grid cells lying within a certain distance are similarly occupied. All other cells are marked as unknown. Once this global map is built, the goal selection phase chooses a location based on certain policies.

3. *Goal Resolution Phase.* Next, this new location is assigned to a waiting node in the goal resolution phase, and a plan for reaching the goal is generated applying a distance transform (also called *flood-fill algorithm*) on the configuration gird, giving rise to a *reachability grid* (see Fig. 9.8d). Thus, the set of reachable cells is a subset of the set of free configuration cells, which in turn is a subset of the set of free occupancy cells. A distance of 0 is assigned to the goal cell (the cell which is chosen to be the optimal location for the next node to be deployed), a distance of 1 to cells adjacent to the goal cell, a distance of 2 to their adjacent cells, and so on. However, distances are not propagated through occupied or unknown cells. Thus, for each node the distance to the goal and whether the goal can be reached is determined.

4. *Execution Phase.* In this phase, the active nodes are deployed sequentially to their respective goal locations. The nodes end up moving in a "conga line"; specifically, as the lead node moves forward, the node immediately behind it steps forward to take its place; this node in turn is replaced by the one behind it, and so on.

9.5.8   Integer Linear Programming Algorithm (ILPA)

Chakrabarty et al. [4] model the optimization problem of coverage with integer linear programming (ILP) and represent the sensor field as a two- or three-dimensional grid. Given a variety of sensors with different ranges and costs, they provide strategies for minimizing the cost, provide coding-theoretic bounds on the number of sensors, and present methods for their placement with desired coverage. Their approach of maximizing coverage in the sensing field is different in the sense that it determines a deployment strategy, such that every gridpoint is covered by a unique subset of sensors. In this way, the set of sensors reporting a target at a particular time uniquely identifies the grid location for the target at that time.

9.5.9 Uncertainty-Aware Sensor Deployment Algorithm (UADA)

In most of the sensor deployment algorithms discussed so far, the optimal positions of the sensors are determined for maximizing coverage. However, there is an inherent uncertainty in sensor locations when sensors are dispersed, scattered, or airdropped. Hence, for every point in the sensing field there is only a certain probability of a sensor being located at that point. Zou and Chakrabarty [51] present two algorithms for efficient placement of sensors when exact locations are not known.

The sensor locations are modeled as random variables following Gaussian distribution. Let the intended sensor locations (x, y) be taken as mean values and σ_x, σ_y as standard deviations in the x and y dimensions, respectively. Assuming that these deviations are independent, the join distribution $p_{xy}(x', y')$ of a sensor's actual location is calculated. Then, the uncertainty in sensor location is modeled by a conditional probability $c_{ij}^*(x, y)$, for a gridpoint (i, j) to be detected by a sensor that is supposed to be deployed at (x, y). Hence, the miss probability (probability of missing) of a gridpoint (i, j) due to a sensor at (x, y) is calculated as $m_{ij}(x, y) = 1 - c_{ij}^*(x, y)$. From this, the collective miss probability of the gridpoint (i, j) due to a set L_s of already deployed sensors is given by $m_{ij} = \prod_{(x,y) \in L_s} (1 - c_{ij}^*(x, y))$. The algorithm then determines the location of the sensors one at a time. It finds out all possible locations that are available on the grid for the next sensor to be deployed and calculates the overall miss probability $m(x, y)$, due to the already deployed sensors and this sensor, assuming that it will be placed at (x, y): $m(x, y) = \sum_{(i,j) \in \text{Grid}} m_{ij}(x, y) m_{ij}$. Based on the $m(x, y)$ values, the current sensor can be placed at gridpoint (i, j) with maximum overall miss probability (worst-case coverage) or minimum overall miss probability (best-case coverage). Once the best location is found, the miss probabilities are updated and the process continues until each gridpoint is covered with a minimum confidence level. The complexity of the first phase of the algorithm where it calculates the conditional and miss probabilities is $O((mn)^2)$, for a $m \times n$ grid. The computational complexity of the second phase where the algorithm deploys the sensors is $O(mn)$.

9.5.10 Comparison of the Deployment Algorithms

The various sensor deployment strategies discussed in the previous sections, have different assumptions and goals depending on the underlying application requirements and the nature of the sensor network. Some of those strategies are applicable to mobile sensor networks, whereas some other ones are applicable only to static sensor networks. Then, a couple of the algorithms work under the scenarios having a mixture of static and mobile nodes. Therefore, the algorithms vary in terms of their applicability, complexity, and several other factors. In this section, we compare these sensor deployment strategies on the basis of their goals, advantages, disadvantages, performance, computation complexity, and applicability. These comparisons are summarized in Table 9.1.

TABLE 9.1 Comparison of Sensor Deployment Strategies

Algorithm	Network and Goals	Advantages	Disadvantages	Performance
IDA [9]	Static; minimize the number of sensors and communication traffic	Minimum confidence level for each grid; allows modeling of obstacles and preferential areas	Complete terrain knowledge assumed and sensor detections assumed to be independent	Outperforms random deployment in case of obstacles and preferential coverage; $O(n^2)$ time complexity, where n is the number of gridpoints
PFA [22,32]	Mobile; redeploy mobile nodes from an initial configuration to maximize coverage while maintaining at least k-connectivity	Good coverage without global maps; does not require centralized control, localization; hence scalable	Computationally expensive and assumes that each node can sense the exact relative range and bearing of its neighbors	Outperforms random deployment but poorer in performance than tiled networks (networks in which nodes are deployed in tiled patterns, e.g., hexagonal, triangular)
VFA [50,52]	Mobile; redeploy mobile nodes from an initial random placement to enhance coverage	One time computation and sensor location determination; allows modeling of obstacles and preferential areas	Centralized and extra computational capability of cluster head; no route plan for repositioning of nodes; efficiency depends on the force parameters; discrete coordinate system	Outperforms random placement. $O(nmk)$ time complexity for a $n \times m$ grid with k sensors deployed
DSSA [16]	Mobile; spread nodes from an initial random deployment to maximize coverage and maintain uniformity	Distributed self-deployment algorithm	Every node should know its own location; obstacles and preferential coverage areas are not modeled	Outperforms simulated annealing [41] (forms the basis of an optimization technique for combinatorial problems) in trms of uniformity, deployment time, and the mean distance traveled by the nodes to reach their final locations.

(continued)

247

TABLE 9.1 (*Continued*)

Algorithm	Network and Goals	Advantages	Disadvantages	Performance
VEC, VOR, min–max, [44]	Mobile; reduce or eliminate coverage holes by relocating mobile sensors	Distributed algorithms; extensible to large deployment because communication and movements are local	Poor performance on initial clustered deployment and lower communication range	Perform better if initial deployment is random rather than clustered; does not perform well in case of insufficient communication range
BIDP [43]	Mixed; reduce or eliminate coverage holes and minimize cost by relocating mobile nodes	Distributed protocol; provides cost balance by using a combination of static and mobile nodes	No obstacle modeling; performance depends on ratio of mobile to static sensors	Coverage increases as the percentage of mobile sensors increases; however, duplicate healing occurs and average movement distance of sensors increases along with it
ISDA [17,19]	Mobile; deploy mobile nodes in an unknown environment to maximize coverage while retaining line of sight	Incremental and greedy algorithm; not dependent on prior environment models; global maps are built from live sensory data	It takes a long time; difficult to make it scalable for large networks	Coverage increases linearly with the number of deployed sensors; $O(n^2)$ worst-case time complexity, where n is the number of deployed sensors
ILPA [4]	Static; provide maximum grid coverage for surveillance and target detection, while minimizing the cost of sensors	Targets can be uniquely identified from the subset of sensors that detect the targets	High computational complexity makes it infeasible for large-scale deployment; relies on perfect binary sensing model	Computationally very expensive
UADA [51]	Static; determine minimum number of sensors and their locations under constraint of imprecise detection and terrain properties	Each gridpoint is covered with a minimum confidence level; models sensor locations as random variables	Sensor detections are assumed to be independent; computationally expensive	Time complexity of first phase of algorithm is $(O(mn)^2)$ and that of second phase is $O(mn)$, for a $m \times n$ gird

9.6 MISCELLANEOUS STRATEGIES

Our discussion so far has concerned mainly algorithms that guarantee optimal coverage of the sensing field. However, as mentioned earlier, a sensor network needs to be connected as well, so that the data sensed by the nodes can be transmitted by multihop communication paths to other nodes and possibly to a basestation where intelligent decisions can be made. Therefore, it is equally important for a coverage algorithm to ensure a connected network. In this section, we will discuss a few techniques that ensure coverage as well as connectivity in a sensing field while at the same time reduce redundancy and increases overall network lifetime.

It is envisioned that a typical wireless sensor network would consist of large numbers of energy-constrained nodes deployed with high density. In such a network, it is sometimes undesirable to have all the nodes in the active state simultaneously, because there would be redundancy in sensing and excessive packet collisions. Also, keeping all the nodes active simultaneously would dissipate energy at a much faster rate and would reduce overall system lifetime. Hence, it is important to turn off the redundant nodes and maximize the time interval of a continuously monitoring, transmitting, or receiving function. Scheduling of nodes that would control the density of active nodes in a sensor network has been the focus of many research works. An optimal scheduling scheme ensures that only a subset of nodes are active at any given point of time, while satisfying the following two requirements relating coverage and connectivity:

1. The area that can be monitored by the working set of nodes is not smaller than the area that can be monitored by the set of all nodes.
2. Network connectivity is maintained even after turning off the redundant nodes.

Zhang and Hou [48] proposed a decentralized and localized density control algorithm [Optimal geographic density control (OGDC)] based on certain optimality conditions of coverage and connectivity for large-scale sensor networks. They investigated the relation between coverage and connectivity and proved that if the communication range is at least twice the sensing range ($R_c \geq 2R_s$), then complete coverage of an area guarantees a connected network. The OGDC algorithm tries to minimize the overlap of sensing areas of all the nodes and finds a node scheduling scheme. It defines the notion of a *crossing point* as an intersection point of the sensing circles of two nodes (see Fig. 9.9a) and proves that to cover one crossing point of two nodes with minimum overlap, only one other node should be used and the centers of the three nodes should form an equilateral triangle with side-length $\sqrt{3}R_s$. As illustrated in Figure 9.9a, nodes A and B have two crossing points. To cover that crossing point optimally, another node C should be placed such that the centers of the three nodes form an equilateral triangle $\triangle ABC$. Furthermore, to cover one crossing point of two nodes whose positions are fixed (i.e., with x_1 fixed), only one disk[3] should be used and $x_2 = x_3 = (\pi - x_1)/2$.

---

[3]Refer to Section 9.3.1 for definition of a *disk*.

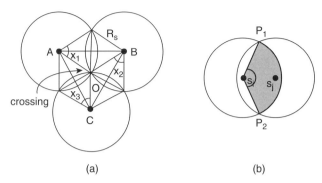

(a)   (b)

Figure 9.9   (a) Optimal positions of sensors to minimize overlap; (b) sponsored sector.

Tian and Georganas [40] proposed a self-scheduling scheme that can reduce overall energy consumption and increase system lifetime by exploiting the redundancy of nodes. Their approach is based on sponsorship criteria, by which each node decides whether to turn itself off or on using only local neighborhood information. We define the notion of sponsorship and the algorithm in the following.

Definition 9.7 (Sponsor Nodes)   Let $N(i)$ denote the set of one-hop neighbors of node s_i. Node s_i is said to be sponsored by its one-hop neighbors if the union of its neighbors' sensing areas is a superset of node s_i's sensing area. If we denote the sensing area of a node s_i as $S(i)$, then the sponsorship criterion is $\bigcup_{j \in N(i)} S(j) \supseteq S(i)$.

Definition 9.8 (Sponsored Sector)   Let the sensing areas of node s_i and one of its one-hop neighbors s_j intersect at points P_1 and P_2, respectively, as shown in Figure 9.9b. The area bounded by radius s_iP_1, radius s_iP_2, and the inner arc $\widehat{P_1P_2}$ is called the *sponsored sector* of node s_i by node s_j. The central angle of the sector is denoted as $Q_{j \to i}$, which lies in the interval $120 \leq \theta_{j \to i} \leq 180$.

Gao et al. [13] proved that at least three and at most five one-hop neighbors are needed to cover the whole sensing area of node s_i.

The algorithm described by Tian and Georganas [40] consists of two phases: self-scheduling phase and sensing phase. In the self scheduling phase, each sensor broadcasts its position and node id, and listens to the advertisement messages from its neighbors to obtain their location information. Then it calculates the sponsored sectors by its neighbors and checks whether the union of their sponsored sectors can cover its own sensing area. If so, it decides to turn itself off. However, if all the nodes make decisions simultaneously, blindspots might appear. To avoid such a situation, each node waits a random period of time and also broadcasts its status message to other nodes. In this way the nodes self-schedule, thus reducing energy consumption while maintaining the original coverage area.

Ye et al. [47] described a distributed localized algorithm for density control based on probing mechanism. In their algorithm, each node can be in one of the three states: sleeping, wakeup, or working. A working node is responsible for sensing and data communication, while nodes in wakeup state prepare themselves for replacing a dying node due to energy depletion or other kinds of failures. A sleeping node wakes up after sleeping for an exponentially distributed period of time (termed as wakeup rate λ) and broadcasts a probing message within a radius of r. If there are any working nodes in the vicinity, they reply back to the wakeup node. If the wakeup node hears such a reply message, it knows that there is a working node within its probing range r and goes back to sleep again. If the wakeup node does not hear a reply message within a certain time, it assumes that there is no working node within its probing range and it starts working. By tuning the parameters λ and r in simulations, the authors show that an optimal node density can be achieved while ensuring that each area is monitored by at least k working nodes. The algorithm is fully distributed and localized, and no neighborhood topology discovery is necessary. The computation and memory overhead per node is also negligible and is independent of the number of neighbors.

Shakkottai et al. [38] considered an unreliable wireless sensor grid network with n nodes placed over a unit area. Defining $r(n)$ as the transmitting radius of each node, and $p(n)$ as the probability that a node is active at some time t, they found that the necessary and sufficient condition for the grid network to cover the unit square region as well as ensure that the active nodes are connected is of the form $p(n)r^2(n) \approx \log(n)/n$. This result indicates that when n is large, each node can be highly unreliable and the transmission power can be small and we can still maintain connectivity with coverage. They have also shown that the diameter of the random grid (i.e., the maximum number of hops required to travel from any active node to another) is of the order of $\sqrt{n/\log(n)}$. A corollary of this is that the shortest-hop path between any pair of nodes is nearly the same as a straight-line path between the nodes. Finally, the authors derived a sufficient condition for connectivity of the active nodes (without necessarily having coverage) and showed that if $p(n)$ is small enough, connectivity does not imply coverage.

9.7 DISCUSSIONS AND CONCLUSIONS

In this chapter, we have discussed the importance of coverage and connectivity, which are two fundamental factors for ensuring efficient resource management in wireless sensor networks, and surveyed various methods and protocols, which optimally cover a sensing field while maintaining global network connectivity at the same time. We have seen that exposure paths can be viewed as a measure of goodness of detectability of a moving target in a sensing field. The notions of min–max exposure paths, breach paths, and support paths provide critical information to the application in terms of identifying sparsely and densely covered areas. We also discussed and compared several node deployment algorithms for static and mobile as well as for mixed-sensor networks, and observed that, depending on the coverage requirements, topological information, presence of obstacles, and other variables, the algorithms vary with respect to their goals, assumptions, and complexities.

The node scheduling schemes that we described under miscellaneous strategies using the notion of sponsored sectors ensure longer network lifetime and guarantee uniform dissipation of battery power throughout the network. This in turn implies better resource management.

However, the works existing in the literature have not addressed some of problems on theoretical bounds related to coverage and connectivity. Although Zhang and Hou [48] provided a theoretical result, proving that if the communication range is at least twice the sensing range, then complete coverage of an area guarantees a connected network, the probabilistic bounds on the number of nodes for a certain percentage of coverage is still unresolved. Future research in this area would provide insights into the probabilistic bounds on the best coverage that one can achieve given a number of nodes. The problem can be formulated as, given a total N number of nodes and a rectangular sensing field $A = a \times b$, with what probability one can guarantee p percentage coverage, while ensuring k degrees of connectivity across the network. This is a combinatorial optimization problem and can be tackled using statistical techniques and integer linear programming.

The deployment of nodes in mixed-sensor networks, which require one to strike a balance between the number of static and mobile sensors, involves the optimization of a cost/performance-based objective function and is therefore challenging. We discussed one approach [43] that initially deploys a fixed number of static and mobile nodes in a sensing field, after which the static nodes are required to find local coverage holes and bid for mobile sensors to relocate to the targeted locations and reduce or eliminate those holes, thus increasing area coverage. However, this approach has a drawback because it deploys a fixed number of mobile nodes.

To overcome this shortcoming, we [14] considered a mixed-sensor network, where initially a fixed number of static nodes are deployed, which deterministically find the exact amount of coverage holes existing in the entire network using the structure of Voronoi diagrams and then dynamically estimate the additional number of mobile nodes needed to be deployed and relocated to the optimal locations of the holes to maximize overall coverage. This approach of deploying a fixed number of static nodes and a varying estimated number of mobile nodes can provide optimal coverage under controlled cost. A mixed sensor approach is a very attractive one, because it allows one to choose the degree of coverage required by the underlying application as well as gives an opportunity to optimize on the number of additional mobile nodes needed to be deployed. We [15] provided distributed algorithms to find suboptimal minimum connected sensor covers, such that the whole sensing field is covered using a suboptimal number of sensors. In another study [5,7] we proposed a novel energy conserving data gathering strategy based on a tradeoff between coverage and data reporting latency with the ultimate goal of maximizing a network's lifetime. The basic idea is to select in each data reporting round only a minimal number of k sensors as data reporters, based on a desired sensing coverage specified by the user or application. Besides conserving energy, such a selection of minimum data reporters also reduces the amount of traffic flow, thus avoiding traffic congestion and channel interference. Simulation results of our proposed schemes demonstrate that the user-specified percentage of the monitored area can be covered using only k sensors. It also shows that the sensors can conserve a significant amount of energy with a small tradeoff and that the higher the network density, the higher is the energy conservation rate without any additional computation cost. In one of our works [6] for efficient resource management in wireless sensor networks, we presented a two-phase clustering scheme for energy saving and delay-adaptive data gathering in order to extend a network's lifetime.

Further research on optimization algorithms in mixed sensor networks and evaluating tradeoffs between latency and data gathering strategies can provide valuable information to optimize resources in a sensing field and help answer questions related to the theoretical bounds on coverage and connectivity.

ACKNOWLEDGMENT

This work is supported by NSF grant IIS-0326505.

REFERENCES

1. S. Adlakha and M. Srivastava, Critical density thresholds for coverage in wireless sensor networks, *Proc. IEEE Wireless Communications and Networking Conf. (WCNC'03)*, New Orleans, LA, March 2003, pp. 1615–1620, Louisiana, Mar. 2003.

2. I. Akyildiz, W. Su, Y. Sankarasubramaniam, and E. Cayirci, Wireless sensor networks: A survey, *Comput. Networks* **38**(2):393–422 (2002).

3. L. Booth, J. Bruck, and R. Meester, Covering algorithms, continuum percolation and the geometry of wireless networks, *Annals Appl. Probability* **13**(2):722–741 (May 2003).

4. K. Chakrabarty, S. S. Iyengar, H. Qi, and E. Cho, Grid coverage for surveillance and target address in distributed sensor networks, *IEEE Trans. Comput.* **51**(12):1448–1453 (Dec. 2002).

5. W. Choi and S. K. Das, in S. Phoha and T. La Porta, eds., *An Energy-conserving Data Gathering Strategy Based on Trade-off between Coverage and Data Reporting Latency in Wireless Sensor Networks*, Sensor Network Operations, IEEE Press, 2004.

6. W. Choi and S. K. Das, A framework for energy-saving data gathering using two-phase clustering in wireless sensor networks, *Proc. Mobile and Ubiquitous Systems: Networking and Services Conf., Mobiquitous'04*, Boston, Aug. 2004, pp. 203–212.

7. W. Choi and S. K. Das, Trade-off between coverage and data reporting latency for energy-conseving data gathering in wireless sensor networks, *Proc. 1st Int. Conf. Mobile Ad Hoc and Sensor Systems, MASS'04*, Ft. Lauderdale, FL, Oct. 2004.

8. T. Clouqueur, V. Phipatanasuphorn, P. Ramanathan, and K. K. Saluja, Sensor deployment strategy for target detection, *Proc. 1st ACM Int. Workshop on Wireless Sensor Networks and Applications (WSNA'02)*, Atlanta, GA, Sept. 2002, pp. 42–48.

9. S. S. Dhilon, K. Chakrabarty, and S. S. Iyengar, Sensor placement for grid coverage under imprecise detections, *Proc. 5th Int. Conf. Information Fusion (FUSION'02)*, Annapolis, MD, July 2002, pp. 1–10.

10. A. Elfes, Occupancy grids: A stochastic spatial representation for active robot perception, *Proc. 6th Conf. Uncertainty in AI*, Cambridge, MA, July 1990, pp. 60–70.

11. S. V. Fomin and I. M. Gelfand, *Calculus of Variations*, Dover Publications, Oct. 2000.

12. D. W. Gage, Command control for many-robot systems, *Proc. 19th Annual AUVS Technical Symp.* Reprinted in *Unmanned Syst. Mag.* **10**(4):28–34 (Jan. 1992).

13. Y. Gao, K. Wu, and F. Li, Analysis on the redundancy of wireless sensor networks, *Proc. 2nd ACM Int. Conf. Wireless Sensor Networks and Applications* (*WSNA'03*), San Diego, CA, Sept. 2003 pp. 108–114.

14. A. Ghosh, Estimating coverage holes and enhancing coverage in mixed sensor networks, *Proc. 29th Annual IEEE Conf. Local Computer Networks* (*LCN'04*), Tampa, FL, Nov. 2004, pp. 68–76.

15. A. Ghosh and S. K. Das, A distributed greedy algorithm for connected sensor cover in dense sensor networks, *Proc. 1st IEEE/ACM Int. Conf. Distributed Computing in Sensor Systems* (*DCOSS'05*), Marina del Rey, CA, June–July 2005, pp. 340–353.

16. N. Heo and P. K. Varshney, A distributed self-spreading algorithm for mobile wireless sensor networks, *Proc. IEEE Wireless Communications and Networking Conf.* (*WCNC'03*), New Orleans, LA, March 2003, pp. 1597–1602.

17. A. Howard, M. J. Matari, and G. S. Sukhatme, An incremental self-deployment algorithm for mobile sensor networks, *Autonomous Robots* special issue on intelligent embedded systems **13**(2):113–126 (2002).

18. A. Howard, M. Mataric, and G. Sukhatme, Mobile sensor network deployment using potential fields: A distributed scalable solution to the area coverage problem, *Proc. 6th Int. Symp. Distributed Autonomous Robotic Systems* (*DARS'02*), Fukuoka, Japan, June 2002, pp. 299–308.

19. A. Howard and M. J. Mataric, Cover me! A self-deployment algorithm for mobile sensor networks, *Proc. IEEE Int. Conf. Robotics and Automation* (*ICRA'02*), Washington DC, May 2002, pp. 80–91.

20. C.-F. Huang and Y.-C. Tseng, The coverage problem in a wireless sensor network, *Proc. 2nd ACM Int. Conf. Wireless Sensor Networks and Applications* (*WSNA'03*), San Diego, CA, Sept. 2003, pp. 115–121.

21. J. M. Kahn, R. H. Katz, and K. S. J. Pister, Next century challenges: Mobile networking for smart dust, *Proc. 5th Annual ACM/IEEE Int. Conf. Mobile Computing and Networking* (*MOBICOM'99*), Seattle, WA, Aug. 1999.

22. O. Khatib, Real-time obstacle avoidance for manipulators and mobile robots, *Int. J. Robotics Res.* **5**(1):90–98 (1986).

23. M. M. Kokar, J. A. Tomasik, and J. Weyman, Data vs. decision fusion in the category theory framework, *Proc. 4th Int. Conf. Information Fusion* (*FUSION'01*), Montreal, Australia, Aug. 2001.

24. X.-Y. Li, P.-J. Wan, and O. Frieder, Coverage in wireless ad-hoc sensor networks, *IEEE Trans. Comput.* **52**:753–763 (2003).

25. A. Mainwaring, J. Polastre, R. Szewczyk, D. Culler, and J. Anderson, Wireless sensor networks for habitat monitoring, *Proc. 1st ACM Int. Workshop on Wireless Sensor Networks and Applications* (*WSNA'02*), Atlanta, GA, Sept. 2002, pp. 88–97.

26. S. Megerian, F. Koushanfar, G. Qu, G. Veltri, and M. Potkonjak, Exposure in wireless sensor networks: Theory and practical solutions, *Wireless Networks* **8**(5):443–454 (2002).

27. S. Meguerdichian, F. Koushanfar, M. Potkonjak, and M. Srivastava, Coverage problems in wireless ad-hoc sensor networks, *Proc. IEEE InfoCom (InfoCom'01)*, Anchorage, AK, April 2001, pp. 115–121.

28. S. Meguerdichian, F. Koushanfar, G. Qu, and M. Potkonjak, Exposure in wireless ad-hoc sensor networks, *Proc. 7th Annual Int. Conf. Mobile Computing and Networking (MobiCom'01)*, Rome, Italy, July 2001, pp. 139–150.

29. A. Okabe, B. Boots, K. Sugihara, and S. N. Chiu, *Spatial Tessellations: Concepts and Applications of Voronoi Diagrams*, 2nd ed., Wiley July 2000.

30. J. O'Rourke, *Art Gallery Theorems and Algorithms*, Oxford Univ. Press, Oxford, UK, 1987.

31. M. Penrose, The longest edge of the random minimal spanning tree, *Annals Appl. Probability* **7**(2):340–361 (May 1997).

32. S. Poduri and G. S. Sukhatme, Constrained coverage in mobile sensor networks, *Proc. IEEE Int. Conf. Robotics and Automation (ICRA'04)*, New Orleans, LA, April–May 2004, pp. 40–50.

33. G. J. Pottie, Wireless sensor networks, *Proc. Information Theory Workshop*, June 1998, pp. 139–140.

34. G. J. Pottie and W. Caiser, Wireless sensor networks, *Commun. ACM* **43**(5):51–58 (May 2000).

35. P. Santi, *Topology Control in Wireless Ad Hoc and Sensor Networks*, Wiley, May 2005.

36. P. Santi and D. M. Blough, The critical transmitting range for connectivity in sparse wireless ad hoc networks, *IEEE Trans. Mobile Comput.* **2**(1):25–39 March 2003.

37. R. C. Shah and J. M. Rabaey, Energy aware routing for low energy ad hoc sensor networks, *Proc. IEEE Wireless Communications and Networking Conf. (WCNC'02)*, Orlando, FL, March 2002.

38. S. Shakkottai, R. Srikant, and N. Shroff, Unreliable sensor grids: Coverage, connectivity and diameter, *Proc. IEEE InfoCom (InfoCom'03)*, pages 1073–1083, San Francisco, CA, March 2003.

39. C. Shurgers and M. B. Srivastava, Energy efficient routing in wireless sensor networks, *Proc. Military Communications Conf. (MilCom'01)*, Vienna, VA, Oct. 2001.

40. D. Tian and N. D. Georganas, A coverage-preserving node scheduling scheme for large wireless sensor networks, *Proc. 1st ACM Int. Workshop on Wireless Sensor Networks and Applications (WSNA'02)*, Atlanta, GA, Sept. 2002, pp. 32–41.

41. P. J. M. van Laarhoven and E. H. L. Aarts, *Simulated Annealing: Theory and Applications*, Reidel Publishing, Kluwer, 1987.

42. G. Veltri, Q. Huang, G. Qu, and M. Potkonjak, Minimal and maximal exposure path algorithms for wireless embedded sensor networks, *Proc. 1st Int. Conf. Embedded Networked Sensor Systems (SenSys'03)*, Los Angeles, Nov. 2003, pp. 40–50.

43. G. Wang, G. Cao, and T. LaPorta, A bidding protocol for deploying mobile sensors, *Proc. 11th IEEE Int. Conf. Network Protocols (ICNP'03)*, Atlanta, GA, Nov. 2003, pp. 80–91.

44. G. Wang, G. Cao, and T. LaPorta, Movement-assisted sensor deployment, *Proc. IEEE InfoCom (InfoCom'04)*, Hong Kong, March 2004, pp. 80–91.

45. D. B. West, *Introduction to Graph Theory*, 2nd ed., Prentice-Hall, Aug. 2003.

46. B. Yamauchi, A frontier-based approach for autonomous exploration, *Proc. IEEE Int. Symp. Computational Intelligence in Robotics and Automation* (*CIRA'97*), Monterey, CA, June 1997, pp. 146–156.

47. F. Ye, G. Zhong, S. Lu, and L. Zhang, Peas: A robust energy conserving protocol for long-lived sensor networks, *Proc. 10th IEEE Int. Conf. Network Protocols* (*ICNP'02*), Paris, Nov. 2002, pp. 200–201.

48. H. Zhang and J. C. Hou, Maintaining sensing coverage and connectivity in large sensor networks, *Proc. Int. Workshop on Theoretical and Algorithmic Aspects of Sensor, Ad Hoc Wireless and Peer-to-Peer Networks* (*AlgoSensors'04*), Florida, Feb. 2004.

49. F. Zhao, J. Liu, J. Liu, et al. Collaborative signal and information processing: An information-directed approach, *Proc. IEEE* **91**(8):1199–1209, (2003).

50. Y. Zou and K. Chakrabarty, Sensor deployment and target localization based on virtual forces, *Proc. IEEE InfoCom* (*InfoCom'03*), San Francisco, CA, April 2003, pp. 1293–1303.

51. Y. Zou and K. Chakrabarty, Uncertainty-aware sensor deployment algorithms for surveillance applications, *Proc. IEEE Global Communications Conf.* (*GLOBECOM'03*), Dec. 2003.

52. Y. Zou and K. Chakrabarty, Sensor deployment and target localization in distributed sensor networks, *Trans. IEEE Embedded Comput. Syst.* **3**(1):61–91 (2004).

CHAPTER 10

Storage Management in Wireless Sensor Networks

SAMEER TILAK and NAEL ABU-GHAZALEH

Department of Computer Science, Watson School of Engineering and Applied Sciences, Binghampton University, Binghampton, New York

WENDI B. HEINZELMAN

Department of Electrical and Computer Engineering, University of Rochester, Rochester, New York

10.1 INTRODUCTION

Wireless sensor networks (WSNs) hold the promise of revolutionizing sensing across a range of civil, scientific, military, and industrial applications. However, many battery-operated sensors have constraints such as limited energy, computational ability, and storage capacity, and thus protocols must be designed to deal efficiently with these limited resources in order to maximize the data coverage and useful lifetime of the network.

Storage management is an area of sensor network research that is starting to attract attention. The need for storage management arises primarily in the class of sensor networks where information collected by the sensors is not relayed to observers in real time. In such applications, the data must be stored, at least temporarily, within the network until it is later collected by an observer (or until it ceases to be useful). An example of this type of application is scientific monitoring, where the sensors are deployed to collect detailed information about a phenomenon for later playback and analysis. Another example is that of sensors that collect data that are later accessed by dynamically generated queries from users. In these types of applications, data must be stored in the network, and thus storage becomes a primary resource that, in addition to energy, determines the useful lifetime and coverage of the network.

Mobile, Wireless, and Sensor Networks: Technology, Applications, and Future Directions
Edited by Rajeev Shorey, Akkihebbal L. Ananda, Mun Choon Chan, and Wei Tsang Ooi
Copyright © 2006 John Wiley & Sons, Inc.

With the knowledge of the relevant application and system characteristics, a set of goals for sensor network storage management can be determined: (1) minimizing storage size to maximize coverage/data retention, (2) minimizing energy, (3) supporting efficient query execution on the stored data (note that in the reachback method where all the data must be sent to the observer, query execution is simply the transfer of the data to the observer), and (4) providing efficient data management under constrained storage. Several approaches to storage management have been proposed to meet these requirements, with most approaches involving a trade-off among these different goals.

One basic storage management approach is to buffer the data locally at the sensors that collect them. However, such an approach does not capitalize on the spatial correlation of data among neighboring sensors to reduce the overall size of the stored data (the property that makes data aggregation possible [11]). Collaborative storage management, on the other hand, can provide the following advantages over a simple buffering technique:

- More efficient storage allows the network to continue storing data for a longer time without exhausting storage space.
- Load balancing is possible. If the rate of data generation is not uniform at the sensors (e.g., in the case where a localized event causes neighboring sensors to collect data more aggressively), some sensors may run out of storage space while space remains available at others. In such a case, it is important for the sensors to collaborate to achieve load balancing for storage to avoid or delay data loss due to insufficient local storage.
- Dynamic, localized reconfiguration of the network (such as adjusting sampling frequencies of sensors based on estimated data redundancy and current resources) is possible.

Thus, collaborative storage is often better able to meet the goals of storage management.

This chapter overviews issues and opportunities in storage management for sensor networks. We first motivate the need for storage management by overviewing sensor network applications that require storage management. We then discuss how the application characteristics and various resource constraints influence the design of storage management techniques. Application characteristics define what data are useful and how the data will eventually be accessed; these features have significant implications on storage protocol design. In addition, hardware characteristics determine the capabilities of the system and the energy efficiency of the storage protocols. We overview the characteristics of flash memory, the prevalent storage technology in sensor networks, and we compare the cost of storage to that of computation and communication to establish the basic data manipulation costs using current technologies for subsequent tradeoff analyses. Then, as a case study, we discuss the design and implementation of the matchbox file system, which supports mote based applications.

10.2 MOTIVATION: APPLICATION CLASSES

In this section, the storage management problem is motivated by describing two application classes that require effective use of storage resources. In general, storage is required whenever data are not relayed to observers outside the network in real time; the data must be stored by the sensors until they are collected or discarded (or cease to be useful, or are sacrificed or compressed to make room for more important data). The following two applications illustrate different situations where such a requirement arises.

10.2.1 Scientific Monitoring: Playback Analysis

In scientific monitoring applications, the sensor network is deployed to collect data. These data are not of real-time interest; they are collected for later analysis to provide an understanding of some ongoing phenomenon. Consider a wildlife tracking sensor network that scientists could use to understand the social behavior and migratory patterns of a species. Suppose sensors are deployed in a forest and collect data about nearby animals. In such an application, the sensors may not have an estimate regarding the observer's schedule for accessing the data. Furthermore, the data generation rate at the sensors may be unpredictable, as it depends on the observed activity, and the sensor density may also be nonuniform [5]. The observer would like the network to preserve the collected data samples, and the collection time should be small since the observer may not be in range of the sensors for very long. An example of such a network is the ZebraNet project [13].

Scientists (observers) collect the data by driving around the monitored habitat, receiving information from sensors as they come within range of them. Alternatively, the data may be relayed by the sensors in a multihop fashion toward the observer. Data collection is not preplanned; it might be unpredictable and infrequent. However, the data "query" model is limited (one-time collection in a known fashion). This information about data access patterns may be exploited; for example, data may be stored at sensors closer to the observer. The long-term availability of the data allows aggregation or compression that can be significantly more efficient than that achieved by real-time data collection sensor networks.

This type of situation also occurs in sensor networks that report their data in near real time when the network becomes partitioned. For example, in the remote ecological microsensor network [22], remote visual surveillance of federally listed rare and endangered plants is conducted. This project aims at providing near-real-time monitoring of important events, such as visitation by pollinators and consumption by herbivores, along with monitoring a number of weather conditions and events. Sensors are placed in different habitats, ranging from scattered low shrubs to dense tropical forests. Environmental conditions can be severe; for instance, some locations frequently freeze. In such applications, network partitioning (relay nodes becoming unavailable) may occur as a result of the extreme physical conditions (e.g., deep freeze). Important events that occur during disconnection periods should be recorded and reported once the connection is reestablished. Effective storage management is needed to maximize the partitioning time that can be tolerated without loss of data.

10.2.2   Augmented Reality: Multiple Observers, Dynamic Queries

Consider a sensor network that is deployed in a military scenario and collects information about nearby activity. The data are queried dynamically by soldiers to help with mission goals or with avoiding sources of danger, and by commanders to assess the progress of the mission. The queried data are real-time as well as long-term data about enemy activity (e.g., to answer a question such as where the supply lines are located). Thus, data must be stored at the sensors to enable queries that span temporally long periods, such as days or even months. One can envision similar applications with sensor networks deployed in other contexts that answer questions about the environment using real-time as well as recent or even historical data. The data are stored at the sensors and used, perhaps collaboratively, to answer queries.

In addition to efficiently using data storage to allow the sensor network to retain more and higher-resolution data, one issue in this application is effective indexing and retrieval of the data. While the data may be addressed by content (e.g., searching for information about a certain type of vehicle), since the observers do not know where the data are stored, or even what data exist, the queries can be quite inefficient. Furthermore, data may be accessed multiple times, by different observers. Efficient indexing and retrieval of the data is desirable.

10.3   PRELIMINARIES: DESIGN CONSIDERATIONS, GOALS, AND STORAGE MANAGEMENT COMPONENTS

The applications described in the previous section provide insight into the range of issues that must be addressed to achieve effective storage management. In this section, these issues are described more directly. We first outline the factors that influence the design of storage management and then discuss the design goals of such a system. We also break the storage management problem into different components that require efficient protocols.

10.3.1   Design Considerations

Energy is a precious resource in all wireless microsensor networks; when the battery energy at a sensor expires, the node ceases to be useful. Therefore, preserving energy is a primary concern that permeates all aspects of sensor network design and operation. For storage-bound applications, there is an additional finite resource: the available storage at the sensors. Once the available storage space is exhausted, a sensor can no longer collect and store data locally; the sensors that have run out of storage space cease to be useful. Thus, a sensor network's utility is bound by two resources: its available energy and its available storage space. Effective storage management protocols must balance these two resources to prolong the network's useful lifetime.

The two limiting resources — storage and energy — are fundamentally different from each other. Specifically, storage is reassignable while energy is not; a node may free up some storage space by deleting or compressing data. Furthermore, storage at other nodes may be utilized, at the cost of transmitting the data. Finally, the use of storage consumes energy. However, storage devices currently consume less energy than do wireless communication devices.

The tradeoff between storage and energy is intricate. Sensors may exchange their data with nearby sensors. Such exchanges allow nearby sensors to take advantage of the spatial correlation in their data to reduce the overall data size. Another positive side effect is that the storage load can be balanced even if the data generation rates or the storage resources are not. However, the exchange of data among the sensors consumes more energy in the data collection phase, as in current technologies, where the cost of storage is significantly smaller than the cost of communication. On the surface, it may appear that locally storing data is the most energy-efficient solution. However, the extra energy spent in exchanging data may be counterbalanced by the energy saved by storing smaller amounts of data and, more importantly, by the smaller energy expenditure when replying to queries or relaying the data back to observers.

The discussion above pertains to data collection and storage. Another aspect of the storage problem is how to support observer queries on the data — the indexing–retrieval problem. In one of the motivating applications (scientific monitoring and partitioning), data will be relayed once to a possibly known observer. For these applications, storage may be optimized if the direction of the observer is known; while conducting collaborative storage, data can be moved toward the observer, making the storage-related data exchange effectively free. However, in the second application type, dynamic queries for the data from unknown observers can occur. This problem is logically similar to indexing and retrieval in peer-to-peer systems.

10.3.2   Storage Management Goals

In light of the preceding discussion, a storage management approach must balance the following goals:

- *Minimize Size of Stored Data.* Since sensors have limited storage available to them (state of the art sensors have approximately 4 Mb (megabits) of flash memory), minimizing the size of data that need to be stored leads to improved coverage/data retention since the network can continue storing data for longer periods of time. Furthermore, query execution becomes more efficient if the data size is small.

- *Minimize Energy Consumption.* Most of the sensors are battery-powered, and thus energy is a scarce resource, requiring that storage management be as energy-efficient as possible.

- *Maximize Data Retention/Coverage*. Collecting data is the primary goal of the network. If storage is constrained, data reallocation must be carried out efficiently; in certain cases, if no more storage space is available to store new data, some of the less important data that are already stored need to be deleted. The management protocol should attempt to retain relevant data at an acceptable resolution, where relevancy and acceptable resolution are application-dependent.

- *Perform Efficient Query Execution*. Whether the queries are dynamically generated or static (e.g., the data relayed to an observer once), storage management can influence the efficiency of query execution. For example, query execution efficiency can be improved by effective data placement and indexing. Query execution efficiency can be measured in terms of the communication overhead and energy consumption required to get the requested data to the observer.

10.3.3   Storage Management Components

We break down the storage management problem into the following components: (1) system support for storage management, (2) collaborative storage, and (3) indexing and retrieval. Each of these components has a unique design space to consider, and some preliminary research has been done to improve each of these three components from the perspective of energy and storage efficiency. Furthermore, each of these components exhibit tradeoffs in the design goals described above. In the following sections, we highlight the problems and issues in each of these components of storage management, and we describe proposed solutions.

10.4   SYSTEM SUPPORT FOR STORAGE

Because of the limited available energy on sensors, energy efficiency is a primary objective in all aspects of sensor design. In this section, the focus is on the system design issues that are most relevant to storage. These include design of hardware components and the sensor network filesystem.

10.4.1   Hardware

Magnetic disks (hard drives) are the most widely used persistent storage devices for desktop and laptop computers. However, power, size, and cost considerations make hard disks unsuitable for microsensor nodes. Compact flash memories are the most promising technology for storage in sensor networks, as they have excellent power dissipation properties compared to magnetic disks. In addition, their prices have been dropping significantly. Finally, they have a smaller form factor than do magnetic disks.

As a case study, we consider the family of Berkeley motes [27], MICA-2 (see Table 10.1), MICA2DOT, and MICA nodes, which feature a 4-Mb serial flash

TABLE 10.1   **Energy Characteristics of MICA-2 Components**

Component	Current	Duty Cycle (%)
Processor		
Fully operational	8 mA	1
Sleep	8 μA	99
Radio		
Receive	8 mA	0.75
Transmit	12 mA	0.25
Sleep	2 μA	99
Logger memory		
Write	15 mA	0
Read	4 mA	0
Sleep	2 μA	100

(nonvolatile) memory for storing data, measurements, and other user-defined information. TinyOS [10] supports a microfile system that manages this flash/data logger component. The serial flash device supports over 100,000 measurement readings, and this device consumes 15 mA of current when writing data. Table 10.1 gives the energy characteristics of important hardware components of MICA-2 motes [27].

10.4.2   Filesystem Case Study: Matchbox

Traditional file systems (such as FFS [17] and the log-structured file system [24]) are not suitable for sensor network environments. They are designed and tuned for magnetic disks that have operational characteristics different from those of flash memories, which are typically used for storage in sensors. For example, traditional filesystems employ clever mapping and scheduling techniques to reduce seek time since this is the primary cost in disk access. In addition, these filesystems support a wide range of sophisticated operations, making them unsuitable for a resource-constrained embedded environment. For example, typically filesystems support hierarchical filing, security (in terms of access privilege and data encryption in some cases), and concurrent read/write support. Finally, sensor storage disk access patterns are not typical of traditional filesystem workloads. New technological constraints, low resources, and different application requirements demand the design of a new filesystem for sensor network applications.

One component of the TinyOS [10] operating system is an evolving microfile-system called *matchbox* [6,7]. Matchbox is designed specifically for mote-based applications and has the following design goals:

1. *Reliability* — matchbox provides reliability in the following two ways:
    a. *Data corruption detection* — matchbox maintains CRCs to detect errors.
    b. Metadata updates are atomic, and therefore the file system is resilient to failures such as powerdown. Data loss is limited only to files being written at the time of failure.

2. *Low resource consumption.*

3. *Meeting technological constraints* — some flash memories have constraints that must be met. For example, the number of times a memory location can be written to may be limited.

These design goals translate into a file system that is very simple. Matchbox stores files in an unstructured fashion (as a bytestream). It supports only sequential reads and append-only writes. The current design does not aim at providing security, hierarchical filesystem support, random access, or concurrent read/write access. Typical clients of matchbox are TinyDB [15], generic sensor kit for data logging applications, and the virtual machine for storing programs.

Matchbox divides the flash memory into sectors [mostly of 128 kB (kilobytes)], and then each sector is divided into pages. Each page is of size 264 B, divided into 256 B of data and 8 B of metadata. Free pages are tracked using a bitmap.

At present, both the filesystem and sensor network applications are evolving. Sensor network applications offer a workload different from that in traditional applications. We believe that as these applications mature and become better understood, the design of the filesystem will evolve to support these canonical applications. The filesystem evolution may include changes in the way the files are stored (extending the bytestream view), and supporting a new set of file operations for real-time streaming data (extending append-only write operations).

Unfortunately, existing collaborative storage management studies have not considered their interaction with the filesystem. It would be interesting to study how different data manipulations can be supported by matchbox. For example, in one of the storage management protocols, old data are gracefully degraded in storage-constrained conditions [4]. With append-only writes, deleting intermediate data might require rewriting much of (even most of) the existing data. This might result either in extending append-only writes to other techniques or modification of the application itself. Still, implications of such operations on lifetime of flash memory (recall that a given memory location can be written only a certain number of times) and energy consumption are worth exploring. Moreover, filesystem designers should consider the requirements of collaborative storage when designing sensor filesystems.

10.5   COLLABORATIVE STORAGE

A primary objective of storage management protocols is to efficiently utilize the available storage space to continue collecting data for the longest possible time without losing samples in an energy-efficient manner. In this section, we describe collaborative storage management protocols that have been proposed and discuss the important design tradeoffs in these different approaches.

10.5.1 Collaborative Storage Design Space

Storage management approaches can be classified as follows:

1. *Local Storage.* This is the simplest solution whereby every sensor stores its data locally. This protocol is energy efficient during the storage phase since it requires no data communication. Even though the storage energy is high (because all the data are stored), the current state of technology is such that storage costs less than communication in terms of energy dissipation. However, this protocol is storage-inefficient since the data are not aggregated and redundant data are stored among neighboring nodes. Furthermore, local storage is unable to balance the storage load if data generation or the available storage varies across sensors.

2. *Collaborative Storage. Collaborative storage* refers to any approach where nodes collabo-rate. It can provide the following advantages over a simple buffering technique: (a) more efficient storage allows the network to continue storing data for a longer time without exhausting storage space, (b) load balancing is possible — if the available storage varies across the network, or if the rate of data generation is not uniform at the sensors (e.g., in the case where a localized event causes neighboring sensors to collect data more aggressively), some sensors may run out of storage space while space remains available at others — in such a case, it is important for the sensors to collaborate to achieve load balancing for storage to avoid or delay data loss due to insufficient local storage; and (c) dynamic, localized reconfiguration of the network (such as adjusting sampling frequencies of sensors on the basis of estimated data redundancy and current resources) is possible.

It is important to consider the energy implications of collaborative storage relative to local storage. Collaborative storage requires sensors to exchange data, causing them to expend energy during the storage phase. However, because they are able to aggregate data, the energy expended in storing this data to a storage device is reduced. In addition, once connectivity with the observer is established, less energy is needed during the collection stage to relay the stored data to the observer.

10.5.2 Collaborative Storage Protocols

Within the space of collaborative storage, a number of protocols have been proposed. One such protocol is the cluster-based collaborative storage (CBCS) protocol [26]. CBCS uses collaboration among nearby sensors only; these have the highest likelihood of correlated data and require the least amount of energy for collaboration. Wider collaboration was not considered because the collaboration cost may become prohibitive, as the energy cost of communication is significantly higher than the energy cost of storage under current technologies. The remainder of this section describes CBCS operation.

In CBCS, clusters are formed in a distributed, connectivity-based or geographically based fashion — almost any one-hop clustering algorithm would suffice. Each sensor sends its observations to the elected cluster head (CH) periodically. The CH then aggregates the observations and stores the aggregated data. Only the CH needs to store aggregated data, thereby resulting in low storage. The clusters are rotated periodically to balance the storage load and energy usage. Note that only the CH needs to keep its radio on during its tenure, while a cluster member can turn off its radio except when it has data to send. This results in high energy efficiency, since idle power consumes significant energy in the long run if radios are kept on. Furthermore, this technique reduces the reception of unnecessary packets by cluster members, which can also consume significant energy.

Operation during CBCS can be viewed as a continuous sequence of rounds until an observer or basestation is present and the reachback stage can begin. Each round consists of two phases. In the first phase, called the *CH election phase*, each sensor advertises its resources to its one-hop neighbors. On the basis of this resource information, a CH is selected according to a clustering protocol. The remaining nodes then attach themselves to a nearby CH. In the second phase, termed the *data exchange phase*, if a node is connected to a CH, it sends its observations to the CH; otherwise, it stores its observations locally.

The CH election approach used in CBCS is based on the characteristics of the sensor nodes such as available storage, available energy, or proximity to the "expected" observer location. The criteria for CH selection can be arbitrarily complex; experiments presented used available storage as the criteria. CH rotation is done by repeating the cluster election phase with every round. The frequency of cluster rotation influences the performance of the protocol, as there is overhead for cluster formation due to the exchange of messages. Thus cluster rotation should be done frequently enough to balance storage or energy yet not so often to render the overhead of cluster rotation prohibitive.

10.5.3 Coordinated Sensor Management

Using a clustering approach such as the one for CBCS enables coordination that can be used for redundancy control to reduce the amount of data actually generated by the sensors. Specifically, each sensor has a local view of the phenomenon, but cannot assess the importance of its information given that other sensors may report correlated information. For example, in an application where 3 sensors are sufficient to triangulate a phenomenon, 10 sensors may be in a position to do so and be storing this information locally or sending it to the cluster head for collaborative storage. Through coordination, the cluster head can inform the nodes of the degree of the redundancy, allowing the sensors to alternate triangulating the phenomenon. Coordination can be carried out periodically at low frequency, with a small overhead (e.g., with CH election). Similar to CH election, the nodes exchange metadata describing their reporting behavior, and it is assumed that some application-specific estimate of redundancy is performed to adjust the sampling rate.

As a result of coordination, it is possible that a significant reduction in the data samples produced by each sensor is achieved. We note that this reduction represents a portion of the reduction that is achieved from aggregation. For example, in a localization application, with 10 nodes in position to detect an intruder, only 3 nodes are needed. Coordination allows the nodes to realize this and adjust their reporting so that only three sensors produce data in every period. However, the three samples can still be aggregated into the estimated location of the intruder once the values are combined at the cluster head.

Coordination can be used in conjunction with local storage or collaborative storage. In *coordinated local storage* (CLS), the sensors coordinate periodically and adjust their sampling schedules to reduce the overall redundancy, thus reducing the amount of data that will be stored. Note that the sensors continue to store their readings locally. Relative to local storage (LS), CLS results in a smaller overall storage requirement and savings in energy in storing the data. This also results in a smaller and more energy-efficient data collection phase.

Similarly, *coordinated collaborative storage* (CCS) uses coordination to adjust the sampling rate locally. Similar to CBCS, the data are still sent to the cluster head where aggregation is applied. However, as a result of coordination, a sensor can adapt its sampling frequency or data resolution to match the application requirements. In this case, the energy in sending the data to the cluster head is reduced because of the smaller size of the generated data, but the overall size of the data is not reduced.

In the next section, we discuss some experimental results for CLS and CCS, and we compare these techniques to their noncoordinated counterparts, LS and CBCS.

10.5.4 Experimental Evaluation

We simulated the proposed storage management protocols using the ns-2 simulator [19]. We use a CSMA-based MAC layer protocol. A sensor field of 350×350 m is used, with each sensor having a transmission range of 100 m. We considered three levels of sensor density: 50 sensors, 100 sensors, and 150 sensors deployed randomly. We divide the field into 25 zones (each zone is 70×70 m to ensure that any sensor in the zone is in range with any other sensor). The simulation time for each scenario was set to 500 s, and each point represents an average over five different topologies. Cluster rotation and coordination are performed every 100 s in the appropriate protocols.

We assume that sensors have a constant sampling rate (set to one sample per second). Unless otherwise indicated, we set the aggregation ratio to a constant value of 0.5. For the coordination protocols, we used a scenario where the available redundancy was on average 30% of the data size — this is the percentage of the data that can be eliminated using coordination. We note that this reduction in data size represents a portion of the reduction possible using aggregation. With aggregation, the full data are available at the cluster head and can be compressed at a higher efficiency.

10.5.4.1   Storage–Energy Tradeoffs

Figure 10.1a shows the average storage used per sensor as a function of the number of sensors (50, 100, and 150 sensors) for the four storage management techniques: (1) local storage (LS), (2) cluster-based collaborative storage (CBCS), (3) coordinated local storage (CLS), and (4) coordinated collaborative storage (CCS). In the case of CBCS, the aggregation ratio was set to 0.5. The storage space consumption is independent of the network density for LS and is greater than the storage space consumption for CBCS and CCS (roughly in proportion to the aggregation ratio). The CLS storage requirement is in between those of the two approaches because it is able to reduce the storage requirement using coordination (we assumed that coordination yields improvement uniformly distributed between 20 and 40%). Note that after data exchange, the storage requirements for CBCS and CCS are roughly the same since aggregation at the cluster head can reduce the data to a minimum size, regardless of whether coordination took place.

Surprisingly, in the case of collaborative storage, the storage space consumption decreases slightly as the density increases. While this is counterintuitive, it is due to the higher packet loss observed during the exchange phase as the density increases; as density increases, the probability of collisions increases. These losses are due to the use of a contention-based unreliable MAC layer protocol. The negligible difference in the storage space consumption between CBCS and CCS is also an artifact of the slight difference in the number of collisions observed in the two protocols. A reliable MAC protocol such as that in IEEE 802.11 (which uses four-way handshaking) or a reservation-based protocol such as the TDMA-based protocol employed by LEACH [9] can be used to reduce or eliminate losses due to collisions (at an increased communication cost). Regardless of the effect of collisions, one can clearly see that collaborative storage achieves significant savings in storage space compared to local storage protocols (in proportion to the aggregation ratio).

We assumed transfer energy/MB 0.055 J for flash memory. In the case of ratio, we assumed transmit power = 0.0552 W, receive power = 0.0591 W, and idle power = 0.00006 W. Figure 10.1b shows the consumed energy for the protocols in joules as a function of network density.

The *x* axis represents protocols for different network densities: L and C stand for local buffering and CBCS, respectively. L-1, L-2, and L-3 represent the results using the local buffering technique for network size 50, 100, and 150, respectively. The energy bars are broken into two parts: *preenergy*, which is the energy consumed during the storage phase, and *postenergy*, which is the energy consumed during data collection (the relaying of the data to the observer). The energy consumed during the storage phase is higher for collaborative storage because of the data communication among neighboring nodes (not present in local storage) and the overhead for cluster rotation. CCS spends less energy than does CBCS because of the reduction in data size that results from coordination. However, CLS has higher expenditure than LS since it requires costly communication for coordination. This cost grows with the density of the network because our coordination implementation has each node broadcasting its update and receiving updates from all other nodes.

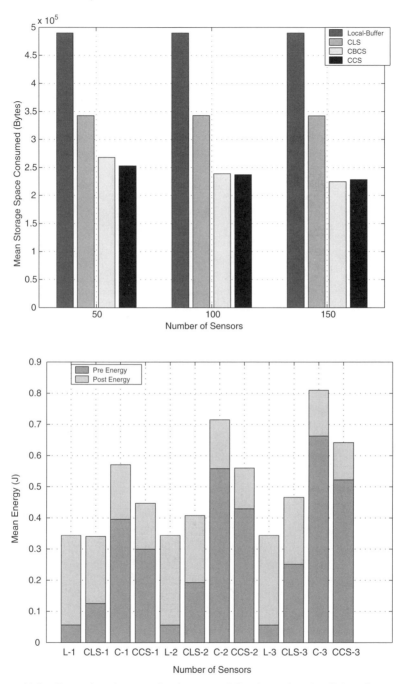

Figure 10.1   Comparison between local storage (LS), cluster based collaborative storage (CBCS), coordinated local storage (CLS), and coordinated collaborative storage (CCS): (a) storage space versus network density; (b) energy consumption versus network density.

For the storage and communication technologies used, the cost of communication dominates that of storage. As a result, the cost of the additional communication during collaborative storage might not be recovered by the reduced energy needed for storage except at very high compression ratios. This tradeoff is a function of the ratio of communication cost to storage cost; if this ratio goes down in the future (e.g., due to the use of infrared communication or ultra-low-power RF radios), collaborative storage becomes more energy-efficient compared to local storage. Conversely, if the ratio goes up, collaborative storage becomes less efficient.

10.5.4.2 Storage Balancing Effect

In this study, we explore the load balancing effect of collaborative storage. More specifically, the sensors are started with a limited storage space, and the time until this space is exhausted is tracked. We consider an application where a subset of the sensors generates data at twice the rate of the others, for example, in response to higher observed activity close to some of the sensors. To model the data correlation, we assume that sensors within a zone have correlated data. Therefore all the sensors within a zone will report their readings with the same frequency. We randomly select zones with high activity; sensors within those zones will report twice as often as will those sensors within low-activity zones.

In Figure 10.2, the x axis denotes time (in multiples of 100 s), whereas the y axis denotes the percentage of sensors that have no storage space left. Using LS, in the even data generation case, all sensors run out of storage space at the same time and all data collected after that are lost. In comparison, CBCS provides longer time without running out of storage because of its more efficient storage.

The uneven data generation case highlights the load balancing capability of CBCS. Using LS, the sensors that generate data at a high rate exhaust their storage quickly; we observe two subsets of sensors getting their storage exhausted at two different times. In comparison, CBCS has much longer mean sensor storage depletion time because of its load balancing properties, with sensors exhausting their resources gradually, extending the network lifetime much longer than LS.

The sensor network coverage from a storage management perspective depends on the event generate rate, the aggregation properties, and the available storage. If the aggregated data size is independent of the number of sensors (or grows slowly with it), the density of the zone correlates with the availability of storage resources. Thus, both the availability of storage resources as well as the consumption of them may vary within a sensor network. This argues for the need of load balancing across zones to provide long network lifetime and effective coverage. This is a topic of future research.

Collaborative storage reduces storage requirements by taking advantage of temporal and spatial correlations in data from nearby sensors. Furthermore, coordinated storage proactively exploits spatial redundancy in sensor data, thereby reducing not only storage requirements but also energy dissipation in communication of the data as well as in the sensing hardware, which can be turned off on nodes that are not assigned to sense data for the current period. Another approach that takes advantage

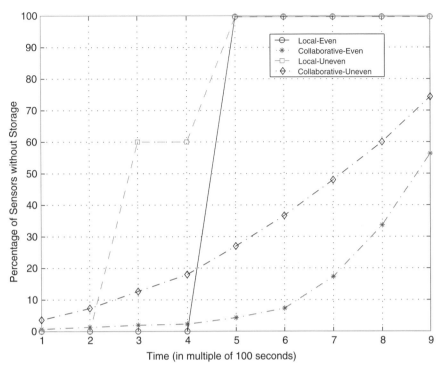

Figure 10.2   Percentage of storage-depleted sensors as a function of time.

of the correlations in the data to manage storage space is the multiresolution technique, described next.

10.5.5   Multiresolution-Based Storage

Much existing research, such as directed diffusion [12], data-centric storage (DCS) [21], TAG [16], and Cougar [2], has focused on in network aggregation and query processing when features of interest are known. For example, DCS stores named events at known locations to reduce query overhead. Since the events are constellations of low-level sensor observations, they are not storage-intensive. Therefore DCS does not address the issue of managing limited storage space. TAG assumes knowledge of aggregation operators. Recently, Ganesan et al. [4] proposed an in-network wavelet-based summarization technique accompanied by progressive aging of these summaries to support data-intensive applications where features of interest are not known in advance and the networks are storage- and communication-constrained. Their system strives to support storage and search for raw sensor data (low-level observations) in data-intensive scientific applications by providing a lossy, progressively degrading storage model. The key factors of this work are wavelet-based spatiotemporal data summarization (construction of multiresolution summaries of data), a hierarchically decomposed distributed storage structure, drill-down queries, and progressive aging of summaries. We briefly describe each of these features in detail in next few sections.

10.5.5.1 Multiresolution Summarization

The authors proposed the use of a wavelet-based data summarization technique that is carried out in two steps: temporal summarization and spatial summarization.

Temporal summarization is done by every sensor node locally using techniques such as time-series analysis to find redundancy in its own signal. Temporal summarization involves just computation — no communication is needed.

Spatial summarization involves constructing a hierarchy and using spatiotemporal summarization techniques to resummarize data at each level. Also, data at lower levels can be summarized at higher levels at larger spatial scale but with higher compression (and thereby are more lossy).

10.5.5.2 Drilldown Queries

The basic idea behind drilldown queries is quite intuitive. Data are summarized at multiple resolutions across the network. Queries are injected at the highest level of hierarchy, which has the coarsest and most highly compressed summary of very large spatiotemporal data. Processing the query over this summary gives an approximate answer or pointer to the part of the network that is very likely to give a more accurate answer, since it has a more detailed view of the subregion. Further queries can then be directed to this region, if needed. This process is applied recursively until the user is satisfied with the accuracy of the result or the leaves of the hierarchy are encountered. Clearly, the accuracy of the result (query quality) improves with more drilldowns since finer data gets queried at lower levels.

Hierarchical summarization and drilldown query together address the challenges in searching data in an efficient manner. After describing how to compute the summaries, we now discuss another challenging issue that deals with storage space allocation or reclaim of these summaries in a distributed fashion.

10.5.5.3 Aging Problem

In storage-constrained networks, a challenging question is to decide how long a summary is to be stored. The length of time for which a summary is stored is called an *age* of the summary. Let $f(t)$ be a monotonically decreasing user-specified aging function that represents the error that the user is willing to accept as the data ages in the network. The authors argue that, typically the domain experts can supply this kind of function. As an example, the user might be willing to accept 90% query accuracy for week-old data but only 50% accuracy for year-old data. Let us denote the instantaneous quality difference as qdiff (t), which represents the user-specified aging function and the achieved query accuracy at a given time t. The aging problem can be defined as follows. Find the ages of summaries, Age_i, at different resolutions such that the maximum instantaneous quality difference is minimized.

$$\text{Min}_{0 \leq t \leq T}(\text{Max}(\text{qdiff}(t))) \tag{10.1}$$

The constraints are

- *Drilldown constraints* — it is not useful to retain a summary at lower levels if its summary at higher levels is not present, since drilldown queries will not be directed to the lower levels in that case.
- *Storage constraints* — each node has a finite storage space available to the summaries of each level.

The authors proposed three aging strategies — omniscient, training-based, and greedy strategy — and evaluated their performance.

10.6 INDEXING AND DATA RETRIEVAL

The final challenge that we discuss in storage management is indexing and retrieval of the data. In the case of the first application type (scientific monitoring/partitioning), the data are relayed back to an observer during reachback one time. Accordingly, indexing and retrieval is not an important issue for this type of application. Nevertheless, reduction of storage size leads to more efficient retrieval in terms of time and energy. Furthermore, it is possible to improve the retrieval performance if the expected location of the observer is known by favoring sensors closer to the observer.

Indexing and retrieval are more important issues in our second application model (augmented reality), where data can be queried dynamically and by multiple observers. Such networks are inherently data-centric; observers often name data in terms of attributes or content that may not be topologically relevant. For example, a commander may be interested in enemy tank movements. This characteristic of sensor networks is similar to many peer-to-peer (P2P) environments [1], which are also often data-centric. Such a model is in contrast with traditional host-centric applications such as telnet, where the end user is communicating with a specific end host at the other end. We first overview peer-to-peer solutions to data indexing and retrieval and then show that the properties of sensor networks significantly change the trade-offs and invite different solutions.

10.6.1 Design Space: Retrieval in P2P Networks

In P2P networks, the problem of data indexing and retrieval has been attacked in several ways. These approaches can be classified in terms of data placement as follows:

1. *Structured* — the data are placed at specific locations (e.g., using hashing on keys) to make retrieval more efficient [20,25]. Using this approach, a node that is searching for data can use this structure to figure out where to look for the data. However, this structure requires extensive communication of data and may not be suitable for sensor networks. Further, related data may be stored at many different locations, making queries inefficient.

2. *Unstructured* — the data are not forced to specific locations [23]. In this case, searching for data is difficult since they may exist anywhere; in the worst case the user must perform a random search. Replication improves the performance of retrieval [3], but is likely to be too expensive for a sensor network environment.

For unstructured networks, P2P solutions differ in terms of indexing support. Decentralized networks provide no indexing. Centralized networks provide a centralized index structure. Finally, hybrid centralized networks provide hierarchical indexing, where supernodes each keep track of the data present at the nodes managed by them.

While the tradeoffs between these approaches are well studied in the P2P community, it is not clear how they apply for sensor networks. Specifically, P2P networks exist on the Internet, where resources are not nearly as limited as those in a sensor network. Solutions requiring large data movement or expensive indexing are likely to be inefficient. It is also unclear what keys or attributes should be indexed to facilitate execution of common queries. Finally, solutions that cause expensive query floods will also be inefficient. In the remainder of this section, we study existing research into indexing and retrieval in sensor networks.

10.6.2 Data-Centric Storage: Geographic Hash Tables

In the context of peer-to-peer systems, distributed hash tables (DHTs) are a decentralized structured P2P implementation. Typically these DHTs provide the following simple yet powerful interface: Put(data d, key k) operation, which stores the given data item d based on its key k; and Get(key k) operation, which can be used to retrieve all the data items matching the given key k.

The geographic hash table (GHT) system implements a structured P2P solution for data centric storage (DCS) in sensor networks [21]. Even though GHT provides an functionality equivalent to that of structured P2P systems, it needs to address several new challenges. Specifically, moving data is costly in sensor networks. Moreover, the authors target keeping related data close such that queries can be more focused and efficient.

In the next few sections we describe how GHT incorporates physical connectivity in data-centric operations in resource-constrained sensor networks. Before delving into the details of GHT-based data centric storage, we overview three canonical data dissemination methods along with their approximate communication costs and demonstrate the usefulness of data-centric storage [21].

10.6.2.1 *Canonical Methods*

The primary objective of a data dissemination method is to extract relevant data from a sensor network in an efficient manner. The following are three fundamentally different approaches to achieve this objective. Let us assume that the sensor network has n nodes.

TABLE 10.2   **Comparison of Different Canonical Methods**

Method	Total	Hostspot
External storage	$D_{\text{total}}\sqrt{n}$	D_{total}
Local storage	$Q_n + D_q\sqrt{n}$	$Q + D_q$
Data-centric (summary)	$Q\sqrt{n} + D_{\text{total}}\sqrt{n} + Q\sqrt{n}$	$2Q$

1. *External storage* (ES) — on detecting an event, the relevant data are sent to the basestation. ES entails $O(\sqrt{n})$ cost for each event to get to the basestation, with zero cost for queries generated at the basestation (external) and $O(\sqrt{n})$ for queries generated within the sensor network (internal).

2. *Local storage* (LS) — a sensor node, on detecting an event, stores the information locally. LS incurs $O(n)$ cost for query dissemination, since a query must be flooded and $O(\sqrt{n})$ cost to report the event.

3. *Data-centric storage* (DCS) — stores named data within the network. It requires $O(\sqrt{n})$ cost to store the event and both querying and event response require $O(\sqrt{n})$ cost.

Let us assume that a sensor network detects T event types, and denote the total number of events detected by D_{total}. Q denotes the number of event types queries and D_Q denotes the number of events detected for each event queried. Further, assume that there are a total of Q queries (one per event type). Table 10.2 shows the approximate communication cost for the three canonical methods. Total cost accounts for the total number of packets sent in the network, whereas the hostspot usage denotes the maximum number of packets sent by any particular sensor node.

From this analysis, its clear that no single method is preferable under all circumstances. If the events are accessed more frequently than they are generated (by external observers), external storage might be a good alternative. On the other hand, local storage is an attractive option when the events are generated more frequently and accessed infrequently. DCS lies in the middle of these two options and is preferable in cases where the network is large and a large number of events are detected but few are queried. The next sections provide the implementation details and a discussion of GHT.

10.6.3 GHT: A Geographic Hash Table

GHT is a structured approach to sensor network storage that makes it possible to index data on the basis of content without requiring query flooding. GHT also provides load balancing of storage usage (assuming fairly uniform sensor deployment). GHT implements a distributed hash table by hashing a key k into geographic coordinates. As mentioned above, GHT supports Put(data d, key k) and Get(key k) operations in the following way. In Put, data (events) are randomly hashed to a geographic location (x, y coordinates), whereas a Get operation on the same key k hashes to the same location. In the case of sensor networks, sensors can be deployed randomly and the geographic hashing function is oblivious to the topology. Therefore a sensor node might not exist at the precise location given by the hash function. Also, it is crucial to ensure consistency between Get and Put operations, namely, that both of them map the same key to the same node. Put achieves that by storing the hashed event at the node *nearest* to the hashed location, and Get operations can also retrieve an event from the node *nearest* to the hashed location. Of course, this policy ensures the desired consistency between Get and Put operations.

GHT uses greedy perimeter stateless routing (GPSR) [14]. GPSR is a geographic routing protocol that just uses location information to route packets to any connected destination. It assumes that every node knows its own location and also locations of all its one-hop neighbors. It has two flavors of routing algorithms: greedy forwarding and perimeter forwarding.

In the case of greedy forwarding, a node X, on receiving a packet for a destination D, forwards it to its neighbor that is closest to D among all its neighbors including X itself. Intuitively, a packet moves closer to the destination each time it is forwarded and eventually reaches the destination. Of course, greedy forwarding does not work in the case when X does not have any neighbor closer to D than itself. GPSR then switches to its perimeter mode.

In the perimeter mode, GPRS uses the following right-hand rule — on arriving on an edge at node X, the packet is forwarded on the next edge counterclockwise about X from the ingress degree. GPSR first computes a planar subgraph of the network connectivity graph and then applies the right-hand rule to this graph.

10.6.3.1 Interaction of GHT with GPSR

GPSR provides the capability to deliver packets to a destination node located at the specified coordinates, whereas GHT requires the ability to forward a packet to a node closest to the destination node. GHT defines a *home node* as the node that is closest to the hashed location; GHT requires an ability to store an event at the Home node.

Let us assume that hashing an event gives location (X_1, Y_1) as the coordinates of the node where the given event needs to be stored. Further, let us assume that no node exists at location (X_1, Y_1). Then GHT would store that event at its home node H. So node H, on receiving a packet for destination (X_1, Y_1), will switch to GPRS's perimeter routing mode (since it will have no neighbor closer to the destination than itself). By applying perimeter routing, the packet will eventually return to H on traversing the entire perimeter. H can then conclude that it is the home node for the given event and then will store the event. The interaction between GHT and GPSR described so far works well for static sensors, ideal radios, and without considering failures of sensor nodes. However, when thousands of sensors are thrown in a hostile environment, one can hardly assume the presence of these ideal conditions.

GHT ensures robustness (resilience to node failure and topology changes) and scalability (by balancing load across the network) as follows. It uses a novel perimeter refresh protocol to achieve robustness in the face of node failures and topology changes, and it uses structured replication to do load balancing, thereby achieving scalability. We now briefly describe these two techniques.

10.6.3.2 Perimeter Refresh Protocol

GHT needs to address two challenges: home node failure and topology changes (e.g., addition of nodes). Periodically, a home node initiates a perimeter refresh message that traverses across the entire perimeter of the specified location. Every node on the perimeter, on receiving this event, first stores it locally and then marks the association between the given event and the home node for a certain duration with the help of a timer. If the node receives any subsequent refresh messages from the same home node, it resets the timer and reassociates the home node with the event. However, if it fails to receive the refresh message, it assumes that the home node for the event has failed and initiates a refresh message on its own for electing a new home node. In this manner, GHT recovers from home node failures.

As a result of topology changes, some new node H_1 might now be closer to the destination than the current home node H. When the current home node H initiates the refresh message, H_1 will receive it. Since H_1 is closer to the destination than H (by definition of the home node), it reinitiates the refresh message, which then passes through all the nodes on the perimeter, including H, and eventually returning to H_1. All the nodes on the perimeter update their associations and timers accordingly to point to H_1. In this way, GHT addresses the issue of topological changes.

10.6.3.3 Structured Replication

If many events are hashed onto same node (location), then this node can become a *hotspot*. Structured replication is GHT's way of balancing the load and reducing the hotspot effect. The basic idea is to decompose the hierarchy geographically and replicate the home node rather then replicating the data itself. For example, instead of storing data at the root of the hierarchy (home node of the event), a node stores the data at the nearest mirror node (one of the replicas of the home node). For retrieving events, queries are still targeted toward the root node; however, the root node then forwards the query to the mirror node.

10.6.4 Graph EMbedding for Sensor Networks (GEM)

GHT uses GPSR, a geographic routing protocol that routes packets based only on location information. GEM (graph *em*bedding for sensor networks) is an infrastructure for node-to-node routing and data-centric storage and information processing in sensor networks without using geographic information [18]. GEM embeds a ringed tree into the network topology and labels nodes in such a way as to construct a VPCS (virtual polar coordinate space). VPCR is a routing algorithm that runs on top of VPCS and requires no geographic information. It provides the ability to create consistent associations without relying on geographic information.

10.6.5 Distributed Index for Features in Sensor Networks (DIFS)

GHT targets retrieval of high-level, precisely defined events. The original GHT implementation is limited to report on whether a specific high-level event occurred. However, it is not able to efficiently locate data in response to more complex queries. The distributed index for features in sensor networks (DIFS) attempts to extend the original architecture to efficiently support range queries. *Range queries* are the queries where only events within a certain range are desired. In DIFS, the authors propose a distributed index that provides low average search and storage communication requirements while balancing the load across the participating nodes [8]. We now describe the notion of high-level events and discuss a few sample range queries that can be posed by the end user.

10.6.5.1 *High-Level Event*

A *high-level event* can be defined in terms of composite measurements of sensor values themselves. For example, a user might query average animal speed or peak temperature in hot regions. Often the user can add timing constraints to these values, for example, by querying the average animal speed within the last hour. Alternatively, the user might incorporate a spatial dimension as well, for example, by looking for an area with average temperature greater than a certain threshold. Some end users might be interested in querying relations among various events as well. An example of a query in the temporal domain would be whether an increase in temperature above a threshold is followed by the detection of an elephant. Of course, all these types can be used in combination, and more complex queries can be posed. DIFS assumes that all sensors store raw sensor readings, while only a subset of sensors serve as an index node to facilitate the search. Note that DIFS runs on top of GHT to support the abovementioned range queries in an efficient way. Since the reader is already familiar with GHT, we first describe a naive quad-tree-based approach for indexing followed by the DIFS architecture.

10.6.5.2 *Simple Quad Tree Approach*

This approach works by constructing a spatially distributed quad tree of histograms that summarize activity within the area they represent. A root node maintains four histograms describing the distribution of data in each of four equisized quadrants (its children). A drilldown query approach can be used on top of this tree; the summaries at children nodes are more detailed. However, in that case, since the root node has to process all the queries (queries move in top–down fashion over the tree), it might prove to be a bottleneck. Also, propagating data up in the hierarchy to update histograms every time a new event is generated might pose problems from an energy perspective. DIFS extends this naive approach to address these problems.

10.6.5.3 *DIFS Architecture*

Similar to the quad tree implementation, DIFS constructs an index hierarchy using histograms. Unlike the quad tree approach, every child has $bfact$ number of parents instead of a single parent, where $bfact = 2^i$, $i \geq 1$. Also, the range of values that a child maintains in its histograms is $bfact$ times the range of values maintained by its parent. The crucial point is that, to ensure energy–storage load balancing, a range of values that an index node knows about is inversely proportional to the spatial extent that it covers. Therefore, rather than having only one query entry point, as in the case of naive quad tree approach, a search may begin at any node in the tree. Query entry points are selected according to both the spatial extent, as well as the range of values mentioned in the query.

Another important feature is *index node selection*, which is done with the help of a geographically bound hash function. A given sensor field is divided into rectangular quadrants recursively proportional to the number of levels in the index hierarchy. Given a source location, a string to hash, and a bounding box, the output of this hash function is a pair of coordinates within the given bounding box. Note that, in the case of GHT, the hash function can result to any location within the sensor field and not to any location within the bounding box. Intuitively, a node forwards an event to the first local index node with the narrowest spatial coverage but widest value range. This node then forwards it to its parent with wider spatial coverage but narrower value range. This forwarding process can be applied recursively.

10.7 CONCLUDING REMARKS

In this chapter, we considered the problem of storage management for sensor networks where data are stored in the network. We discussed two classes of applications where such storage is needed: (1) *offline scientific monitoring*; where data are collected offline and periodically gathered by an observer for later playback and analysis; and (2) *augmented reality applications*, in which data are stored in the network and used to answer dynamically generated queries from multiple observers. In such storage-bound networks, the sensors are limited not only in terms of available energy but also in terms of storage. Furthermore, efficient data-centric indexing and retrieval of the data are desirable, especially for the second application type.

We have identified the goals, challenges, and design considerations present in storage-bound sensor networks. We organized the challenges into three areas, each incorporating a unique design/tradeoff space with respect to the identified goals and possessing preliminary existing research and protocols:

1. *System support for storage* — storage for sensors is different from traditional nonvolatile storage in terms of the hardware, applications, and resource limitations. As such, the hardware and required file system support differ significantly from those of traditional systems.
2. *Collaborative storage* — storage can be minimized by exploiting spatial correlation between nearby sensors and reducing redundancy/adjusting sampling periods. Such techniques require data exchange among nearby sensors (and the associated energy costs). An intricate tradeoff in terms of energy exists; local buffering may save energy during the data collection phase but not during indexing and data retrieval. Furthermore, storage and energy may need to be balanced depending on the resource that is scarcer.
3. *Indexing and retrieval* — the final problem is similar to the indexing and retrieval problem in P2P networks. However, it differs in the type of data and the high costs for communication (requiring optimized queries and limiting of unnecessary data motion). We discussed existing solutions in this area.

Most existing work addresses these different aspects of the problem independent of the other aspects. We believe that there remain large unexplored areas of the design space in each area. However, issues at the intersection of these areas appear most challenging and have not been addressed to our knowledge.

REFERENCES

1. S. Androutsellis-Theotokis, A survey of peer-to-peer file sharing technologies, 2002; available on the Web at `http://www.eltrun.aueb.gr/whitepapers/p2p_2002.pdf`.

2. P. Bonnet, J. Gehrke, and P. Seshadri, Towards sensor database systems, *Lecture Notes in Computer Science*, 1987, 2001.

3. E. Cohen and S. Shenker, Replication strategies in unstructured peer-to-peer networks, *Proc. 2002. ACM SigComm'02 Conf.*, Aug. 2002.

4. D. Ganesan, B. Greenstein, D. Perelyubskiy, D. Estrin, and J. Heidemann, An evaluation of multi-resolution storage for sensor networks, *Proc. 1st Int. Conf. Embedded Networked Sensor Systems*, ACM Press, 2003, pp. 89–102.

5. D. Ganesan, S. Ratnasamy, H. Wang, and D. Estrin, Coping with irregular spatio-temporal sampling in sensor networks, *Proc. ACM Computer Communication Review (CCR) Conf.*, 2004.

6. D. Gay, Design of matchbox, the simple filing system for motes, 2003; available on the Web at `http://www.tinyos.net/tinyos-1.x/doc/matchbox.pdf`.

7. D. Gay, Matchbox: A simple filing system for motes, 2003; available on the Web at `www.tinyos.net/tinyos-1.x/doc/matchbox-design.pdf`.

8. B. Greenstein, D. Estrin, R. Govindan, S. Ratnasamy, and S. Shenker, Difs: A distributed index for features in sensor networks, *Proc. 1st IEEE Int. Workshop on Sensor Network Protocols an Applications (SNPA 2003)*, 2003.

9. W. Heinzelman, *Application-Specific Protocol Architectures for Wireless Networks*, PhD thesis, Massachusetts Institute of Technology, 2000.

10. J. Hill, R. Szewczyk, A. Woo, S. Hollar, D. E. Culler, and K. S. J. Pister, System architecture directions for networked sensors. *Proc. 8th Int. Conf. Architectural Support for Programming Languages and Operating Systems*, ACM Press, 2000, pp. 93–104.

11. C. Intanagonwiwat, D. Estrin, R. Govindan, and J. Heidemann, *Impact of Network Density on Data Aggregation in Wireless Sensor Networks*, Technical Report TR-01-750, Univ. Southern California, Los Angeles, Nov. 2001.

12. C. Intanagonwiwat, R. Govindan, and D. Estrin, Directed diffusion: A scalable and robust communication paradigm for sensor networks, *Proc. 6th ACM Int. Conf. Mobile Computing and Networking (MobiCom'00)*, Aug. 2000.

13. P. Juang, H. Oki, Y. Wang, M. Martonosi, L. S. Peh, and D. Rubenstein, Energy-efficient computing for wildlife tracking: Design tradeoffs and early experiences with ZebraNet, *Proc. 10th Int. Conf. Architectural Support for Programming Languages and Operating Systems*, ACM Press, 2002, pp. 96–107.

14. B. Karp and H. T. Kung, Gpsr: Greedy perimeter stateless routing for wireless networks, *Proc. 6th Annual Int. Conf. Mobile Computing and Networking*, ACM Press, 2000, pp. 243–254.

15. S. Madden, *The Design and Evaluation of a Query Processing Architecture for Sensor Networks*, PhD thesis, Univ. California Berkeley, 2003.

16. S. Madden, M. Franklin, J. Hellerstein, and W. Hong, Tag: A tiny aggregation service for ad-hoc sensor networks *Proc. OSDI*, 2002.

17. M. K. McKusick, W. N. Joy, S. J. Leffler, and R. S. Fabry, A fast file system for UNIX, *Comput. Syst.* **2**(3):181–197(1984).

18. J. Newsome and D. Song, Gem: Graph embedding for routing and data-centric storage in sensor networks without geographic information, *Proc. 1st ACM Conf. Embedded Networked Sensor Systems (SenSys 2003)*, 2003.

19. Network simulator; available on the Web at http://isi.edu/nsnam/ns.

20. S. Ratnasamy, P. Francis, M. Handley, R. Karp, and S. Shenker, A scalable content addressable network, *Proc. ACM SigComm Conf., 2001*, 2001.

21. S. Ratnasamy, B. Karp, S. Shenker, D. Estrin, R. Govindan, L. Yin, and F. Yu, Data-centric storage in sensornets with ght, a geographic hash table, *Mobile Network Appl.* **8**(4):427–442(2003).

22. A remote ecological micro-sensor network, 2000; available on the Web at http://www.botany.hawaii.edu/pods/overview.htm.

23. M. Ripeanu, I. Foster, and A. Iamnitchi, Mapping the gnutella network: Properties of large-scale peer-to-peer systems and implications for system design, *IEEE Internet Comput. J.* **6**(1) (XXXX).

24. M. Rosenblum and J. K. Ousterhout, The design and implementation of a log-structured file system, *ACM Trans. Comput. Syst.* **10**(1):26–52(1992).

25. I. Stoica, R. Morris, D. Liben-Nowell, D. R. Karger, M. F. Kaashoek, F. Dabek, and H. Balakrishnan, Chord: A scalable peer-to-peer lookup protocol for internet applications, *IEEE/ACM Trans. Networking* **11**(1):17–32(2003).

26. S. Tilak, N. Abu-Ghazaleh, and W. Heinzelman, Storage management issues for sensor networks; available on the Web at www.cs.binghamton.edu/~sameer/CS-TR-04-NA01.

27. Crossbow Technology Inc; http://www.xbow.com.

CHAPTER 11

Security in Sensor Networks

FAROOQ ANJUM and SASWATI SARKAR

Applied Research, Telcordia Technologies, Piscataway, New Jersey and
Department of Electrical and Systems Engineering, University of Pennsylvania, Philadelphia

11.1 INTRODUCTION

Sensor networks have increasingly become the subject of intense scientific interest. These networks can vary in size from tens to thousands of inexpensive wireless sensor nodes depending on the application. The sensors are characterized mainly by low cost, small size, dense deployment, limited mobility, and a lifetime constrained by battery power. Additionally, sensor nodes have limited resources in terms of storage, computational, memory, and communication capabilities. Typical architectures of sensor networks involve the sensor nodes forwarding all generated data to a single collection point called the *sink node*.

Potential areas of application for sensor networks include health monitoring, data acquisition in hazardous environments, military operations, and homeland defense. Consider the example where researchers from Syracuse University are working on a project that involves installing a network of sensors in the Seneca River to create the largest underwater monitoring system. The sensor nodes are expected to collect data as frequently as every 10 min on temperature, oxygen, turbidity, light, salt content, phosphorus, iron, nitrates, nitrites, ammonia, and other substances. The collected data are eventually delivered to a main computer at Syracuse and posted on the Web. This information enables scientists to assess whether the water is suitable for consumption, aquatic life, and recreation. Several similar projects are under way by researchers and companies to hone sensor networks for use in homeland security.

It is clear, however, that attention has to be paid to the security aspects of such networks. Sensors are subjected to numerous threats due to operation in a hostile environment. Addressing these threats requires various forms of physical,

Mobile, Wireless, and Sensor Networks: Technology, Applications, and Future Directions
Edited by Rajeev Shorey, Akkihebbal L. Ananda, Mun Choon Chan, and Wei Tsang Ooi
Copyright © 2006 John Wiley & Sons, Inc.

communications, and cryptographic protection. For example, terrorist organizations seeking to attack the Seneca River system would try to compromise the sensor network before launching any attacks. Protecting these sensor networks would have to be a multipronged strategy. It would be necessary to focus on aspects such as protecting the information sent out by the sensor nodes by encrypting it, detection of malicious events in the system, and recovery if possible after detecting malicious events. For example, the intruders might compromise the existing sensors and then use these compromised sensor nodes to send viruses targeting the main computer in the example above. The "bad" sensors can also try to overwhelm the network with unnecessary updates so as to deplete the resources such as bandwidth, and power in the sensor network.

Given the similarity between sensor networks and ad hoc networks, such as the use of wireless links as well as the presence of multihop communication, an issue that logically arises is the applicability of the security schemes proposed for wireless ad hoc networks to sensor networks. One very important reason why these schemes are not suitable for sensor networks is the severe energy constraints associated with sensor nodes. There are many other differences between these two networks as listed below — as a result of these differences, the solutions proposed to provide secure communication in ad hoc networks might not always be applicable to sensor networks:

- Energy is the fundamental resource constraint as mentioned earlier. This is because advances in battery technology still lag behind. Their lifetime is generally limited by the lifetime of a tiny battery. For example, the Berkeley MICA mote lasts for only about 2 weeks when running at full power. Combine this along with the fact that in many cases the sensors are deployed in areas where their batteries cannot be replaced, and we see the importance of conserving power. Hence, if the sensor networks are to last for years, then it is essential that radios operate in the sleep mode (which implies being inactive, thereby saving on energy consumption) much of the time.

- The nodes in a sensor network are densely deployed; that is the distance between two nodes is often less then a few meters. Thus, sensor networks might consist of tens to thousands of nodes. This is also made possible since nodes are small and cheap to build, and therefore a large number of these networks can be used to cover an extended geographic area.

- Sensors have limited computational power and memory resources. Each bit transmitted consumes as much power as does executing 800–1000 instructions [54]. The range of such devices is also limited by power and antennae constraints. Low-sitting sensors sometimes have limited line of sight because of rocks, grass, trees, bushes, and other obstacles.

- Public-key-algorithm-based primitives are ruled out for such networks because of the computational complexity associated with such algorithms.

- Sensor nodes are more prone to failure.

- These networks may typically be deployed in hostile areas where the sensor nodes are more susceptible to capture and exploitation by the enemy. The fact that they are typically unattended might also imply that compromising them might be easier. Mandating tamperproof nodes to address this problem would result in increased costs.

- While a majority of sensor node deployments are expected to be static, mobility is not ruled out in such networks. For example, data collection and control nodes placed on humans, vehicles, and other subjects or objects inherit mobility from the platform where the nodes are deployed.

- These networks also admit specialized communication patterns. Most communication in a sensor network involves a sink, specifically, between the node and the sink (requests for data or responses) or from the sink to all nodes (beacons, queries, or reprogramming) [3]. In some cases the sink might need to correspond with multiple nodes simultaneously. In addition, sensor networks might also allow for local communication in order to allow neighbors to discover and coordinate with each other.

- Although their location is fixed in many applications, the network topology can change frequently because of node failures and objects passing through the sensor field.

In this chapter we focus on the research problems associated with ensuring security in sensor networks. The chapter is organized as follows. In Section 11.2 we comment on the resources available in sensor networks. In Section 11.3 we investigate the various issues associated with providing secure communication in sensor networks. Finally we present our conclusions in Section 11.4.

11.2 RESOURCES

The resources associated with sensors are mainly the processing and power resources. Processing capabilities of microprocessors are improving at an exponential rate, but the battery and energy storage technologies are improving at a much slower pace. So energy efficiency is a critical issue for sensor networks, especially given that it is difficult to change or recharge batteries on nodes in such networks. For example, motes feature an 8-bit central processing unit (CPU) running at 4 MHz, with 128 kB of program memory, 4 kB of random access memory (RAM), 512 kB of serial flash memory, and two AA batteries. The processor provides support only for a minimal reduced instruction set computer (RISC)-like instruction set. No support is provided either for variable-length shifts/rotates or for multiplication. Communication is at a peak rate of 40 kbps (kilobits per second) with a range of up to 100 ft. Sensors are usually built with limited processing, communication, and memory capabilities in order to prolong their lifetime under the limited energy budget.

In the case of a sensor device, the radio transceiver is the dominant energy consumer [36]. The energy consumed when idle is not negligible and has been shown to represent 50–100% of the energy used while transmitting and receiving messages

[37,38,46]. Note that a wireless transceiver has to be powered to receive each incoming packet to determine whether the packet should be accepted or forwarded. Although many of these packets are simply discarded, they consume a significant amount of energy. The energy consumed during communication is several orders of magnitude greater than the energy consumed by computation [41]. For idle power consumption, the current generation of radios (RFM Inc. http://www.rfm.com — used in Berkeley motes) use one or more orders of magnitude lesser power than for transmit power. This can be expected only to improve in the future.

Sensor node capabilities span a range varying from the smart dust sensors [48] that have only 8 kb of program and 512 bytes of memory, and processors with thirty-two 8-bit general registers with a speed of 4 MHz and a voltage of 3.0 V to sensors such as the MIPS R4000 processors, which are over an order of magnitude more capable. But use of asymmetric primitives for cryptographic purposes is ruled out on account of power, energy, and related communicational as well as computational constraints associated with such devices. For example, Deng et al. [27] show that on a processor such as the Motorola MC68328 (which can be considered midrange in terms of capabilities), the energy consumption for a 1024-bit RSA encryption (signature) operation is about 42 mJ (840 mJ) which is much higher than that for a 128-bit AES encryption operation, which is about 0.104 mJ. In order to transmit a 1024-bit block over a distance of 900 m, it was found that RSA encryption required about 21.5 mJ and half this for AES-based encryption. This is for a 10-kbps system that requires $\sim$10 mW of power. It was also shown that symmetric-key ciphers and hash functions are between two to four orders of magnitude faster to process than are digital signatures.

Sensor network applications are mainly in the area of information acquisition [10,11]. Given this application area, two types of communication models have been considered for sensor networks: cluster-based and peer-to-peer. In the cluster-based model [10], multiple clusters are formed statically and/or dynamically. A cluster head (CH) exists for each cluster in order to manage or control the cluster. The CH election and maintenance requires the use of secure message exchanges. In the peer-to-peer model [12] all nodes are homogeneous, thereby having the same capabilities. Thus, each sensor communicates with any of the other sensors without relying on dedicated devices to act as a CH. In each of these models we could assume the presence of a single or multiple destinations to which data are delivered.

11.3 SECURITY

Sensor networks with limited processing power, storage, bandwidth, and energy require special security approaches. The hardware and energy constraints of the sensors add difficulty to the security requirements of ad hoc networks concerning availability, confidentiality, authentication, integrity, and nonrepudiation. Availability ensures the survivability of network services despite denial-of-service attacks. Confidentiality ensures that certain information is never disclosed to unauthorized entities. Integrity guarantees that a message being transferred is never corrupted. Authentication enables a node to ensure the identity of the peer node that it is communicating with. Finally, nonrepudiation ensures that the origin of a message cannot deny having sent the message [11].

In this section we plan to look at the various security-related problems in sensor networks. We first focus on the different attacks that are possible in sensor networks. Then we consider the topic of data encryption and authentication. The topic of key management in sensor networks is addressed next. Following that, the topic of intrusion detection in sensor networks is covered. Routing, secure aggregation, and implementation are the final three topics addressed in this section, in that order.

11.3.1 Attacks on Sensor Networks

Wireless sensor networks may be deployed in hostile environments where attackers can (1) eavesdrop and replay transmitted information, (2) compromise network entities and force them to misbehave, and (3) impersonate multiple network entities by obtaining their identities. In this section we consider the possible problems that can occur in sensor networks. Wood and Stankovic [15] consider denial-of-service attacks in sensor networks. The authors emphasize an important and well-known fact, namely, that security has to be built into the system during design instead of slapping it on later. The latter approach fails in many cases, and examples have been provided to illustrate this. Some of the attacks considered by those authors [15] are as follows:

1. *Tampering*. This might be the result of physical capture and resulting compromise. It is recommended that nodes react to tampering in a fail-complete manner by erasing the cryptographic and program memory. This implies that the sensor nodes are tamperproof. Tamperproofing, though, increases the costs of such nodes. Camouflaging or hiding nodes are other recommended defenses to protect against physical capture and resultant tampering. Remote management of the sensors can also help if an authenticated user can erase the cryptographic keys and information on the sensor if they feel it has been compromised.

2. *Jamming*. This can be detected easily since the sensor nodes will not be able to communicate. A recommended defense is spread-spectrum communication, but that increases the costs, power, and design complexity. So this might not be feasible for sensor networks. In this case we might have different strategies based on the type of jamming. If jamming is permanent, then nodes can use a lower duty cycle and try to outlive the adversary by spending energy frugally. For intermittent jamming, other schemes such as using priority-based schemes to send important messages, cooperation among system nodes to store and forward such important messages, or notification to the base-station of such an event can be used — and if jamming is localized, then nodes surrounding the affected region can cooperate in mapping and reporting about

the region. This will allow communication to bypass the affected region. Using redundancy in terms of paths (by avoiding the jammed region), interfaces (by using different technologies such as optical, infrared, etc.), and frequencies (by transmitting on different frequencies) seems to be a useful approach to address jamming. Of course, this does imply that the attacker is not able to block the redundant aspects such as by blocking all the frequencies.

3. *Link Layer Attacks.* In this case schemes could try to break the medium access control protocol being followed. For example, causing a collision in only a few bits of the transmission can result in a very efficient way of disrupting an entire packet in terms of the energy expended by the attacker. This can be partially addressed by the use of error correcting codes (and protocols that attempt "best-effort delivery"). Note, though, that these codes work best when faced with accidental errors as opposed to malicious errors. Further, the use of these codes results in additional processing and communication overhead. A variation on this could result in unfairness whereby some nodes do not follow the recommended rules to access the channel. For example, if the MAC protocol provides for random backoff after detecting a collision, a malicious node might choose to ignore this and transmit immediately. This will result in unfairness to the other nodes.

Newsome et al. [55] investigated the Sybil attack [56] in sensor networks. In this case a single malicious node assumes the identity of a large number of nodes. It does this either by impersonating other nodes or by claiming false identities. This could be done while using a single physical device. The authors start by providing a taxonomy of the different types of Sybil attack. Possible Sybil attacks on various aspects of sensor networks such as aggregation, voting, fair resource allocation, and misbehavior detection have been explained. In case of aggregation a malicious node using the Sybil attack can send in multiple false sensor readings in order to significantly affect the computed aggregate. Similarly, by using multiple false identities, schemes that depend on voting such as malicious behavior detection schemes can be influenced strongly. Fair allocation of resources can also be circumvented by these attacks.

An approach to defend against Sybil attack is resource testing [56]. This assumes that each physical entity is limited in some resource. Verification of identity involves testing that each identity has as much of the tested resource as is expected from each physical device. Some of the possible resources here are computation, storage, and communication. Unfortunately, this method is unsuitable for sensor networks [55] because of the resource limitations. For example, the attacker can deploy a physical device with capabilities that are several orders of magnitude larger than those of the normal sensor nodes. Hence new defenses [55] would be needed for sensor networks. The defenses proposed by Newsome et al. [55] include radio resource testing, verification of key sets for random key predistribution, registration, and position verification. In the case of radio resource testing, a node can assign each of its neighbours a different channel to transmit on. The node can then choose a channel randomly to listen. If the neighbor that was assigned this channel is genuine, then the node can hear this message. Of course, an attacker with multiple interfaces can foil this defense. Position verification whereby the network verifies the physical position of each device can also be used to foil the Sybil attack. This can be done by checking that different identities are present at different physical locations. In case of random key predistribution, a node's keys are associated with its identity. A one-way hash of the claimed ID combined with a pseudorandom function provides the indices of the keys that the node is expected to posses. If this is not verified, then the claimed ID is rejected. This is a promising approach to address the Sybil attack while also leveraging the presence of keys on sensor nodes. We will look at the key management as well as encryption/authentication aspects in later sections. Registration of a node's identity with a central authority could be another approach although it suffers from drawbacks such as vulnerability to an attacker adding false identities to the list maintained with the central authority. Karlof and Wagner [4] also consider attacks in sensor networks at the network layer in the context of routing. We will look at these later in Section 11.3.5, on routing.

11.3.2   Data Encryption/Authentication

Sensor network deployments result in communication that is typically geared toward data collection. In this case the nodes might periodically collect data that are then forwarded to a basestation or sink. Environmental monitoring is an example of such an activity, including monitoring the temperature, radiation, or chemical activity within a certain field or plant. In addition, communication in such networks might also result from an observer querying the network about information of interest. In such a case, the source node in possession of this information replies with an answer. An example of this is an object tracking system, in which an observer queries the network about the occurrence or behavior of an object. Note that these queries are typically flooded unless the observer knows the exact sensor that has the answer it is looking for.

Now, given the use of wireless links for communication, which makes eavesdropping quite easy, it is vital to secure communication between any two sensor nodes. If this is not done, then it would be very easy for the intruder to mount various attacks on the network such as injecting false data, snooping and modifying transmitted data, or replaying old messages. The communication mechanisms would have to ensure that the communicated data are secured. The standard approach to achieving this is to encrypt the data transmitted by the sensors. In addition, it might also be necessary to provide authentication mechanisms in order to guarantee that the data source is the claimed sender. Another security attribute that is needed is one related to data freshness. Data freshness provides a guarantee that the data are recent and ensures that an adversary cannot replay old messages. Given that sensor networks send measurements over time, if this guarantee is missing, then an adversary can replay old measurements, thereby causing confusion for networks whose focus is on measuring various values over time. We would like to point out that the different proposals discussed here are based on the use of symmetric cryptographic primitives.

SPINS [3] considers the case where most communication involves the basestation. A central basestation acts as the only point of trust, where all nodes trust only the basestation and themselves. Thus two sensor nodes do not communicate directly. It focuses on energy-efficient encryption and authentication mechanisms based on the RC5 block cipher. They selected RC5 because of its small code size and high efficiency. The other ciphers considered were AES, DES, and TEA [26]. SPINS addresses secure communication in resource-constrained sensor networks, introducing two low-level secure building blocks, SNEP and μTESLA. SNEP provides data confidentiality, two-party data authentication, integrity, and freshness between nodes and the sink. μTESLA provides authentication for data broadcast. All cryptographic primitives are constructed out of a single block cipher for code reuse and also because of the limited program memory. But following the recommendation of not reusing the same cryptographic key for different cryptographic primitives (since this might cause weaknesses on any potential interaction between the primitives), different keys are used for encryption and for MAC operations. The main idea of SPINS is to demonstrate the feasibility of security with very limited computing resources, by using symmetric cryptography alone, without emphasis on general applicability. The target wireless network is homogeneous and static.

SNEP uses the block cipher in the counter mode. In order to save energy, the counter is not sent with every message. This can cause problems when the corresponding parties are not synchronized. To address this issue, the authors also propose a counter exchange protocol that will be used by the parties concerned when the counters are unsynchronized. Note that the counter mode provides advantages such as semantic security, which ensures that the intruder has no information about the plaintext, even when multiple encryptions of the same plaintext are transmitted. Further in this case the cryptographic algorithm need be used only in the encryption mode on both the receiver and the transmitter, which is an advantage in terms of performance when considering algorithms like AES.

The communicating parties share a master secret key from which independent keys are derived using a pseudorandom function. Four keys are derived, two encryption keys and two MAC keys, with separate keys for each direction of communication. The data to be transmitted are then encrypted using the encryption key in the counter mode. A MAC is also associated with each message with the MAC computed over the encrypted data and the counter. The counter value in the MAC prevents replay of old messages. In addition, partial message ordering is also provided by the use of the counter. The MAC algorithm used is CBC-MAC, since this allows reuse of the RC5 block cipher.

The other component of SPINS is μTESLA. The focus of μTESLA is to provide authentication for broadcast communication in sensor networks. Using a single MAC key to secure such communication is not possible since all the receivers

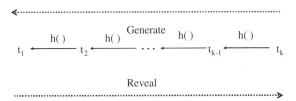

Figure 11.1 TESLA: Keys are generated by using a one way function, such as, hashing in the order shown. They are then used in the reverse order of generation with one key corresponding to each time interval. h() denotes the one way functions.

and the sender will have to share the MAC key, and hence a receiver cannot have the guarantee about the claimed sender.

Else nodes have to make multiple transmissions of messages each time they are encrypted with a different key, which is an inefficient solution. The authors therefore build on the TESLA [40] protocol to make it amenable to sensor networks. TESLA applies the concept of one-way key chain whereby the keys are related using a one-way hash function as shown in Figure 11.1. The time is divided into intervals, and a key used in interval i is revealed in interval $i + 1$. Further, the keys revealed in intervals i and $i + 1$ are connected via the one-way hash function whereby a hash of the key revealed in interval $i + 1$ equals the key revealed in interval i. This delayed disclosure of symmetric keys provides the necessary asymmetry needed to authenticate broadcast packets. The initial packet is authenticated in TESLA using a digital signature. Standard TESLA has an overhead of approximately 24 bytes per packet since the TESLA key for the previous interval is revealed in each packet of the current interval. TESLA uses too much communication and memory on resource-starved sensor nodes.

μTESLA addresses the problems of TESLA for sensor networks and uses only symmetric mechanisms. The key is also disclosed once per interval. Given the fact that it is expensive to store a one-way key chain in a sensor node, μTESLA restricts the number of authenticated senders. Of course, a requirement for sensor nodes is that of loose time synchronization because the division of time into intervals as intervals are crucial in the use and revelation of the symmetric keys. The second disadvantage is the need to store packets in a buffer until the next time interval when the keys are revealed. The node on receiving a key revelation checks on the correctness of the key (using the hash relationship with the key revealed in the previous interval) and then uses the key to authenticate the packets stored in the buffer. If the packet revealing the key in an interval is lost, then the sensor node would have to wait and continue buffering packets until the key for the next interval is revealed.

SPINS [3] assumes that individual sensors are untrusted. Hence, the key setup is designed so that a compromise of a node does not spread to other nodes. Every node shares a master key with the basestation. Any node-to-node communication necessitates authentication via the basestation.

11.3.3 Key Management

When employing cryptographic schemes, such as encryption or digital signatures, to protect both routing information and data traffic, a key management service is always required. For secure communication between any two entities, both the entities should possess a secret value or key. The possible ways in which secure communication can be established are for the entities concerned to share a key (symmetric-key system) or for the entities concerned to possess different keys (asymmetric-key system). Key management is the process by which those keys are distributed to nodes on the network and how they are further updated if required, erased, and so on. But at the same time we need to consider factors such as

- Limited power supply, which limits the lifetime of keys. Battery replacement might cause the device to reinitialize and zeros out keys.
- Working memory of sensor nodes possibly insufficient to even hold variables for asymmetric cryptographic algorithms (e.g., RSA has a key size of at least 1024 bits).
- Use of computationally efficient techniques.
- Inability to predetermine the neighbors after deployment as well as inability to put absolute trust in neighbors.
- Few and expensive basestations.

It is not possible to set up an infrastructure to manage keys used for encryption in the traditional Internet style because of factors such as unknown and dynamic topology of such networks, vagaries of the wireless links used for communications, and lack of physical protection. A practical solution given these constraints is to distribute the keys on the sensors before the sensor nodes are deployed. Thus the nodes have some secret information on them before being deployed, and using this, they need to set up secure communicaton infrastructure for use during operation. In sensor network security, an important challenge is the design of protocols to bootstrap the establishment of a secure communications infrastructure from a collection of sensor nodes that may have been preinitialized with some secret information but have had no prior direct contact with each other. A bootstrapping protocol must not only enable a newly deployed sensor network to initiate a secure infrastructure but also allow nodes deployed at a later time to join the network securely. At present, the most practical approach for bootstrapping secret keys in sensor networks is to use predeployed keying, in which keys are loaded into sensor nodes before they are deployed. Several solutions based on pre-deployed keying have been proposed, including approaches based on the use of a global key shared by all nodes [5], approaches in which every node shares a unique key with the basestation [3], and approaches based on probabilistic key sharing [6,14]. Ignoring the last approach of probabilistic key sharing, we have two possible options:

1. Using a single key for the entire network
2. Pairwise sharing, whereby each node has a separate key for every other node in the network

Solution 1 is problematic from a security standpoint since the compromise of a single sensor will break the security of the entire network. To ensure security in such cases would involve revocation and rekeying, but this is not an attractive option for such networks because of the energy consumption associated with communication (in addition to requiring authentication for the rekeying process in order to prevent an attacker from rekeying or zeroing your keys remotely). Pairwise secret sharing (solution 2) circumvents avoids this problem but then places great demands on the amount of storage needed on each sensor node, which renders it an impractical solution for large networks. This solution also makes it difficult to add more nodes to a deployed system since this involves rekeying with all the deployed nodes. The procedure of loading keys into the sensor must also be factored in.

Basagni et al. [5] consider sensor networks consisting of tamperproof nodes (these nodes are also called "pebbles"). Thus, even when the nodes fall into enemy hands, the keys are not compromised. All nodes before deployment are initialized with a single symmetric key, which thereby saves on storage and search time. This single key is then used to derive the keys used to protect data traffic. The derivation occurs in stages; in the first stage the nodes are organized into clusters. In the second stage the cluster heads (CHs) organize into a backbone, and from these a key manager is chosen probabilistically. The key manager decides the key to be used to encrypt data in the next cycle and distributes this key in a secure fashion (based on a one-way chain derived from the single systemwide key) to the nodes in the network via their CHs. The protocol considers factors such as the capabilities of the sensor nodes in deciding the CHs as well as the key manager. A key manager is selected each time the protocol is run. The fittest node becomes the key manager, and the key manager changes every time, increasing the security. Having a single key shared by all nodes is efficient in terms of storage requirements. Further, as no communication is required among nodes to establish additional keys, this also offers efficient resource usage. The problem with this approach is that if a single node (especially if it is the CH) is compromised, then the security of the entire network is disrupted. Refreshing the key is also expensive because of the communication overhead. In SPINS [3] also, each sensor node shares a secret key with the basestation. Hence a compromised sensor can be used by the intruder to break the security of the network. (A concern with determining the keys from preplaced keys is the efficiency of the function used to determine the new keys. Is it truly random, or is there a hint of probabilistic nature there that could help an attacker determine a key that will be used later on? How much trust can you place in your key manager?)

To address these shortcomings associated with a single key being shared by all the nodes in a network, a probabilistic key sharing approach was first proposed by Eschenaueer and Gligor [14]. The proposals that followed were improvements on the basic scheme in terms of security. The proposed improvements focus on three aspects: key pool structure [24,29], key selection threshold [6], and path key establishment protocol [6,42]. In this case a set of keys is randomly chosen from a large pool and installed on each sensor node prior to deployment. Thus, a pair of nodes would share keys with some probability. As a result, these schemes offer network resilience against capture of nodes since a node has very few keys deployed on it. Thus, there is no need to consider tamperproof nodes (but note that nontamperproof nodes might not be acceptable in some situations such as military deployments). This is an advantage given the problems such as cost, and complexity associated with tamperproof nodes.

In case of probabilistic key sharing, each sensor is loaded with one or more keys before deployment. These keys are randomly chosen from a pool of keys. After deployment, a secure link can be established between a pair of sensors, provided a key happens to be common to both sensor nodes. The drawbacks of this scheme are

- Communication over multiple hops requires decryption and reencryption for each hop, and thus a sensor node cannot just relay packets but will have to cryptographically process all incoming packets as well. This increases the sensor's workload and the latency.
- Setting up a secure channel between sensors that do not share common keys or that are distant is cumbersome.

When two nodes need to discover any common keys, they need to execute a shared-key discovery protocol. The simplest way for two nodes to discover if they share a common key is to have an identifier associated with every key. A node can then transmit the list of identifiers representing the set of keys present on that node. Of course, as a result the adversary might not know the actual keys but would know the identifiers of keys on a given node by eavesdropping. If this is of concern, then alternate methods that encrypt a random number using each key on the node and transmit the random number along with the encrypted data can be used. On receiving this transmission, nodes can search their own set of keys for common keys. Verification of these claims can then be based on a challenge–response protocol. Thus, this phase involves a communication overhead. The communication overhead is considered for key pathlengths of less than 3 in. [6,29].

The number of keys in the key pool is chosen such that any two random chosen nodes will share at least one key with some degree of probability. However, because of the random selection of keys on each node, it is very possible that a shared common key might not exist between any two given nodes. In such a case, a path of nodes that share common pairwise keys between the nodes of interest is found. This is possible if the graph representing the network is connected. The pair of nodes can then use this path of nodes to exchange a key that establishes the direct link. This achieves the same effect as that provided by pairwise sharing between every pair of nodes in the system but at lower cost. Of course, given that the models of connectivity are probabilistic, it is possible that the graph may not be fully connected, thereby preventing some pairs of nodes from communicating with each other securely. The proper choice of parameters is important in minimizing the occurrence of this event. Eschenaueer and Gligor [14] show that to establish connectivity between any two pairs of nodes with a very high probability for a 10,000-node network, it is necessary to distribute only 250 keys on each sensor node, with the keys drawn out of a pool of 100,000 keys. The network expansion by adding more nodes is also not a problem in this case.

Eschenaueer and Gligor [14] handle revocation by employing special nodes called *controller nodes* that contain the list of key identifiers present on a given sensor node. Now, when a sensor node is detected to be compromised, the entire set of keys present on that node will have to be revoked. This is done by the controller node broadcasting securely a single revocation message containing a list of the key identifiers corresponding to the keys to be revoked. On receiving the message, each node will have to delete the corresponding keys after verifying the authenticity of the message.

Chan et al. [6] the authors propose three random key predistribution schemes:

1. *q-Composite random key predistribution scheme* — achieves better security under small-scale attack and trades off increased vulnerability in the face of large-scale attack
2. *Multipath key reinforcement scheme* — attacker has to compromise many more nodes to achieve a high probability of compromising any communication
3. *Random pairwise key scheme* — ensures that network is secure even when some nodes have been compromised; also enables node-to-node mutual authentication between neighbors and quorum-based node revocation without involving a basestation.

In scheme 1, the q-composite scheme, in order to form a secure link, the scheme requires at least a threshold number of shared common keys between any two sensor nodes as opposed to the scheme due to Eschenaueer and Gligor [14], where a single common key is needed for any two sensor nodes to communicate securely. Increasing the threshold makes it exponentially harder for an attacker with a given set of keys to break a link. At the same time, to enable two nodes to establish a secure link with some probability, it is necessary to reduce the size of the key pool. But this is problematic since it allows the attacker to gain a larger proportion of keys from the key pool by compromising fewer nodes. The actual key used during communication can be a hash of all the shared keys. Other operation details are similar to the scheme due to Eschenaueer and Gligor [14]. The authors do address the problem of selecting an appropriate size of the key pool. They also consider the effects of compromised nodes on ensuring secure communications between noncompromised nodes.

Scheme 2, proposed by the present authors, is a multipath reinforcement scheme. This is a method for strengthening the security of an established link by establishing the link key through multiple disjoint paths. In this case, a node generates j random values and routes each random value over a different path to the correspondent node. The link key computed is a XOR (exclusive OR) of all these j random values. The authors show that this scheme provides a significant improvement over the basic scheme due to Eschenaueer and Gligor [14] but has little effect on the q-composite scheme.

Scheme 3, also proposed by the present authors, is the random pairwise scheme. In this case each node is provided with its own ID. Each node is also initialized with a certain number of keys such that each key is shared with only one other node. Each node's key ring contains not only the key but also the ID of the other node that shares the same key. After deployment a node broadcasts its node ID. This information is used by the node's neighbors to search in their key rings to verify whether they share a common pairwise key with the broadcasting node. A cryptographic handshake can then be performed so to enable the nodes to mutually verify knowledge of the key. This scheme is quite resilient against node capture. Node-based revocation and resistance to node replication are also provided by this scheme. A drawback is that this scheme does not scale to large network sizes.

The key predistribution scheme (KPS) [29] and grid-based scheme [24] change the unstructured key pool to a structured key pool. The structured key pool is formed by multiple key spaces. Within each key space, the key space structure uses the group key scheme proposed by Blundo et al. [43]. Chan et al. [6] and Zhu et al. [42] use k key paths to set up a pairwise key. The latter group [42] uses a secret sharing scheme.

Zhu et al. describe LEAP [7] (localized encryption and authentication protocol), a key management protocol for sensor networks that is designed to support in-network processing, while at the same time providing security properties similar to those provided by pairwise key sharing schemes. In their study [7], every node creates a cluster key and distributes this key to its immediate neighbors using pairwise keys that it shares with each of the its neighbors. Given that the different clusters highly overlap, each node has to apply a different cryptographic key before forwarding the message. LEAP also includes an efficient protocol for internode traffic authentication based on the use of one-way key chains. In this case each node simultaneously maintains an individual key, a pairwise key, and a cluster key to support in-network processing. While this scheme offers deterministic security and broadcast of encrypted message, the bootstrapping phase is quite expensive. Further, the storage requirements on each node are also not trivial since every node must set up and store a number of pairwise and cluster keys, and this number is proportional to the number of actual neighbors that the node has. Finally, this scheme is robust against outsider attacks but susceptible to insider attacks in which an adversary needs to compromise only a single node to inject false data.

Another approach employs a cluster-based communication model. In this case a cluster-based key management [5] scheme can be used according to which the cluster head (CH) maintains and distributes the keys within its own cluster. The problems with this scheme are

- Each CH is a single point of failure for the cluster.
- The energy of the CH will be drained faster relative to the others.
- Key management among different clusters can incur a large overhead.

The role of the CH [13] can be rotated in order to balance the energy consumption among sensors. Note that this also incurs an extra overhead.

Slijepcevic et al. [35] consider three different security levels because three types of data are considered in the networks in question. Each node in the network has a set of initial master keys, one of which is active at any given time. The keys needed for each of the security levels are then derived from the master key. For the first level, where the messages are infrequent, the master key is used. For the second level of security the network is divided into hexagonal cells.

All the members in a cell share a unique location-based key. Nodes belonging to the bordering region between cells store the keys of the cells to which they belong to allow traffic to pass through. The model requires that the sensor nodes be able to discover their exact locations, which allows them to organize into cells and produce a location-based key. The third level of security uses a weaker encryption with a focus on computational overhead. The authors assume that the sensor nodes are tamperproof. Thus, the set of master keys and the pseudorandom generator, preloaded to all sensor nodes, cannot be revealed by compromising a node.

11.3.4 Intrusion Detection

It is necessary to focus not only on the design of efficient mechanisms that can minimize intrusions but also on mechanisms that attempt to detect intrusions when they occur. This problem of detection is not specific to sensor networks. In fact, it has been studied thoroughly for wireline networks and is attracting significant attention for wireless mobile networks. Comparing the actual behavior of the system with the normal behavior in the absence of any intrusions is the typical approach taken to detect intrusions. Thus, a basic assumption is that the normal and abnormal behaviors of the system can be characterized.

As a result, there are two main techniques for detecting intrusions: anomaly detection and signature (misuse) detection. Anomaly detection essentially deals with the uncovering of abnormal patterns of behavior given the normal behavior pattern. This might imply the use of extensive "training sets" in order to characterize normal behavior. If the normal behavior can be accurately characterized, then this technique would be able to detect previously unknown attacks. This technique is used in a limited form in current commercial systems (designed for wireline systems) because of the high probability of false alarms.

Signature detection relies on the use of specifically known patterns of unauthorized behavior. In the context of network communications, these techniques rely on sniffing packets and using the sniffed packets for determining whether the traffic consists of malicious packets. For example, the detection system can watch out for malformed packets or viruses that are directed toward the end computer system. So, after sniffing the packets, if any such packet or set of packets is found during analysis, it can be concluded that the destination is under attack. Signature-based techniques are effective at detecting attacks without too many false alarms. Because of its low false alarm rate and maturity, signature-based detection is the mainstay of operational intrusion detection systems today. A drawback of this technique is that it is unable to detect novel attacks whose signatures are unknown. Further, examining the packet details would entail resource usage, which, as mentioned earlier, is a significant factor in sensor networks. Hence, we have to minimize the number of locations where packets are sniffed and analyzed. So a related question that arises is

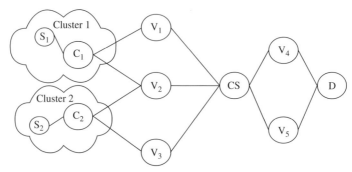

Figure 11.2 The sensor nodes are divided into clusters and a cutset determined using the cluster heads. The IDS modules are placed on the nodes constituting the cutset. In this figure, nodes S_1 and S_2 are intruders, C_1 and C_2 are clusterheads of cluster 1 and cluster 2, respectively, and D is the sink. The clusterheads can forward malicious packets through multiple paths to the sink. All the paths, however, have node CS in common. The custets are $\{CS\}$, $\{V_1, V_2, V_3\}$ and $\{V_4, V_5\}$, etc. Here, CS is the minimum cutset. Hence, it suffices to activate IDS on CS in order to achieve 100% detection.

where to place the sniffing and analysis software modules that we refer to as the *intrusion detection system* (IDS) modules in the sequel. This question is addressed by the present authors and others in another study [60]. Monitoring promiscuously might not be practical in such networks, and hence that study [60] does propose schemes that do not require promiscuous monitoring. The idea here is to divide the entire sensor nework into clusters. The CHs and the single destination are then considered to determine the minimal number of nodes, called the *minimal cutset*, through which all communication has to occur. The IDS modules are then expected to be placed on the nodes in the cutset as shown in Figure 11.2.

Efficient tracing of failed nodes in sensor networks is the focus in another study [18]. All nodes are assumed to have powerful and adjustable radios that can transmit at extended distances. Algorithms to trace failed nodes in a trusted environment are given. This allows the basestation to determine whether measurements from a region of nodes have stopped because all the nodes in that region have been destroyed or whether the reports have ceased as a result of the failure of a few nodes. The algorithms require communicating the topology of the network to the basestation, which is then responsible for tracing the identities of the failed nodes. The authors do not address the issue of compromised nodes.

Kumar et al. [44] consider the problem where, given an area to be protected, the number of sensors to be deployed must be determined to ensure that every point in the region is covered by k sensors, given that the network must last for a specified length of time. This is to facilitate the classification and tracking of intruders.

11.3.5 Routing

Karlof and Wagner [4] focus on routing security in wireless sensor networks, and demonstrate that currently proposed routing protocols for these networks are insecure because these protocols have not been designed with security as a goal. The authors propose threat models and security goals for secure routing in wireless sensor networks, present detailed security analysis of major routing protocols describing practical attacks against them that would defeat any reasonable security goals, and discuss countermeasures and design considerations for secure routing protocols in sensor networks. They considered various attacks on sensor networks such as spoofed, altered, or replayed routing information; selective forwarding; sinkhole attacks; Sybil attacks; wormholes; "Hello" flood; and acknowledgment spoofing. Note that these attacks are straightforward, and hence we will not explain them. They then study the effects of these attacks on routing protocols such as TinyOS beaconing; directed diffusion; geographic routing as used by GPSR or GEAR; minimum-cost forwarding; clustering-based protocols such as LEACH, TEEN, and PEGASIS; and rumor routing, as well as energy conserving topology maintenance protocols.

Deng et al. [16,17] propose an intrusion-tolerant routing protocol for sensor networks. This approach does not focus on detecting intrusions but rather attempts to design a routing protocol that can tolerate intrusions. This is done by taking advantage of the redundant paths possible in such networks from any sensor node to the sink. As a result, the effect of a malicious node is shown to be restricted to a small number of nodes in its immediate vicinity. The authors make use of various mechanisms such as encrypted communication, allowing only the basestation to broadcast information (which is authenticated using a one-way hash chain) to protect the routing protocols from attack. The resource constraints are addressed by allowing every sensor node to share a key only with the basestation and by performing all computations related to the building of routing tables at the basestation. Those authors [17] also consider implementation of their proposed protocol. We will consider this in detail in Section 11.3.7.

Tanachiwiwat et al. [33] monitor the behavior of static sensor nodes. They propose a routing protocol called TRANS (Trust Routing for location aware sensor networks). The authors apply the concept of trust to select a secure path and avoid insecure locations. Each node is responsible for calculating the trust values of its neighbors. Areas of misbehaving nodes are then bypassed in the route.

Deng et al. [59] consider a multipath routing strategy for sensor networks. They consider two set of attack, in which (1) the intruders aim to isolate the basestation and (2) the basestation location is deduced by snooping on data traffic destined to the basestation. The multipath routing strategy is designed to protect against the first set of attacks. Strategies such as hop-by-hop cluster key encryption/decryption and control of the sending rate are then proposed to prevent traffic analysis from determining the location of the basestations.

11.3.6 Aggregation

Sensor networks consist of thousands of sensors capable of generating a substantial amount of data. However, in many cases it may be unnecessary and inefficient to return all raw data collected from each sensor. Instead, information can be

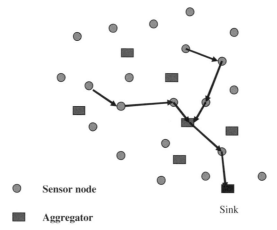

Figure 11.3   Aggregation using sensor nodes.

processed and summarized within the network and only the cumulative information sent to the sink as shown in Figure 11.3. This would ensure efficient usage of the communication resources, which includes the low-bandwidth links as well as the energy associated with transmitting and receiving data. But many factors such as development of efficient information processing and aggregation techniques and decisions about placement of the nodes that will perform this aggregation task have to be addressed.

Some work has been done in this area [49–53], but all these except Hu and Evans [51] assume that every node is honest. Given the security threats that sensor networks face, it is vital to study the performance of such techniques under various conditions such as compromise of sensor nodes, compromise of the nodes that perform the algorithm, and malicious circumvention of the processing and aggregation techniques. The security aspect during aggregation in such networks has also been considered [30–32,51]. We consider these next.

Pryzdatek et al. [30] address the problem of secure computation and aggregation in sensor networks. They assume that every sensor has a unique ID and shares a key with the aggregator and the sink. The aggregator and the sink can derive this from a single key each. The authors assume that only the aggregators are compromised. The compromised aggregator may cheat by forging results that are inconsistent with the raw data provided by the various sensors to the aggregator. The proposed schemes address this. Cases where the raw data themselves are forged or fabricated are not considered. In this work the authors consider only stealthy attacks in which the goal of the attacker is to make the home server accept false aggregation results that are significantly different from the true results. Hence, this also rules out attacks such as denial of service, which are considered detectable attacks. The approach followed by those authors [30] is that of aggregate–commit–prove. In the first step the aggregator collects the data from sensors with the aggregator verifying the authenticity of each sensor reading. The authenticator then commits to the collected data. Finally the aggregator transmits the aggregation result and the committment to the server and proves to the server that the reported results are correct according to interactive proof protocols. Using this framework, the authors then provide protocols for securely computing quantities such as the median, minimum, and maximum values; number of distinct elements; and average. Thus, the problem is addressed using statistical techniques and interactive proofs to ensure that the aggregated result is a good approximation to the true value. As a result this approach is not applicable to information such as nonnumeric data, which are not amenable to statistical techniques.

Protection against event fabrication attacks when a small number of nodes are compromised has also been considered, by probabilistic key sharing over a partitioned key pool and interleaved per hop authentication, respectively [31,32]. When the attacker succeeds in compromising a number of nodes greater than a threshold, the security guarantees for the entire network are lost.

Ye et al. [31] propose a scheme called *statistical en route filtering* (SEF). The authors consider a scenario where multiple sensors detect a signal. The attacker can both suppress reports on events that have occurred (false negatives) and report on events that have not really occurred (false negatives). In this paper the authors focus only on false negatives. Any sensor except the sink can be compromised and thus raw data can be manipulated. The multiple detecting sensors elect a leader among themselves. Each node is initialized with a certain number of keys drawn randomly from a global key pool. Note, however, that this global key pool is divided into groups, and a node gets all its keys randomly from a single group. The sink is assumed to have knowledge of all the keys in all the groups. Other nodes are initialized with a small number of keys drawn randomly from this key pool.

Multiple detecting sensors process a signal. One of these is elected the master, called the *center of stimulus* (CoS) and is responsible for summarizing and producing a synthesized report on behalf of the group. Each detecting node produces a keyed MAC for the report using one of its keys. Each report has multiple MACs associated with it. Intermediate forwarding nodes can detect incorrect MACs and filter out false reports en route. The sink verifies the correctness of each MAC and eliminates remaining false reports. The center of stinulus has one key from each of the group given a threshold number T of groups and collects the MACs and attaches them to the report, and these multiple MACs act as proof that the report is legitimate. A report with an insufficient number of MACs will not be forwarded. A forwarding node checks whether the packet has T MAC from different groups and if not, it drops the packet. The node further checks whether it has any of the keys, by using the key ids. If it has one of the keys, it checks to see if the MAC matches. If the MAC does not match, the packet is dropped. If it does not have the key, it forwards the packet. So it is possible that some packets might be forwarded even without having proper MAC since the forwarding nodes do not have the corresponding keys. Thus a final check is done by the sink, which has all the keys. SEF has an overhead of about 14 bytes per report and is able to drop 80–90% of the injected false reports within 10 forwarding hops.

Zhu et al. [32] focus on detecting and filtering out false data packets either at or en route to the basestation. Nodes are initialized with a unique id as well as keying material that allow it to establish pairwise keys with other nodes. Each node has other nodes both upstream and downstream associated with it with respect to the direction of communication between the source of reports and the basestation. A report is generated by each of a certain number of nodes that form a cluster. Each of these reports is directed to the cluster head and contains two MACs, one using the key that the node shares with the basestation and the other using the key shared with the associated nodes. The clusterhead collects all reports from members of its cluster and forwards them along with the various MACs to the basestation. Every forwarding node verifies the MAC computed by its associated node, removes the MAC, computes a new MAC, and forwards the report to the basestation. If the MAC verification fails, then the packet is dropped. Finally, when the report reaches the basestation, it verifies the MACs appended by the original sources of the report before accepting the report.

Hu and Evans [51] assume a routing hierarchy where each node has a parent to which the node transmits its reading. The report contains the data reading, node id, and a MAC. The MAC is calculated using a key known only to the node and the basestation. The parent node will store the message and its MAC until the key is revealed by the basestation. If the MAC does not match, then the parent node will initiate an alarm. In addition to storing the message, the parent node also waits for a specified time before sending a report to its parent (the grandparent of the first set of nodes). The parent must wait to collect reports from all its children. After a stage of messages arrives at the basestation, the basestation reveals the keys used by various nodes to generate the MACs. To enable this broadcast message to be authenticated, the basestation makes use of µTESLA. This protocol has quite a few shortcomings such as the need for time synchronization (loose) between the various nodes, the requirement for a basestation powerful enough to reach the various nodes, inability to defend against attacks that compromise both a child and a parent node, the extra communication entailed, and the need for a routing hierarchy.

11.3.7   Implementation

In this section we investigate implementation-related details of sensor nodes. A MICA mote is representative of sensor nodes. This is a small sensor/actuator unit with a CPU, power source, radio, and several optional sensing elements. The CPU is a 4-MHz 8-bit Atmel ATMEGA103 CPU with an instruction memory of 128 kB, data memory of 4 kB, and flash memory of 512 kB. When active, the CPU consumes 5.5 mA of power at 3 V. This power consumed is about two orders of magnitude less when the CPU is sleeping. The radio operates at 916 MHz with a bandwidth of 40 kbps on a single shared channel. The range is about a few dozen meters. The radio consumes a power of 4.8 mA (at 3 V) in receive mode, 12 mA in transmit mode, and 5 µA in sleep mode. A temperature sensor, magnetometer, accelerometer, microphone, sounder, and other sensing elements can be mounted on an optional sensor board. The whole device is powered by two AA batteries, which provide approximately 2850 mAh (milliamperehours) at 3 V [4]. Individual motes can last only 100–120 h (4–5 days) on a pair of AA batteries in active mode [45]. On the other hand, when the motes are in sleeping mode, they consume only 0.1% of the energy consumed in active mode. In fact, experimental data show that motes can last for more than a year on a 1% duty cycle [45].

Slijepcevic et al. [35] provide a proof-of-concept implementation using Rockwell WINS sensor nodes [57]. Each node has an Intel StrongARM 1100 processor running at 133 MHz, 128 kB SRAM, 1 MB flash memory, a Conexant DCT RDSSS9M radio, a Mark IV geophone, and an RS-232 external interface. The radio has a bandwidth of 100 kbps using a transmission power of either 1, or 10, or 100 mW. To send a block of 128 bits, the radio consumes 1.28 µJ. The authors consider RC6 and show that 3.9 µJ is consumed to encrypt the block using 32 rounds while 2.7 µJ is consumed to encrypt the block using 22 rounds.

In SPINS [3] applications are implemented on sensor nodes using the mechanisms SNEP and µTESLA. It is shown that the energy spent for security is negligible compared to the energy spent on sending or receiving messages [3]. Thus, this implies that it is possible to encrypt and authenticate all sensor readings on a per packet basis. Of course, this is based on the 10-kbps links that they use. The experimental platform that they use also influences these values.

Deng et al. [17] implement an intrusion-tolerant routing protocol, INSENS, on MICA sensor motes. These motes use a Atmel ATMEGA128 microcontroller. The radio used has a bandwidth of 19.2 kbps. The operating system used is TinyOS 1.0. The default packet size is 30 bytes, although this can be changed. The authors experimented with the use of RC4, RC5, and Rijndael (AES) as candidate cryptographic algorithms to perform encryption and MAC calculation. RC5 was implemented with 5 rounds and 12 rounds, while a standard version of AES (that does not use the 4-kB lookup tables required by a fast version) was used. They compared the performance and found that RC5 uses less memory (in both code size and data size) and is also efficient [17]. AES was found to be slow. RC4 was found to be efficient in terms of code size and also had the best performance. They show that the average time for computing 128 bits of data by AES was 102.4 ms, by RC5 (5 rounds) 5.4 ms, by RC5 (12 rounds) 12.4 ms, and by RC4 1.299 ms. Note that a typical packet size is 240 bits. RSA was also implemented, and the delay for decryption with a 1024-bit RSA key was found to be approximately 15 s considering 64 bytes of data. This reemphasizes the impracticality of using public key cryptography in this domain.

Carman et al. [27] focus on performance comparison of different cryptographic algorithms on various sensor hardware. They show that energy consumption due to communication in sensors is several orders of magnitude higher than that due to computation overhead. Anderson and Kuhn [34] show that building tamperproofing into sensor nodes can substantially increase their cost. Additionally, trusting such nodes can be problematic. The ultra-low-power wireless sensor project has an objective of designing and fabricating sensor systems capable of transmitting data up to 1 Mbps with an average transmission power in the range from 10 µW to 10 mW [58]. The next generation of sensor nodes incorporates a radio conforming to the new IEEE 802.15.4 standard, operating at 2.4GHz with 250 kbps bandwidth (IEEE Inc. IEEE 802.15.4 draft standard, `http://grouper.ieee.org/groups/802/15/pub/TG4.html`).

11.4 SUMMARY

In this chapter we have considered the problem of security in sensor networks and reported on various problems and corresponding solutions proposed in the literature. We have investigated the different attacks that are possible in sensor networks. Then we considered the topic of data encryption and authentication and looked at the state of the art in this area. The various proposals for key management in sensor networks were given next. We have also considered intrusion detection in sensor networks. Routing, secure aggregation, and implementation were the final three topics, addressed in that order.

REFERENCES

1. I. F. Akyildiz, W. Su, Y. Sankarasubramaniam, and E. Cayirci, Wireless sensor networks: A survey, *Comput. Networks* **38**:393–422 (March 2002).

2. C.-Y. Chong and S. Kumar, Sensor networks: Evolution, opportunities and challenges, *Proc. IEEE* **91**:1247–1256(Aug. 2003).

3. A. Perrig, R. Szewczyk, V. Wen, D. Culler, and J. D. Tygar, SPINS: Security protocols for sensor networks, *Proc. Mobile Computing and Networking, Conf.*, Rome, Italy, 2001.

4. C. Karlof and D. Wagner, Secure routing in wireless sensor networks: Attacks and countermeasures. *Proc. 1st Int. Workshop on Sensor Network Protocols and Applications* (*SNPA'03*), May 2003.

5. S. Basagni, K. Herrin, E. Rosti, and D. Bruschi, *Secure pebblenets, Proc. MobiHoc*, 2001.

6. H. Chan, A. Perrig, and D. Song, Random key predistribution schemes for sensor networks, *Proc. IEEE Sympo. Security and Privacy* (*SP*), May 11–14, 2003.

7. S. Zhu, S. Setia, and S. Jajodia, LEAP: Efficient security mechanisms for large-scale distributed sensor networks, *Proc. CCS'03*, Washington, DC, Oct. 27–31, 2003.

8. TinyPK project; available on the Web at `http://www.is.bbn.com/projects/lws-nest`, BBN Technologies.

9. Crossbow Co., MICA, MICA2 motes and sensors; available on the Web at `http://www.xbow.com`

10. H. Wang, D. Estrin, and L. Girod, Preprocessing in a tiered sensor network for habitat monitoring, *EURASIP JASP Special Issue on Sensor Networks* **4**: 392–401 (March 2003).

11. A. Mainwaring, J. Polastre, R. Szewczyk, D. Culler, and J. Anderson, Wireless sensor networks for habitat monitoring, *Proc. ACM WSNA 2002*, Sept. 2002.

12. F. Ye, H. Luo, J. Cheng, S. Lu, and L. Zhang, A two-tier data dissemination model for large-scale wireless sensor networks, *Proc. IEEE/ACM MobiCom 2002*, 2002.

13. B. Chen, K. Jamieson, H. Balakrishnan, and R. Morris, Span: An energy-efficient cooridination algorithm for topology maintenance in ad-hoc wireless networks, *Proc. IEEE/ACM MobiCom 2001*, 2001.

14. L. Eschenaueer and V. Gligor, A key-management scheme for distributed sensor networks, *Proc. ACM CCS 2002*, Nov. 2002.

15. A. Wood and J. Stankovic, Denial of service in sensor networks, *IEEE Comput.*, 54–62 (Oct. 2002).

16. J. Deng, R. Han, and S. Mishra, *INSENS: Intrusion-Tolerant Routing in Wireless Sensor Networks*, Technical Report CU-CS-939-02. Dept. Computer Science, Univ. Colorado, Nov. 2002.

17. J. Deng, R. Han, and S. Mishra, A performance evaluation of intrusion-tolerant routing in wireless sensor networks, *Proc. IEEE Int. Workshop on Information Processing in Sensor Networks* (*IPSN'03*), April 2003, pp. 349–364.

18. J. Staddon, D. Balfanz, and G. Durfee, Efficient tracing of failed nodes in sensor networks, *Proc. WSNA 2002*, Atlanta, GA, 2002.

19. B. Deb, S. Bhatnagar, and B. Nath, Information assurance in sensor networks, *Proc. 2nd ACM Int. Conf. WSNA*, Sept. 2003, pp. 160–168.

20. B. Deb, S. Bhatnagar, and B. Nath, ReInForM: Reliable information forwarding using multiple paths in sensor networks, *Proc. 28th IEEE Conf. Local Computer Networks* (LCN'03), Oct. 2003.

21. D. Ganesan, R. Govindan, S. Shenker, and D. Estrin, Highly resilient, energy-efficient multipath routing in wireless sensor networks, *Mobile Comput. Commun. Rev.*, **5**(4): 10–24 (2002).

22. D. Ganesan, B. Krishnamachari, A. Woo, D. Culler, D. Estrin, and S. Wicker, *Complex Behavior at Scale: An Experimental Study of Low-Power Wireless Sensor Networks*, Technical Report 02-0013, Computer Science Dept., UCLA, July 2002.

23. D. Liu and P. Ning, Location-based pairwise key establishment for relatively static sensor networks, *Proc. ACM Workshop on Security of Adhoc and Sensor Networks* (SASN 2003), Oct. 2003.

24. D. Liu and P. Ning, Establishing pairwise keys in distributed sensor networks, *Proc. ACM CCS*, 2003.

25. J. Hill, R. Szewczyk, A.Woo, S. Hollar, D. Culler, and K. Pister, System architecture directions for networked sensors, *Proc. Int. Conf. Architectural Support for Programming Languages and Operating Systems* (ASPLOS 2000), Cambridge, Nov. 2000.

26. D. Wheeler and R. Needham, TEA, a tiny encryption algorithm (1994); available on the Web at `http://www.ftp.cl.cam.ac.uk/ftp/papers/djw-rmn/djw-rmn-tea.html`.

27. D. W. Carman, P. S. Kruus, and B. J. Matt, *Constraints and Approaches for Distributed Sensor Network Security*, NAI Labs Technical Report 00-010, 2002.

28. W. Du, J. Deng, Y. Han, S. Chen, and P. Varshney, A key management scheme for wireless sensor networks using deployment knowledge, *Proc. IEEE InfoCom*, 2004.

29. W. Du, J. Deng, Y. Han, and P. Varshney, A pairwise key pre-distribution scheme for wireless sensor networks, *Proc. ACM CCS*, 2003.

30. A. B. Pryzdatek, D. Song, and A. Perrig, SIA: Secure information aggregation in sensor neworks, *Proc. ACM SenSys*, 2003.

31. F. Ye, H. Luo, S. Lu, and L. Zhang, Statistical en-route filtering of injected false data in sensor networks, *Proc. IEEE InfoCom*, 2004.

32. S. Zhu, S. Setia, S. Jajodia, and P. Ning, An interleaved hop-by-hop authentication scheme for filtering false data in sensor networks, *Proc. IEEE Symp. Security and Privacy*, 2004.

33. S. Tanachiwiwat, P. Dave, R. Bhindwale, and A. Helmy, Secure locations: Routing on trust and isolating compromised sensors in location-aware sensor networks, *Proc. ACM SenSys*, 2003.

34. R. Anderson and M. Kuhn, Tamper resistance — a cautionary note, *Proc. 2nd Usenix Workshop on Electronic Commerce*, Nov. 1996, pp. 1–11.

35. S. Slijepcevic, M. Potkonjak, V. Tsiatsis, S. Zimbeck, and M. Srivastava, On communication security in wireless ad-hoc sensor networks, *Proc. 11th IEEE Int. Workshop on Enabling Technologies: Infrastructure for Collaborative Enterprises*, June 2002, pp. 139–144.

36. ASH transceiver designer's guide, 2002; available on the Web at `http://www.rfm.com`.

37. O. Kasten, Energy consumption; available on the Web at `http://www.inf.ethz.ch/~kasten/research/bathtub/energy_consumption.html`.

38. M. Stemm and R. H. Katz, Measuring and reducing energy consumption of network interfaces in hand-held devices, *IEICE Trans. Commun.* **E80-B**(8):1125–1131(Aug. 1997).

39. `http://www.cs.berkeley.edu/~awoo/smartdust`.

40. R. Canetti, D. Song, and D. Tygar, Efficient authentication and signing of multicast streams over lossy channels, *Proc. IEEE Security and Privacy Symp.*, May 2000.

41. E. J. Riedy and R. Szewczyk, Power and control in networked sensors; available on the Web at `http://today.cs.berkeley.edu/tos/`.

42. S. Zhu, S. Xu, S. Setia, and S. Jajodia, Establishing pairwise keys for secure communication in ad-hoc networks: A probabilistic approach, *Proc. 11th IEEE Int. Conf. Network Protocols (ICNP'03)*, Nov. 2003.

43. C. Blundo, A. D. Santis, A. Herzberg, S. Kutten, U. Vaccaro, and M. Yung, Perfectly-secure key distribution for dynamic conferences, *Inform. Comput.* **146**(1):1–23(1998).

44. S. Kumar, T.-H. Lai, and J. Balogh (Ohio State Univ.), On k-Coverage in a mostly sleeping sensor network, *Proc. IEEE/ACM MobiCom*, 2004.

45. Crossbow, Power management and batteries, application notes, 2004; available on the Web at `http://www.xbow.com/support/appnotes.htm`, 2004.

46. S. Singh and C. S. Raghavendra, Pamas: Power aware multi-access protocol with signalling for ad-hoc networks, *ACM Comput. Commun. Rev.* **28**(3):5–26(July 1998).

47. N. Sastry, U. Shankar, and D. Wagner, Secure verification of location claims, *Proc. WISE'03*, San Diego, CA, Sept. 2003.

48. J. M. Kahn, R. H. Katz, and K. S. J. Pister, Mobile networking for smart dust, *Proc. ACM/IEEE MobiCom*, 1999, pp. 271–278.

49. A. Deshpande, S. Nath, P. B. Gibbons, and S. Seshan, Cache-and-query for wide area sensor databases, *Proc. SIGMOD 2003*, 2003

50. D. Estrin, R. Govindan, J. Heidemann, and S. Kumar, Next century challenges: Scalable coordination in sensor networks, *Proc. IEEE/ACM MobiCom 99*, Aug. 1999.

51. L. Hu and D. Evans, Secure aggregation for wireless networks, *Proc. Workshop on Security and Assurance in Ad-hoc Networks*, Jan. 2003.

52. C. Intanagonwiwat, D. Estrin, R. Govindan, and J. Heidemann, Impact of network density on data aggregation in wireless sensor networks, *Proc. Int. Conf. Distributed Computing Systems*, Nov. 2001.

53. S. R. Madden, M. J. Franklin, J. M. Hellerstein, and W. Hong, TAG: A Tiny Aggregation service for ad-hoc sensor networks, *Proc. OSDI*, Dec. 2002.

54. J. Hill, R. Szewczyk, A. Woo, S. Hollar, D. Culler, and K. Pister, System architecture directions for networked sensors, *Proc. ACM ASPLOS IX*, Nov. 2000.

55. J. Newsome, E. Shi, D. Song, and A. Perrig, The Sybil attack in sensor networks: Analysis and defenses, *Proc. IPSN'04*, April 2004.

56. J. R. Douceur, The Sybil attack, *Proc. IPTPS'02*, March 2002.

57. J. Agre, L. Clare, G. Pottie, and N. Romanov, Development Platform for self-organizing wireless sensor networks, *Proc. SPIE AeroSense'99 Conf. Digital Wireless Communication*, Orlando, FL, April 1999.

58. `http://www-mtl.mit.edu/~jimg/project_top.html`.

59. J. Deng, R. Han, and S. Mishra, Intrusion tolerance and anti-traffic analysis strategies for wireless sensor networks, *2004 IEEE International Conf. Dependable Systems and Networks* (DSN2004), Florence, Italy, June 2004.

60. F. Anjum, D. Subhadrabandhu, S. Sarkar, and R. Shetty, On optimal placement of intrusion detection modules in sensor networks, *Broadnets*(Oct. 2004).

PART III

MIDDLEWARE, APPLICATIONS, AND NEW PARADIGMS

In distributed systems, middleware bridges the gap between the operating system and the application to facilitate the development of distributed applications. Traditional computer-network-related middleware such as CORBA, PVM, DCOM, and GLOBUS are generally heavyweight in terms of memory and computation requirements.

As described in Part II, wireless sensor networks have unique characteristics such as small size, limited memory and processing power, limited battery life, low communication bandwidth and range, and a mix of heterogeneous devices that may or may not talk to each other. These characteristics, coupled with node mobility, node failures, and environmental obstruction, result in frequent topology changes and network partitions. Hence, it is important to provide middleware support for sensor-based applications.

The main purpose of middleware for sensor networks is to support the development, maintenance, deployment, and execution of sensor-based applications, and in particular, provide services in the areas of sensor network control, monitoring, management, network agent support, data query, data mining, and application-specific services.

In Chapter 12, Prabhu, et al. present a middleware architecture designed for applications using radiofrequency identification (RFID) technology. RFID middleware is a specialized software that sits between the RFID hardware (readers) and the enterprise applications or conventional middleware.

The authors first introduce different types of RFID tags, including their characteristics. The architecture of WinRFID, a multilayered middleware developed in .NET framework at WINMEC, UCLA, is then presented and some of the main modules are discussed. The XML-based Web services are briefly presented as a convenient way to not only exchange data but also integrate an RFID application into an existing application. The challenges in collecting/filtering/categorizing large volume of data, without false reading, duplicate reading, or misreading are also

Mobile, Wireless, and Sensor Networks: Technology, Applications, and Future Directions
Edited by Rajeev Shorey, Akkihebbal L. Ananda, Mun Choon Chan, and Wei Tsang Ooi
Copyright © 2006 John Wiley & Sons, Inc.

highlighted. To add domain-specific rules, a rule engine is used in WinRFID to analyze the data from low layers.

The vision of pervasive computing concerns the creation and management of largely invisible digital or smart spaces where many devices interact seamlessly in wired as well as wireless environments. To highlight this aspect, the next chapter describes the design of a smart environment using sensors and distributed agents for learning and prediction. In this chapter, Chapter 13, Das and Cook present a sensor-enabled environment model. The "MavHome" is an example implementation of this model.

In the MavHome environment, intelligent agents perceive the state of the home through sensors and act on the environment through device controllers with the aim of maximizing comfort of its inhabitants while minimizing the cost of running the home. The authors discuss in detail on how this objective is met through automatic learning and prediction. A demonstration of the smart environment model is presented as an example. The chapter also discusses some practical considerations.

In such a smart sensor network environment, information has to be managed and accessed quickly, securely, efficiently, and effortlessly anywhere, anytime. A key technology in realizing the vision of pervasive computing is mobile networking that supports secure access to sensor data.

In Chapter 14, Bao et al. address the security issues in mobile networks. Although the example network used is IPv6-based, the concepts and issues relating to security in a mobile network discussed in the chapter are equally applicable to the case of ad hoc and sensor networks.

In this chapter, the authors have chosen mobile IPv6 as a vehicle to provide a good understanding of the security issues in a mobile environment, and discuss the redirect attacks in particular. They present and analyze three different protocols: (1) the return routability protocol, (2) the cryptographically generated address protocol, and (3) the home agent proxy protocol. These protocols are designed to secure correspondent binding updates in order to prevent redirect attacks. The comparison of these three protocols in terms of security, performance, and scalability are discussed.

An on-demand business is an enterprise whose business processes — integrated end-to-end across the company and with key partners, suppliers and customers — can respond with speed to any customer demand, market opportunity, or external threat. On-demand business concerns the integration of services, in ways unlike those the past world of distributed computing. It is, in many ways, the realizable result of an evolution of many networked technologies, in conjunction with some key business process transformation activities.

The need for wireless ad hoc and sensor networks from an on-demand business perspective, in the global pervasive ecosystem of today, is discussed in the final chapter of Part III. In this chapter, Chpater 15, Fellenstein et al. provide an in-depth treatment of on-demand business. The authors describe how an on-demand business needs an intelligent network with quick response attributes to meet the global challenges. Technologies such as wireless, mobile, and sensor networks are helping businesses respond faster and more effectively to planned and unplanned situations in the markets. The chapter focuses on issues ranging from advanced forms of on-demand business services architectures, on-demand business operating environments, a variety of networking protocols in a pervasive computing ecosystem, business modeling, and security issues, and concludes with an analysis of economic, cultural, and market trends. The authors also discuss on-demand business solutions integration challenges of mobile, wireless, and sensor networks.

CHAPTER 12

WinRFID: A Middleware for the Enablement of Radiofrequency Identification (RFID)-Based Applications

B. S. PRABHU, XIAOYONG SU, HARISH RAMAMURTHY, CHI-CHENG CHU, and RAJIT GADH

University of California at Los Angeles, Wireless Internet for the Mobile Enterprise Consortium (WINMEC)

12.1 INTRODUCTION

Globalization and accelerated innovation cycles are forcing industry to adopt new technological improvements in manufacturing automation, process execution, engineering practices, and control applications. At the same time it is desired that these improvements inject flexibility into the system to enable it to respond quickly to alterations and disruptions in real time. The key enabler for this to happen is *ubiquitous information flow.*

Toward this end, industry is looking for a new paradigm, which can provide real time visibility for most of the activities to the collaborating partners involved; enabling quick decisions and shorten process times. Activities that are in the forefront of this plan include supply chain management, inventory control, asset management and tracking, theft and counterfeiting prevention, access restrictions and security, and hazardous-materials management.

A number of mobile wireless technologies have all been significant catalysts for this transformation and have also spurred the development of inventive business process facilitators such as speed, quality, timeliness, adaptability, and depth of information. Among them, however, radiofrequency identification (RFID) has drawn a lot of attention from the industry as it has already demonstrated the

Mobile, Wireless, and Sensor Networks: Technology, Applications, and Future Directions
Edited by Rajeev Shorey, Akkihebbal L. Ananda, Mun Choon Chan, and Wei Tsang Ooi
Copyright © 2006 John Wiley & Sons, Inc.

potential to enhance efficiencies of activities across business processes by providing a means to affix unique identification and related information to individual items and enable the items to travel with the information, which can be utilized as the items pass through the different process stages, increasing productivity, minimizing errors, improving accuracy, and potentially reducing labor costs [27,39,68].

RFID does not require line-of-sight access to communicate; without requiring physical contact, multiple tags can be identified, and tags mounted on products or consignments can survive in harsh environments such as extreme temperatures, moisture, and rough handling. The technology significantly enhances the velocity of information flow by overcoming the limitations of other manual, data collection methods [59]. With proper deployment, it is deemed to enable industry to focus on real-time optimization of the activities by providing accurate, timely visibility of the various stages of the activities and make intelligent strategic decisions [33,34,55,56,66,69].

However, there are also some major weaknesses associated with the currently available RFID technology, which has hampered the technology from going prime-time to some extent. In some of the more recent industry pilots and studies done by researchers and industry, it was found that a number of systems had a lot of problems — they either failed or provided erroneous reads, had problems in handling large amounts of data generated by the tags, were found costly for some applications, lacked mature standards, were burdened with collisions due to multiple-tag reads, failed in the presence of metal and liquid based products, and so on [14,15,28,42,46].

Nonetheless, the industry is very interested in RFID technology because it is expected to provide a means of bringing passive objects online and integrate physical assets into the overall IT infrastructure, subject to intelligent decisionmaking directly from the information available on these physical objects, thereby increasing efficiency, reducing losses, providing superior quality control, and providing other benefits [5,49].

This situation provides an excellent research opportunity and the ongoing research work under the aegis of Wireless Internet for the Mobile Enterprise Consortium (WINMEC) at UCLA attempts to propose a RFID ecosystem to mitigate a number of the abovementioned problems by architecting a new generic paradigm using remoting, Windows, and Web services technologies and frameworks-based distributed middleware.

Before we begin discussing the middleware for RFID applications, we introduce the concepts of RFID technology, middleware architectures, and enterprise class distributed systems in the sections that follow.

12.1.1 Radiofrequency Identification (RFID) System

RFID is an automatic identification technology that can be used to provide electronic identity to an item or object. A typical RFID system consists of transponders (tags), reader(s), antennas, and a host (computer to process the data) as shown in Figure 12.1.

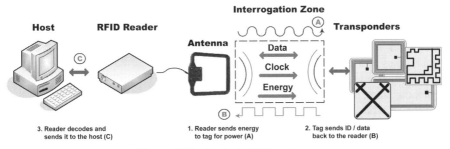

Figure 12.1 Typical RFID system.

The history of RFID can be traced back to "friend or foe" long-range transponder systems fitted to aircraft in World War II, followed by a number of reported scientific works prior to the 1970s; in the 1980s and 1990s they appeared in commercial applications such as animal tracking, vehicle tracking, factory automation, and toll collection, and finally, now in the twenty-first century, the technology is on the verge of exploding by virtue of its capability to facilitate "reality mining" mainly in retail, food, and drug supply chains, as well as in security and trade [12,17,25,44].

Communication in RFID occurs through radiowaves, where information from a tag to a reader or vice versa is passed via an antenna. Unique identification or electronic data are stored in RFID tags, which can consist of serial numbers, security codes, product codes, and other object-specific data. Using an RFID reader, the data on the tag can be read wirelessly, even without line-of-sight access, even when tagged objects are embedded inside packaging or even when the tag is embedded inside an object itself. RFID reader can read multiple RFID tags simultaneously [6,24,27,70]. RFID technologies that work at different frequencies are currently available, and their selection depends mainly on the requirements of the end application.

The strong interest shown by industry in RFID is due mainly to the following features, which potentially would lead to better business and workflow processes:

- Ability of tags to store or archive data that can be modified and updated during various stages of the processes
- Automation of data collection at a high rate, eliminating the need for manual scanning
- Accurate data collection and hence less problems due to erroneous data for decisionmaking
- Less handling of goods and hence less labor required
- Simultaneous identification of multiple tagged items in the read area

To better understand RFID, a discussion of the types and protocols of the tags is warranted. The following section briefly describes the salient points of the currently available tags and supporting protocols.

12.1.1.1 Types of RFID Tags

RFID tags come in a wide variety of shapes, sizes, capabilities, and materials — ranging from as small as the tip of a pencil or a grain of rice to as large as a 6-in. ruler. They are available in a variety of shapes, such as key fobs, credit cards, capsules, bands, disks, and pads. Tags can have metal external antennas or embedded antennas; the latest are printed antennas.

Tags can be either "passive" — working without a power source, or "active" — equipped with an embedded power source. The tags can be either read-only or read/write-capable. The range for sensing the tags from a reader can vary from a few centimeters to a few meters depending on the power output, radiofrequency used, and the type and size of tag antennas.

The frequency bands and emission power for the RFID systems are limited according to governmental regulations [26]. The choice of a particular frequency depends on application requirements such as absorption in liquids, the reflection on surfaces, tag densities, power demand, size and location of tags, exposure to temperature range, data transmission speed, and data (processing) rates. Even within the same class of tags, say, EPC Class 0 and Class 1, tags working with a particular reader may pose problems related to variations in antenna designs and tag sizes. Some of these factors are contradictory, and hence an optimal combination will have to be identified for each application [43]. Table 12.1 lists the most popular RFID technologies currently in use and their typical characteristics, highlighting the range of applications where tag technology is best suited.

As RFID is embraced by industry and the technology is an internalized in-house processes, the infrastructure is expected to support different RFID technologies to provide optimal benefits to a number of business processes, by utilizing the appropriate features of different RFID technologies such as read/write range, data rate, and interference, at different stages of the business or workflow processes.

12.1.2 Middleware Technologies

Middleware is multidisciplinary and attempts to merge features and knowledge from diverse areas such as distributed systems, networks, and even embedded systems [18,21,41,61]. The term *middleware* refers to the software layer that resides between the physical layer components (hardware), firmware, or operating systems, which deal with low-level system calls and communication protocols, and the upper layer standalone or even distributed enterprise applications generally interacting via the network. The boundaries between layers is not very sharp, and as software evolves, the features of middleware become part of operating systems, firmware, application frameworks, and other layers of the information technology (IT) infrastructure [8].

However, middleware components are an essential part of the latest-generation distributed systems — both new developments or involving integration of existing applications and services. The anatomy of a typical middleware is shown in Figure 12.2. The success of the architecture depends on how well the different pieces in different layers fit together or are made to fit together by modifying some of the modules.

TABLE 12.1 Current RFID Technologies

Frequency Band	Read Range	Characteristics	Typical Applications	RFID Protocols
Low (100–500 kHz)	≤4–6 in.	Short to medium read range Inexpensive Low reading speed Can read through liquids	Access control Animal identification Inventory control Vehicle immobilizer	ISO/IEC 18000-2
High (10–15 MHz)	≤8 ft	Short to medium read range Potentially inexpensive Medium read speed Can read through liquids	Access control Smart cards Item tracking Electronic surveillance	ISO/IEC 18000-3 EPC HF Class 1 ISO/IEC 15693 ISO 14443 (A/B) I-Code, Tag-It, Hitag, MiFare
Ultrahigh (850–950 MHz)	10–20 ft	Long read range High reading speed Reduced chance of signal collision Problems with liquids and metal	Railroad asset monitoring Toll collection systems Supply chain Item tracking	ISO 18000-6 EPC Class 0, Class 1
Microwave (2.4–5.8 GHz)	<3 ft	Moderate read range Chance of signal collisions Very high data rates Problems with liquids and metal	Railroad asset monitoring Toll collection systems Airline baggage tracking	ISO/IEC 18000-4

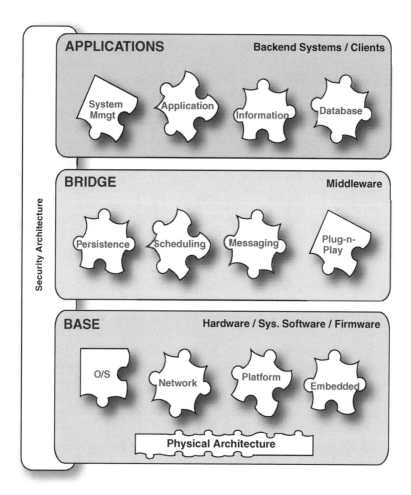

Figure 12.2 Anatomy of a middleware-supported IT infrastructure.

Middleware systems generally support the interaction of disparate application programs, collaborative groupware, and other federated workflow systems. They seek primarily to hide the underlying networked environment's complexity by insulating applications from heterogeneous hardware, explicit protocol handling, distributed data repositories, networking technologies, and so on and provide quality-of-service guarantees, security, scalability, ubiquity, and ease of integration of applications and systems [16,36,38,48,52,57].

Thus, in developing a successful middleware, a number of features must be considered. Some of the important features are network, language and operating system independence; architecture interoperability (object-oriented, client/server, push/ pull, Web services); plug-n-play operation of different modules and components; service location, and message and data routing; scheduling transactions through

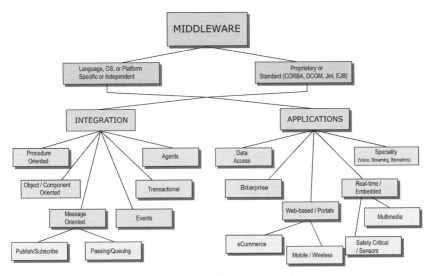

Figure 12.3   Middleware categories.

publish/subscribe schemes; mechanisms for fault tolerance and recovery from failures; and end-application-specific features such as events, persistence, and adapters [18,21,37,50].

Middleware functions fall broadly into three main categories: application-specific, information exchange, and management and support functions. Figure 12.3 shows an accepted categorization of the middleware systems currently in use [9,23,61]. Each of these categories caters to a principal requirement of an enterprise IT infrastructure. However, as the infrastructure matures, features of other categories will have to be integrated and a fully working middleware will have features of many of the categories shown in the figure. In Section 12.3 we will see how some of these features are relevant to RFID middleware, how can they be integrated, and what value they add to the infrastructure.

12.1.3   Web Services

Web services are middleware components using Extensible Markup Language (XML). They are reusable components that can be accessed by multiple clients at the same time, allow two or more Web applications to communicate with each other, or can be used as a glue to patch new applications and/or services with legacy applications. They are self-contained units of functionality exposing well-defined and precise interface to receive or generate messages. They are registered with a directory service or a registry and discovered by users [40,62].

Web Services architecture allows a decentralized computing model in which interactions between the Web service components can occur over distributed domains (different host machines in the network) using existing Internet technologies without much reengineering in comparison to earlier component models such as CORBA, DCOM, and COM+, where the component ownership generally resides in a single trusted domain such as an enterprise intranet [35].

It is this decentralized deployment option which can be exploited in an RFID infrastructure, where the capabilities of subdomains of different value chain partners of an enterprise can be utilized to provide real-time access to data as an integrated seamless environment. The decentralized model also bestows the architecture a unique advantage of maintaining, updating, and adding the Web Services individually without disrupting the existing infrastructure, which may prove to be very valuable in RFID networks as new technology, protocols, and standards will be introduced, and also the architecture may be very dynamic.

Thus, it is expected that Web services will play a pivotal role in helping RFID adopters integrate RFID-based applications into existing enterprise applications such as logistics, warehouse management, and inventory management, and supply chain management, and enable sharing of up-to-date, if not real-time, data about the tagged objects' location and shipping events and history, enabling quick decision-making. Another important feature of a Web service is its capability to be deployed to communicate between machines and to react to event triggers automatically in the background while performing such tasks as event logging and event verification, seldom requiring the immediate attention of the operators — again, this is a feature that will add substantial advantage to RFID-based solutions.

It is this feature of Web services, in addition to remoting and Windows services, that will be examined and investigated in WinRFID — RFID middleware at WINMEC. These services will be used as building blocks of the WinRFID.

12.2   RFID MIDDLEWARE

RFID middleware is a new breed of specialized software that sits between the RFID hardware (readers) and the enterprise applications or conventional middleware. The main goal of this middleware is to process data from tags collected by the readers deployed in the RFID infrastructure, or to write ID numbers and/or business process data to the tags while commissioning these tags for assignment to individual items. In addition, it deals with a number of important issues related to avoidance of data duplication, mitigating errors, and proper presentation of data. RFID middleware is being developed and made available by some software vendors on a service basis to suppliers of large retailers such as Wal-Mart and Target, and to the U.S. Department of Defense, and pharmaceutical companies who have to meet mandated deadlines in 2005, requiring tagging at pallet and carton levels. These vendors also conduct pilots and proof-of-concept projects for the suppliers.

According to the latest research report by Venture Development Corporation, companies planning to implement RFID are worried mainly about data quality and data synchronization. Many of the research survey respondents indicated that they were experiencing difficulty extending their RFID pilots because their legacy systems were not able to process the vast amount of information generated. In addition, there were problems such as large number of missed tags producing a high volume of false negatives and readers reading the tags multiple times and thus generating duplicate data [60,64].

RFID middleware will play a large role in reducing these problems and eventually in mitigating them. This is vouched for by the VDC survey, and it is expected that the RFID middleware market will grow by 162% in the year 2005, from $16.4 million in 2004 to $43.1 million in 2005. Middleware will account for roughly 3% of RFID system revenues in 2007, or $135 million [51,60]. This shows that RFID middleware will be a significant software suite in times to come.

All these issues will be dealt with in the sections describing the middleware research being undertaken at WINMEC.

12.2.1   Benefits of RFID

The benefits of deploying RFID will begin to accrue in phases. This is because of the scope of various mandates and the degree of compliance required. A clear return on investment (ROI) is still disputed, but obvious payback is expected in the form of a decrease in labor, facility/equipment productivity, process improvements throughout the supply chain, reduction in theft and reduced inventory, and other benefits. The accompanying chart in Figure 12.4 illustrates the various stages and the timeline during which a number of enterprise activities will benefit [1,2,13,19, 20,22,32,45].

A good middleware solution will be greatly impacted by the technology adoption rate. Benefits will increase as tagging goes to the item level, and at the same time it is expected to burden the middleware design. The design adopted by us in this work demonstrates that the solution framework can evolve with the increase in degree of difficulty of the adoption.

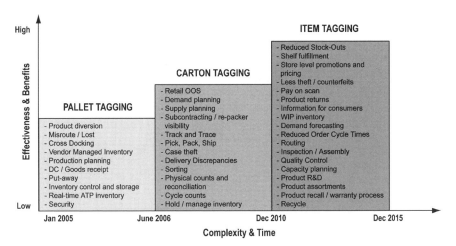

Figure 12.4   RFID benefits across supply chain activities.

So, RFID will undoubtedly be a very disruptive technology with the potential to drastically change a number of business practices affecting a large number of industry verticals, and some of these verticals will see exceptional benefits. And middleware architecture will be most appropriate as only such a distributed architecture would successfully wrap the differences and nuances of the technology and allow the dynamic character of this infrastructure to flourish as against a monolithic application. However, there are a number of challenges, as laid out in the next section.

12.2.2 Challenges in Adopting RFID

It is anticipated that the impact of introducing RFID in an enterprise IT infrastructure would be felt on many fronts, but two main ones would be on the network and sharing of data generated by the RFID systems. The challenge posed will be from the point of integration of the RFID infrastructure with the existing IT network, and the potential business process transformation that enterprises would have to entertain to gainfully exploit the benefits of RFID [29,31].

The demand progresses from the readers, which may be deployed at various locations connected to the edge host. The edge host will sieve through all the input data, filter them, and glean information from them. The edge hosts will relay the information to intermediate aggregator servers which in turn will update the enterprise repository from where the value chain partners can utilize it for decisionmaking.

Given this scenario, understanding and designing the data infrastructure would require resolution of three key issues: (1) the volume of data that the system would generate, (2) the locations from which the data will be generated, and (3) where and how long the data need to be maintained.

The answers vary by the industry verticals. Many issues such as the number of partners who will be using the data and the format in which they will require the data, regulatory mandates, granularity of tagging, and the degree of distributed nature of the RFID network would further impact the answers to the abovementioned needs.

In a typical RFID network it is anticipated that a few hundred (some may require thousands) readers, tens of edge hosts, and a few aggregators would constitute the RFID network infrastructure. At each of these nodes in the network, data in different formats and different quantities would reside at any given point. The amount of data at each node would probably be fairly limited. However, it is speculated that the aggregation of data from many of these nodes would create large data volumes. Thus, the challenge lies in developing a distributed network to gather data from a large number of independent and fairly tiny data sources. This may also require an intelligent framework to aggregate and cross-index the tiny data sources to enable the users to apply only the information of interest to them. The tiny distributed data sources would also be significant sources of failure due to possible outages, requiring sufficient data redundancy to be built in the network architecture or to implement a mechanism to recover data after failure for seamless data visibility at all times.

   In addition, any software solution that is developed today will at best be a stop-gap arrangement as the technology itself is evolving. New technologies are being introduced with different RF physics and transmission schemes, supporting different frequency bands, new protocols, new standards, multiprotocol support, changing governmental regulations, and so on. Thus, any software solution developed now will have to evolve with the RFID technology — and should be extendible and adaptable with minimum disruption to the deployed infrastructure.

   Section 12.3 describes the architecture, different frameworks, data layers, and integrating strategies being developed and implemented at WINMEC with the goal of overcoming some of these challenges.

12.3   RFID ECOSYSTEM RESEARCH AT WINMEC

The WinRFID–RFID middleware research work involves developing new algorithms and data structures, exploring the option of employing new paradigms, namely, remote, Windows, and Web services, which have been used in the case of large distributed applications requiring a high degree of autonomy and flexibility. Autonomous services can facilitate the incorporation of reasoning capabilities within the application logic, which we think would be an ideal feature for large-scale RFID-based systems, which require effective use of interoperability between diverse business processes and diverse information and data required to achieve cooperation over the Internet and collaborating business partners.

   In the following sections WinRFID architecture and some of the main modules will be discussed, highlighting the technological aspects that benefit a large distributed RFID ecosystem, and how it may alleviate the challenges delineated in Section 12.2.2.

12.3.1   Architecture of WinRFID

WinRFID is a multilayered middleware developed using .NET framework. There are five main layers. The first layer deals with the hardware — readers, tags, and other sensors. The second layer abstracts the reader-tag protocols. Above that lies the data processing layer which deals with processing the data streams generated by the reader network. The fourth layer constitutes the XML framework for data and information representation. The top layer deals with the data presentation as per the requirements of the end users or different enterprise applications.

   The communication, management, aggregation, formatting, and customization of data, messages, and information between these layers is marshaled by supporting services and modules such as the business rule engine, intelligent remote objects and coordinators, and some libraries. Figure 12.5 shows the overall architecture of the middleware.

   The functional and operational features of each layer will be described in the following sections.

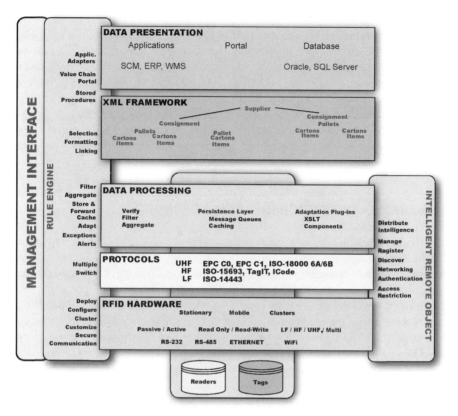

Figure 12.5 WinRFID middleware stack.

12.3.2 Physical Layer: RFID Hardware

This layer deals with the abstraction of three elements of the RFID infrastructure, namely, readers, tags, and the I/O module of the readers. The abstraction is designed to greatly facilitate the derivation of any new specific reader, tag, or I/O class extending the middleware capabilities in the advent of introduction of new RFID technology.

The reader object assists in management, configuration, location assignment, associate tag protocol(s), security, and the interface for command engine or the dedicated API/SDK provided by the vendor. The reader object supports the requirements of physical readers which can be stationary, mobile, handheld, or even clusters of readers and facilitates their integration into the infrastructure agnostically. This feature addresses the idiosyncrasies of the readers such as read mode, support for a number of antennas, and command structures to interact with the tags, by providing a common interface with high-level methods to carry out these tasks. The reader object also provides a wizard interface to manage and configure the reader both during and after deployment. Currently readers operating on low, high, and ultrahigh frequencies, both stationary and mobile, are supported for passive RFID, and active technology operating at 415 MHz and 802.11b-based RFID has been implemented.

The tag object abstracts the operations and payload formats such as access to designated areas of memory, read, write, ID structure, and the syntax of the commands or the API calls. It also provides a common interface to tags that are read-only, read/write, passive, or active and working on different frequencies. To the user of this object the top-level operational methods wrap around the individual commands or the API calls specific to the tag type just as in the case of the readers. Support is built in not only for reading and writing tag IDs but also for read/write operations or processing data on to tags that have additional memory. With each type of tag is associated a protocol, and the tag object maps the tag operations to the protocol syntax and semantics. Currently support is available to tags of all frequencies and protocols listed in Table 12.1.

The next component in this layer is the input/output (I/O) object. This object abstracts the functionality of different I/O protocols such as RS-232, RS-485, transistor–transistor logic (TTL) and Ethernet, which are currently used for communications between the RFID readers and the edge hosts. Support is also built in for communication protocols such as HTTP, Telnet, and TCP. So, using the reader wizard the I/O module of the reader can be switched from one to the other according to the physical I/O connection employed.

In the next section we will see how components in this layer interface with the protocol component in the next layer of the middleware.

12.3.3 Protocol Layer

In the case of a comprehensive RFID middleware, support for multiple tag protocols and the capability to add new ones as they become available is imperative. To facilitate this, in WinRFID, the protocol component is also abstracted to wrap the command syntax and semantics of a variety of published protocols such as ISO 15693, ISO 14443, ISO 18000-6 A/B, ICode, EPC Class 0, EPC Class 1, EPC Class 1, Gen. 2. It deals with protocol specifics such as byte-based, block, or even page reading and writing, structure and length of the command frames, partitioning of the tag memory space, and checksums.

The essence of this layer is the protocol engine which will parse and process the raw data from the tags in accordance with any particular standard protocol as mentioned above. The physical layer or the reader object subscribes to the protocol parser service with the type of protocol that the reader will have to communicate with for the tags that it will negotiate. This is done using the configuration wizard of the reader object.

When the data are parsed using the selected protocol by the reader, they are still in a raw format and are passed on to the data processing layer for further processing.

12.3.4   Data Processing Layer

Given the state of the current RFID technology, the read and write operations in a reading area are influenced mainly by the tag density, read/write distance from the reader antenna(s), orientation of the tags, material of the item that is tagged, and spatial resolution between tags (closeness of tags to each other).

Many of these characteristics introduce inconsistencies in reading or writing such as multiple reads of the same tag, failure to read some of the tags, or erroneous reads. These issues are addressed by establishing processing rules that will weed out duplicate reads and verify the tag reads, and when advanced records are available such as advanced shipping notices, this layer reconciles the records with the tag reads. Any discrepancy is processed as an exception, and a variety of alerting systems are available for resolution — emails, messages, or user-defined triggers.

At the same time, due to the business process requirements, the reading of the tags may have to be intelligently processed. Demands such as consolidation of carton and pallet information from a particular supplier or vendor, information on a particular product, data about items passing through a dock or warehouse door, items passing during a particular timeframe, reads from a particular reader or cluster of readers, and so on are quite commonplace in warehousing, supply chain management, and logistics. This capability is quite challenging as individual tagged items (cartons or pallets) from a consignment may arrive at the designated warehouse or distribution center in a staggered manner over a period of time through multiple inward doors by different transport media, but the middleware will have to keep tabs on all the items received and outstanding, and reconcile the consignment contents and provide a status view when queried. In this case specific rules can be incorporated to aggregate and classify the data accordingly and make them available in a variety of formats which will be discussed in the next section. In this layer provision is also made to log the activities on the basis of user-selectable criteria.

In addition to these features, there is a data persistence component which provides a local data store. This component is based on message queues. This design facilitates asynchronous processing of the data streams coming from the lower layers of the middleware and gives sufficient time for the abovementioned rules to work on the raw data and covert or adapt them for the upper layers or subscription services.

The intelligence needed to define these requirements and process them accordingly is built into this layer. This is achieved by way of a customizable business rule engine and a framework for adding custom data adaptation plugins. The features of these modules are discussed later.

12.3.5   Extensible Markup Language (XML) Framework

The raw cleaned (verified and filtered) tag data from the physical layer data streams are formatted in a variety of ways to a high-level XML-based representation. The information is filtered, cleaned, aggregated, and adapted as per the custom plugins, which can be added to the middleware services. The objective is to provide data in a format amenable to decisionmaking at the application layer as shown in Figure 12.5.

The layer is supported by default templates and tag libraries, using which the raw data are pulled from the message queues in the data processing layer. The data from the queues are corralled according to the rules or plugins as described in various criteria (particular supplier or vendor, particular product, etc.) in the previous section. The data from the framework can be published to registered connectors, or connectors can subscribe to specific data. The data are designed to facilitate search of containers on the basis of key fields such as suppliers, vendors, order number, and consignment type. The search pointers are required to set up the correct link between data sources and the connectors.

We expect the XML-based representation to facilitate data consumption by enterprise applications such as warehouse management, supply chain management, and enterprise resource planning. This is because most of these systems have adapters to export XML-based data and wizards used to design templates to parse the XML tree within these systems are also quite mature.

This framework can be deployed as an in-memory database or a native XML database.

12.3.6 Data Presentation Layer

This is the application layer, and it obtains the data for visualization and decisionmaking from the XML framework. Currently, we have considered only the portal and the database connectors. We are attempting to link the XML framework with SharePoint server to provide a portal interface. The main feature of this portal would be capability to set up secure subscriber (e.g., a value chain partner in a supply chain) accounts with complete authentication and access control. Each subscriber can then subscribe to the information of his or her interest. The data delivery format can be by default as provided by the middleware, or the subscriber can register data adapter plugins as Web parts in SharePoint. Each supplier can also provide access rights to these Web parts for other subscribers to share from the community. All such Web parts would be available through a library.

Other features of the portal are plugging the RFID data into graphic visual widgets for presentation. This would be extended to provide data in other denominations such as charts and graphs. From each of these widgets the portal will allow the subscribers to make decisions such as trigger events for rerouting, reassignment, billing, and alerts.

The other connector is the database connector. Currently the middleware can populate SQL server and Oracle RDBMS. The databases become populated in an asynchronous fashion in a trickle-down mode — a process with least priority so as to prevent the edge hosts from being locked up. The priority of the resources is skewed toward processing the activities of the lower three layers as shown in Figure 12.5 and the upper layers being catered to in the background at lower priority.

12.3.7   Services in WinRFID

Using the modules based on Window services, Web services, and remote objects, middleware modules are deployed as independent components that run as self-contained modules in dispersed machines. These services run unattended with or without a user interface, run within their own process space, and can start up during the operating system (OS) boot process. They can be configured and managed over the Internet and be set up in the polling mode (service monitors applications) or event mode (application sends events).

The following sections provide a glimpse of the main services of WinRFID, which impart dynamic distributive control to the middleware.

12.3.7.1   Reader Windows Service

This service is hosted by the edge host in the WinRFID network. We have used Windows service here because of its better performance using the TCP (Transmission Control Protocol) channel for communication in comparison to Web services. It also provides the container services for the remote-object-based reader coordinator.

It is used in monitoring the physical connection of the readers to the edge hosts and the health of the readers, which is monitored at predefined intervals; it authenticates the readers per the approved reader repository; and above all, it provides a means for other applications to discover and interact with it. Windows services' feature of remote activation during the OS bootup is also beneficial in an RFID network (with a large number of edge hosts with readers) as the service boots up when the edge host is remotely booted in case of any service breakage, or else the system needs to be reset.

12.3.7.2   Remote Object Based Service

This service is built on the .NET remoting framework. It acts as the coordinator at the edge hosts for managing the readers physically connected to the edge host. This service allows the coordinator to directly interact with applications or services running remotely on other machines. It provides features such as activation and trans-actional lifetime support and communication channels (TCP or HTTP) for passing data and messages. It allows formatters for encoding and decoding data and messages, and these can be used to provide security to the content before they are transported over the communication channel. Thus, the main advantage of this framework is that it allows secure custom binary encoding for the payload, which reduces the size of the content transported over the network, in contrast to the bulkiness of the payload in a Web services–based solution as the content and messages are encoded in XML.

These features of the service are exploited for a variety of functions of WinRFID such as reader deployment, configuration, and management; they also allow clients to subscribe to different formatters for events and data; and enable consolidated functions such as management of read and write cycles using multiple antennas connected to a particular reader, clustering of readers, and aggregating the data stream, and providing plugin hooks for data subscription from edge hosts, portals, or applications. This concept is illustrated in Figures 12.6 and 12.7.

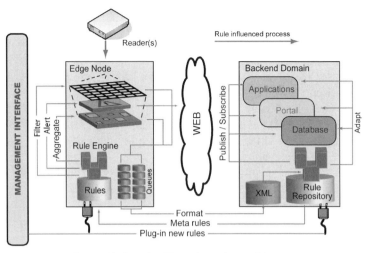

Figure 12.6   WinRFID rule engine architecture.

12.3.7.3 Reader Web Service

Web service technologies are based on standards and use XML-based languages for message and data passing, and platform support services framework based on universal description, discovery, and integration (UDDI) for subscription, discovery, transactions, and other functions. They provide platform, OS, and programming language transparency to the distributed system.

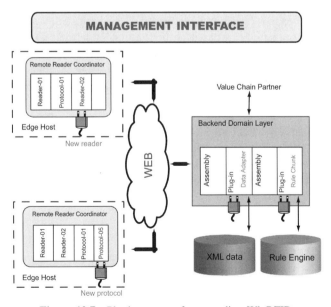

Figure 12.7   Plugin concept for extending WinRFID.

In WinRFID the reader Web service is functionally equivalent to the two services described above put together. This service allows readers connected to any platform (UNIX, Linux, etc. with, e.g., a Java programming environment) that support the standard Web services framework. The Simple Open Access Protocol (SOAP) is used for payload transactions; service request and response, and Web Service Description Language (WSDL) [63,67] is used for service description, discovery, and query responses.

Currently, in WinRFID we have integrated mobile readers attached to WIN CE/Pocket PC PDAs and use the reader Web service to transact with the reader over 802.11b and Bluetooth wireless connectivity. Since PDAs are computationally challenged and do not support remote object references, in WinRFID we use the Web services and pass on all the intensive processing from the PDA.

12.3.8 Rule Engine

The philosophy of a rule engine is to solve a problem using a set of logical rules specific to the problem domain. They have been very popular in solutions requiring processing of large sets of rapidly changing data, as in an RFID network. The other main advantage is that the rules can be flexibly updated and modified even by end users as and when the system requirements change, without requiring the services of system developers.

Driven by these benefits, and the nature (large sets of data from multiple sources) and status (changing standards, protocols) of the RFID technology, we have designed a tightly coupled rule engine into the architecture of WinRFID. An attempt is made to craft the architecture in a flexible way to the extent that it provides a means to even the end users to incorporate their own rule chunks by way of "plugins," the concept of which is described in the next section. The "inference engine" of the rule engine is based on forward chaining (data driven) as defined by RuleML [10,11,65].

As per the design, the rule engine influences a number of processes and activities of WinRFID. As shown in Figure 12.6, a number of processes such as raw data filtering, aggregating, exception handling and alerting at the edge node, and data adaptation, options to publish or subscribe the data to and from the enterprise applications and others are all driven by these rules. The main objective of using a rules system in WinRFID is to convert data and messages from lower layers to *actionable information* for the upper layers, based on the business or process semantics as perceived by the user of the information.

This module is being refined with different sets of rules, and is also being tested for accuracy, reliability, integrity, and performance.

12.3.9 Extendibility Through Plugins

The .NET framework facilitates the addition of functionality to the applications through runtime plugins. This feature is exploited in WinRFID for adding new reader modules to the remote object service, protocol modules to the system repository, and data adaptation plugins in the data processing layer, and can be extended to add other sensors or auto-ID technologies at the physical layer. We expect that this feature will help refine and reinforce WinRFID with minimal rework as and when readers, new protocols, and standards become available. The major benefit of this feature is that these modules or assemblies, as they are called in .NET, can be added to existing infrastructure with minimal or no disruption. Figure 12.7 illustrates the concept of WinRFID plugins as assemblies of .NET framework.

For this feature to work, the added assembly will have to be discovered, which can take place at runtime. The discovery can be facilitated in a number of ways, but we employ two methods. First, we use an XML configurations file with the information of the assembly to be used — its name, location, and activation methods to are registered in a registry. This method is used in adding new reader, tag protocol, or business rules. This is the method currently (as of 2005) employed in WinRFID. The second method employs reflection (explanations of reflection can be found in the reports by Liberty [47] and MSDN [53]), where the discovery method can be automated by storing the plugin in a publicized location relative to the application directory.

In WinRFID, plugin framework allows external value chain partners to add data adaptation plugins to transform the XML data to the required format, add new reader objects, and add business rule chunks.

With this feature of WinRFID, we are working on simulating an EPC network with the functionality of each modules of the network — savant, naming, discovery, information, and trust embedded in WinRFID's services. Considering the success of this exercise, it would be possible to experiment with EPC and non-EPC technologies working together and exploiting their synergies.

12.4 SUMMARY

WinRFID is an RFID-technology-agnostic middleware and a holistic distributed application. The design of the architecture is federated with the functional, system, business, and process knowledge residing in self-contained software units — the different services providing a variety of independent and complementary capabilities.

Our experience has been encouraging in terms of system-specific features such as reliability, extendibility, scalability, and ease of use. This has been possible because of the architecture providing access to end users right up to the remote-object-based reader coordinator on the edge of the network, and the ability to customize it even during runtime by injecting new process rules, adding new hardware, and supporting new protocols and standards. In our view this feature is very important in order to assimilate and sustain the number of changes across the entire gamut of RFID technology that is expected, as well as effortlessly deploy a solution exploiting the best-of-breed RFID technology, catering to the needs of desired solutions for a variety of enterprise verticals. As a result, it is our opinion that RFID technology rollouts and internalization will be a journey, not a destination, for quite some time.

From the perspective of incorporating business process knowledge and semantics, we feel confident that the rule-based engine will prove its worth as it is very flexible and facilitates an apt description of the process activity syntax and semantics that can blend with the RFID data to assist quick decisionmaking.

We have been successful in supporting a substantial number of useful reader-tag technologies, protocols, and standards, and providing a transparent tier from which enterprises can build solutions, focusing only on the end application. The middleware has been deployed on an RFID testbed at WINMEC and various experiments conducted (http://www.wireless.ucla.edu/rfid/research/).

In the next phase support for a reconfigurable sensor platform will be added [54] (see also http://www.winmec.ucla.edu/rewins/). We believe that RFID in conjunction with a variety of sensors (temperature, pressure, chemical, motion, etc.) will provide a big value-added benefit to a number of supply chain, security, and logistics activities, enabling new effective business models and improving existing practices multifold.

REFERENCES

1. A. T. Kearney, *Meeting the Retail RFID Mandate*, Research analysis report, A. T. Kearney, Inc. Chicago, Nov. 2003 (a discussion of the issue facing CPG companies).

2. A. T. Kearney, *RFID/EPC: Managing the Transition (2004–2007)*, Research analysis report, A. T. Kearney, Inc., Chicago, Aug. 1, 2004.

3. G. Agha, Adaptive middleware, *Commun. ACM* **45**(6):30–32 (2002).

4. G. Agha, S. Frølund, W. Kim, R. Panwar, A. Patterson, and D. Sturman, Abstraction and modularity mechanisms for concurrent computing, *Parallel Distrib. Technol.* **1**(2):3–15 (1993).

5. AIMGlobal, *A Study of Data Carrier Issues for the Next Generation of Integrated AIDC Technology*, research study report, AIM, Inc., Warrendale, PA; available on the Web at http://www.aimglobal.org/technologies/rfid/resources/dcstudy/datacarrier_study.htm.

6. AIMGlobal, *Draft Paper on the Characteristics of RFID-Systems*, version 1.0, Aim Inc., Warrendale, PA, July 2000.

7. M. Astley, D. Sturman, and G. Agha, Customizable middleware for modular distributed software, *Commun. ACM* **44**(5):99–107 (2001).

8. A. C. P. Barbosa and F. A. M. Porto, Configurable data integration middleware system, *Proc. Int. Workshop Information Integration on the Web — Technologies and Applications (WiiW 2001)*, Rio de Janeiro, Brazil, April 9–11, 2001.

9. T. Bishop and R. Karne, A survey of middleware, *Proc. 18th Int. Conf. Computers and Their Applications*, Honolulu, March 26–28, 2003, pp. 254–258.

10. H. Boley, The Rule Markup Language: RDF-XML data model, XML schema hierarchy, and XSL transformations, *Web Knowledge Management and Decision Support at 14th Int. Conf. Applications of Prolog (INAP 2001)*, Tokyo, Oct. 20–22, 2001.

11. H. Boley, S. Tabet, and G. Wagner, Design rationale of RuleML: A markup language for semantic Web rules, *Proc. Int. Semantic Web Working Symp. (SWWS'01), Infrastructure and Applications for the Semantic Web*, Stanford, CA, July 30–Aug. 1, pp. 381–401.

12. G. Boone, Reality mining: Browsing reality with sensor networks, *Sensors* **21**:9 (Sept. 2004).

13. M. Boushka, L. Ginsburg, J. Haberstroh, T. Haffey, J. Richard, and J. Tobolski, *Auto-ID on the Move — The Value of Auto-ID Technology in Freight Transportation*, Research report, Accenture, Feb. 2003.

14. M. Brandel, Smart tags, high costs, *ComputerWorld* (Dec. 15, 2003); available on the Web at http://www.computerworld.com/softwaretopics/erp/story/ 0,10801,88130,00.html.

15. B. Brewin, RFID users differ on standards, *ComputerWorld* (Oct. 27, 2003); available on the Web at http://www.computerworld.com/softwaretopics/erp/ story/0,10801,86486,00.html.

16. C. Britton, Classifying middleware, *Business Integrator J.* 27–30 (Winter 2001).

17. M. W. Cardullo, Genesis of the versatile RFID tag, *RFID J.* (April 24, 2004); available on the Web at http://www.rfidjournal.com/article/articleview/ 392/1/2/.

18. H. S. Carvalho, A. L. Murphy, W. B. Heinzelma, and C. J. N. Coelho, Network-based distributed systems middleware, *Proc. Int. Middleware Conf.*, Rio de Janeiro, Brazil, June 16–20, 2003 pp. 13–20.

19. G. Chappell, D. Durdan, G. Gilbert, L. Ginsburg, J. Smith, and J. Tobolski, *Auto-ID on Delivery: The Value of Auto-ID Technology in the Retail Supply Chain*, Research report, Accenture, Feb. 2003.

20. G. Chappell, L. Ginsburg, P. Schmidt, J. Smith, and J. Tobolski, *Auto-ID on Demand: The Value of Auto-ID Technology in Consumer Packaged Goods Demand Planning*, Research report, Accenture, Feb. 2003.

21. A. Colyer, G. Blair, and A. Rashid, Managing complexity in middleware. *Proc. 2nd AOSD Workshop on Aspects, Components, and Patterns for Infrastructure Software (ACP4IS)*, Boston, 2003, pp. 21–26.

22. eForce, RFID Solution Lifecycle Management, presentation by eForce IT team, 2004; available on the Web at http://www.eforceglobal.com/ppt/eF RFID Assesment Overview.ppt.

23. W. Emmerich, Software engineering and middleware: A roadmap, *Proc. Conf. Futures of Software Engineering*, Limerick, Ireland, June 4–11, 2000, pp. 117–129.

24. G. Enciu, Applications of RFID in electromechanical systems, *Proc. Int. Symp. Electrical Engineering*, Valahia Univ. Targoviste, Romania, June 3–4, 2002.

25. J. Evilsizer, B. Patel, D. Robles, and M. Mintz, *Radio Frequency Identification (RFID)*, INFO3229 Business Data Communications Technical Briefs, Belk College of Business, Univ. N. Carolina, Charlotte, Spring 2004.

26. Federal Communications Commission (FCC), *Part 15 — Radio Frequency Devices*, The Office of Engineering and Technology, Federal Communications Commission, U.S. government report, Oct. 1, 2001.

27. K. Finkenzeller, *RFID Handbook: Fundamentals and Applications in Contactless Smart Cards and Identification*, 2nd ed., Wiley, Chichester, UK, 2003.

28. C. Floerkemeier and M. Lampe, *Issues with RFID Usage in Ubiquitous Computing Applications*, draft technical paper submitted for publication, Computer Science Dept., ETH Univ., Zurich, 2004; available on the Web at http://www.vs.inf.ethz.ch/ publ/papers/RFIDIssues.pdf.

29. H. Fornicio, Middleware filters RFID data, *Managing Automation Mag.* **19**:2 (March. 2004); available on the Web at http://www.managingautomation.com/maonline/magazine/read.jspx?id=196620.

30. S. Frølund, *Coordinated Distributed Objects: An Actor-Based Approach to Synchronization*, MIT Press, Cambridge, MA, 1996.

31. Frontline, Confusion still common for RFID users, *Frontline Solutions* **5**:8 (Aug. 2004); available on the Web at http://www.frontlinetoday.com/frontline/article/articleDetail.jsp?id=110305.

32. Frontline, RFID benefits will come in phases, *Frontline Solutions* **5**:9 (Sept. 2004); available on the Web at http://www.frontlinetoday.com/frontline/article/articleDetail.jsp?id=122209.

33. R. Gadh, The state of RFID: Heading toward a wireless Internet of artifacts, *Computerworld* (Aug. 11, 2004).

34. R. Gadh, Oct. 2004, RFID: Getting from mandates to a wireless Internet of artifacts, *Computerworld*, (Oct. 4, 2004); available on the Web at http://www.computerworld.com/mobiletopics/mobile/story/0,10801,96416,00.html.

35. J. Ganesh, S. Padmabhuni, and D. Moitra, Web services and multi-channel integration: A proposed framework, *Proc. IEEE Int. Conf. Web Services (ICWS'04)*, San Diego, CA, June 6–9, 2004, pp. 70–79.

36. K. Geihs, Middleware challenges ahead, *IEEE Comput.* **34**(6):24–31 (2001).

37. G. Gimenez and K. H. Kim, A Windows CE implementation of a middleware architecture supporting time-triggered message–triggered objects, *Proc. 25th Annual Int. Computer Software and Applications Conf. (COMPSAC'01)*, Chicago, Oct. 8–12, 2001, pp. 181–189.

38. C. Hartwich, *A Middleware Architecture for Transactional, Object-Oriented Applications*, PhD thesis, Fachbereich Mathematik u. Informatik, Freie Univ. Berlin, Nov. 2003.

39. IBM Wireless e-Business Group, *RFID in Business Processes, A Thought Leadership Technical Paper*, 2003; available on the Web at http://www.ibm.com/industries/wireless/doc/content/bin/SmartTagsInformationv.1.0.pdf.

40. D. Karastoyanova and A. Buchmann, Components, middleware and Web services, *Proc. IADIS Int. Conf. WWW/Internet (ICWI2003)*, Algarve, Portugal, Nov. 5–8, 2003.

41. A. Kelkar and R. F. Gamble, Understanding the architectural characteristics behind middleware choices, in S. Rubin, ed., *Proc. 1st Int. Conf. Information Reuse and Integration*, Nov. 1999.

42. L. Kellam, P&G rethinks supply chain, *Optimize Mag.* **24** (Oct. 2003); available on the Web at http://www.optimizemag.com/printer/024/pr_supplychain.html.

43. C. Kern, RFID-technology — recent development and future requirements, *Proc. European Conf. Circuit Theory and Design (ECCTD '99)*, Stresa, Italy, Aug. 29–Sept. 2, 1999.

44. J. Landt, *Shrouds of Time — the History of RFID*, an AIM publication, Oct. 2001; available on the Web at http://www.aimglobal.org/technologies/rfid/resources/shrouds_of_time.pdf.

45. Y. M. Lee, F. Cheng and Y. T. Leung, Exploring the impact of RFID on supply chain dynamics, *Proc. 2004 Winter Simulation Conf. (WSC'04)*, Washington DC, Dec. 5–8, 2004, Vol. 2, pp. 90–97.

46. J. Lewis, *RFID: Small Package, Big Problem*, course paper: on information security and cryptography, Information Security Laboratory, Dept. Electrical Engineering and Computer Science, Oregon State Univ., Dec. 2003; available on the Web at `http://islab.oregonstate.edu/koc/ece399/f03/final/lewis2.pdf`.

47. J. Liberty, *Programming C#*, 3rd ed., O'Reilly Media, Inc., Sebastopol, CA, May 2003.

48. W. Lowe and M. L. Noga, A lightweight XML-based middleware architecture, *Proc. 20th IASTED Int. Conf. Applied Informatics (IASTED AI 2002)*, Innsbruck, Austria, ACTA Press, Feb. 18–21, 2002, pp. 131–136.

49. D. McFarlane, *Auto-ID Based Control: An Overview*, White Paper, AUTO-ID Center, Feb. 2002.

50. S. van der Meer, *Middleware and Application Management Architecture*, PhD thesis, Electrotechnik und Informatik der Technischen Univ., Berlin, Sept. 2002.

51. E. Michielsen, RFID Middleware market competition heats up, *RFID Int. Newsl.* II(6) (March 15, 2004).

52. Middleware Domain Team, 2001, *A Middleware Framework for Delivering Business Solutions*, Middleware Architecture Report version 1.0, The COTS Enterprise Architecture Workgroup, The Council on Technology Services Commonwealth of Virginia. May 2001.

53. MSDN, *Reflection*; available on MSDN's .NET Framework Technology Samples Website: `http://msdn.microsoft.com/library/default.asp?url=/library/en-us/cpsamples/html/reflection.asp`.

54. H. Ramamurthy, B. S. Prabhu, and R. Gadh, in I. Niemegeers and S. Heemstra de Groot, eds., Reconfigurable wireless interface for networking sensors (ReWINS), *Proc. 9th Int. Conf. Personal Wireless Communications (IFIP TC6) (PWC 2004) (LNCS 3260 / 2004])*, Delft, The Netherlands, Sept. 21–23, 2004.

55. K. R. Sharp, A sense of the real world, *Supply Chain Syst. Mag.* **20**:9 (Sept. 2000).

56. K. R. Sharp, Channel tuning in the RFID market, *Supply Chain Syst. Mag.* **23**:5 (May 2003).

57. S. M. Sutton, Middleware selection, in W. Emmerich and S. Tai, eds., *Proc. 2nd Int. Workshop on Engineering Distributed Objects (EDO 00)*, LNCS 1999/2001, Davis, CA, Nov. 2–3, 2000, pp. 2–7.

58. A. Tripathi, Challenges designing next-generation middleware systems, *Commun. ACM* **45**(6):39–42 (2002).

59. P. Välkkynen, I. Korhonen, J. Plomp, T. Tuomisto, L. Cluitmans, H. Ailisto, and H. Seppä, A user interaction paradigm for physical browsing and near-object control based on tags, *Proc. Physical Interaction (PI03) Workshop on Real World User Interfaces*, held during MobileHCI 2003, Udine, Italy, Sept. 8–11, 2003.

60. VDC — Venture Development Corporation, *Radio Frequency Identification (RFID) Middleware Solutions: Global Market Opportunity*, a market research report, Natick, MA, Aug. 2004.

61. S. Vinoski, Where is middleware? *IEEE Internet Comput.* **6**(2):83–85 (March–April 2002).

62. B. Violino, Linking RFID with Web services, *RFID J.* (Oct. 6, 2003); available on the Web at `http://www.globeranger.com/papers/Article%20%20RFID%20Journal%20RFID%20&%20Web%20Services.pdf`.

63. W3C, *Web Services Activity*, World Wide Web Consortium's Web Services Working Groups and Interest Groups Website: `http://www.w3.org/2002/ws/`.

64. J. Walker, T. Mendelsohn, and C. S. Overby, *Vendors Race to fill a New Void: RFID Middleware*, TechStrategy Research brief, Forrester Research, Inc., Cambridge, MA, Jan. 26, 2004.

65. G. Wagner, How to design a general rule markup language, *Proc. Workshop XML Technologies for the Semantic Web* (*XSW 2002*), Institut fur Informatik, HU Berlin, March 24–25, 2002.

66. R. Want, K. P. Fishkin, A. Gujar, and B. L. Harrison, Bridging physical and virtual worlds with electronic tags, *Proc. ACM CHI'99, Conf. Human Factors in Computing Systems*, Pittsburgh, PA, May 15–20, 1999, pp. 370–377.

67. MS-WSDC, Microsoft's Web Services Developer Center, *Microsoft's .NET Web Services Portal*; available on the Web at `http://msdn.microsoft.com/webservices/`.

68. L. H. Zawada and P. O'Kelly, 2003, Assess RFID's transformational potential, *.NET Mag.* **3**:11 (Oct. 2003).

69. L. Zhekun, R. Gadh, and B. S. Prabhu, *A Study of RFID Smart Parts*, Technical Report UCLA-WINMEC-2003-202, Wireless Internet for the Mobile Enterprise Consortium, School of Engineering and Applied Science, Univ. California Los Angeles, 2003.

70. L. Zhekun, R. Gadh, and B. S. Prabhu, Applications of RFID technology and smart parts in manufacturing, *Proc. DETC'04: ASME 2004 Design Engineering Technical Conf. and Computers and Information in Engineering Conf.*, Salt Lake City, UT, Sept. 28–Oct. 2, 2004.

CHAPTER 13

Designing Smart Environments: A Paradigm Based on Learning and Prediction

SAJAL K. DAS and DIANE COOK

Department of Computer Science and Engineering, University of Texas at Arlington

13.1 INTRODUCTION

We live in an increasingly connected and automated society. Smart environments embody this trend by linking computers and other devices to everyday settings and commonplace tasks. Although the desire to create smart environments has existed for decades, research on this multidisciplinary topic has become increasingly intense since the early 1990s or so. Indeed, tremendous advances in such areas as smart (portable) devices and appliances, wireless mobile communications, pervasive computing, wireless sensor networking, machine learning and decision-making, robotics, middleware and agent technologies, and human computer interfaces have made the dream of smart environments become a reality. As depicted in Figure 13.1, a smart environment is a small world where sensor-enabled and networked devices work continuously and collaboratively to make lives of inhabitants more comfortable. A definition of "smart" or "intelligent" is "the ability to autonomously acquire and apply knowledge," while an "environment" refers to our surroundings. We therefore define a "smart environment" as *one that is able to acquire and apply knowledge about an environment and to adapt to its inhabitants in order to improve their experience in that environment* [8].

The type of experience that individuals wish from their environment varies with the individual and the type of environment considered. For example, they may wish the environment to ensure the safety of its inhabitants, they may want to reduce the cost or overhead of maintaining the environment, they may wish to optimize the

Mobile, Wireless, and Sensor Networks: Technology, Applications, and Future Directions
Edited by Rajeev Shorey, Akkihebbal L. Ananda, Mun Choon Chan, and Wei Tsang Ooi
Copyright © 2006 John Wiley & Sons, Inc.

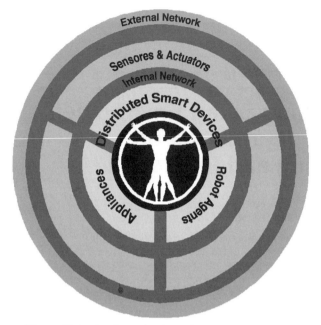

Figure 13.1 A schematic view of a smart environment.

resource (e.g., utility/energy bills or communication bandwidth) usage, or they may want to automate tasks that they typically perform in the environment. The expectations of such environments have evolved with the history of the field. We [8] have introduced the necessary technologies, architectures, algorithms, and protocols to build a smart environment along with a variety of existing applications. In this chapter, we will demonstrate that wireless mobile and sensor networks play a significant role in this domain.

Reflecting the increased interest in smart environments, research labs in academia and industry are picking up the theme and creating environments with their own individual spin and market appeal. For example, the Georgia Tech Aware Home [1,22], the Adaptive House at the University of Colorado at Boulder [26], and the MavHome smart home at the University of Texas at Arlington [10] use sensors to learn models of the inhabitants and automate activities accordingly. Other types of smart environments, including smart offices, classrooms, kindergartens, tables, and cars, have been designed by MIT [4,33], Stanford University [14], the University of California at Los Angeles (UCLA) [31,32], INRIA in France [23], and Ambiente, Nissan, and Intel. Connected homes with device communications capability have become the focus of companies such as Philips, Cisco [6], GTE, Sun, Ericsson, and Microsoft [5]. Still other groups have focused on smart environments to assist individuals with health challenges. These projects include the Gloucester Smart Home [15], the Edinvar Assisted Interactive Dwelling House [13], the Intel Proactive Health Project [21], agent-based smart health monitoring in MavHome [11], and MALITDA smart house for individuals with special needs

[18]. It is easy to see that such environments are results of phenomenal advancements in wireless mobile communications infrastructures and sensor networking technologies, among others.

This chapter presents our research experience in developing smart environments through a project called *MavHome* [10], funded by the U.S. National Science Foundation. In particular, we propose "learning and prediction" as an overarching framework or paradigm for designing efficient algorithms and smart protocols in such environments. The foundation of this paradigm lies in information theory as it manages inhabitants' uncertainties in mobility and activities in their daily lives. The underlying idea is to build intelligent (compressed) dictionaries of mobility and activity profiles (or histories) of inhabitants, collected from sensor data, learn from this information, and then predict future mobility and action. The prediction in turn helps automate device operations and manage resources efficiently, thus optimizing the goals of the smart environment.

The chapter is organized as follows. Section 13.2 describes salient features of smart environments. Section 13.3 presents the architectural details of our Mav-Home smart home project. Section 13.4 deals with the proposed paradigm for inhabitant's (indoor) location and activity prediction and automated decisionmaking capability. Section 13.5 discusses MavHome implementation issues, while Section 13.6 highlights practical considerations. Finally, Section 13.7 concludes the chapter.

13.2 FEATURES OF SMART ENVIRONMENTS

Important features of smart environments are that they possess a degree of autonomy, adapt themselves to changing environments, and communicate with humans in a natural way. Intelligent automation can reduce the amount of interaction required by the inhabitants, as well as reducing utility consumption and other potential wastages. These capabilities can also provide important features, such as detection of unusual or anomalous behaviors for health monitoring and home security.

The benefits of automation can influence every environment that we interact with in our daily lives. For example, consider operations in a smart home and illustrate with the help of the following scenario. To minimize energy consumption, the home keeps the temperature cool throughout the night. At 6:45 A.M., the home turns up the heat because it has learned that it needs 15 min to warm to the inhabitant's favorite waking temperature. The alarm sounds at 7:00 A.M., which signals the bedroom light to go on as well as the coffeemaker in the kitchen. The inhabitant, Bob, steps into the bathroom and turns on the light. The home records this manual interaction, displays the morning news on the bathroom videoscreen, and turns on the shower. While Bob is shaving, the home senses that Bob is 4 lb over his ideal weight and adjusts his suggested daily menu and displays in the kitchen. When Bob finishes grooming, the bathroom light turns off while the kitchen light and display turn on. During breakfast, Bob requests the janitor robot to clean the house. When Bob leaves for work, the home secures all doors behind him and starts the lawn sprinklers despite knowledge of the 30% predicted chance of rain. To reduce energy costs, the house turns down the heat until 15 min before Bob is due home. Because the refrigerator is low on milk and cheese, the home places a grocery order. When Bob arrives home, his grocery order has arrived, the house is back at Bob's desired temperature, and the hot tub is waiting for him.

This scenario highlights a number of desired features in a smart environment such as a home. In the following, let us look at some of these features in more detail [8].

13.2.1 Remote Control of Devices

The most basic feature of smart environments is the ability to control devices remotely or automatically. Powerline control systems have been available for decades, and basic controls offered by X10 can be easily installed. By plugging devices into such a controller, inhabitants of an environment can turn lights, coffeemakers, and other appliances on or off in much the same way as couch potatoes switch television stations with a remote control (Fig. 13.2). Computer software can

Figure 13.2 Device control in smart environments.

additionally be employed to program sequences of device activities and to capture device events executed by the powerline controllers.

With this capability, inhabitants are free from the requirement of physical access to devices. Individuals with disabilities can control devices from a distance, as can the person who realized when he got to work that he left the sprinklers on. Automated lighting sequences can give the impression that an environment is occupied while the inhabitants are gone, thus handling basic routine procedures by the environment with almost no human intervention.

13.2.2 Device Communications

With the maturity of wireless mobile communications and middleware technology, smart environment designers and inhabitants have been able to raise their standards and expectations. In particular, devices use these technologies to communicate with each other, share data to build a more informed model of the state of the environment and/or inhabitants, and retrieve information from outside sources over the Internet or wireless communication infrastructure. This allows better response to the current state and needs.

As mentioned earlier, such "connected environments" have become the focus of many industry-developed smart homes and offices. With these capabilities, for example, the environment can access the weather page to determine the forecast and query the moisture sensor in the lawn to determine how long the sprinklers should run. Devices can access information from the Internet such as menus, operational manuals, or software upgrades, and can post information such as a grocery store list generated from monitoring inventory with an intelligent refrigerator or trash bin.

Activation of one device can also trigger other sequences, such as turning on the bedroom radio, kitchen coffeemaker, and bathroom towel warmer when the alarm goes off. Inhabitants can benefit from the interaction between devices by muting the television sound when the telephone or doorbell rings; temperature as well as motion sensors can interact with other devices to ensure that the temperature is kept at a desired level wherever the inhabitants are located within the environment. Moreover, a smart environment will provide a neat service forwarding capability with the help of individual smart devices that communicate with each other without human intervention. For example, in a smart environment, calls on a mobile phone can be automatically forwarded to a nearby landline phone while emails can be received in the mobile phone instead of an outdoor cellular network.

13.2.3 Sensory Information Acquisition/Dissemination

The recent past has observed tremendous advancements in sensor technology and the ability of sensors to share information and make low-level decisions. As a result, environments can provide constant adjustments based on sensor readings and can better customize behaviors to the nuances of the inhabitants' surroundings. Motion detectors or force sensors can detect the presence of individuals in the environment and accordingly adjust lights, music, or climate control. Water and gas sensors can monitor potential leaks and force the valves, thus closing them when a danger arises. Low-level control of devices offers fine-tuning in response to changing conditions, such as adjusting window blinds as the amount of daylight coming into a room changes. Networks composed of these sensors can share data and offer information to the environment at speeds and complexity not experienced in the earlier versions of smart environments. For example, the Smart Sofa [30] developed at Trinity College in Dublin, Ireland can identify an individual by the person's weight and can theoretically use this information to customize the settings of devices around the house.

13.2.4 Enhanced Services by Intelligent Devices

Smart environments are usually equipped with numerous smart devices and appliances that provide varied and impressive capabilities. Networked together and tied to intelligent sensors and the outside world, the impact of these devices becomes even more powerful. Such devices are becoming the focus of a number of manufacturers, including Electrolux, Whirlpool, and a collection of startup companies.

As examples of such devices, Frigidaire and Whirlpool offer intelligent refrigerators with features that include Web cameras to monitor inventory, barcode scanners, and Internet-ready interactive screens. Through interactive cameras, inhabitants away from home can view the location of security or fire alerts and remote caregivers can check on the status of their patients or family. Merloni's Margherita 2000 washing machine is similarly Internet-controlled, and uses sensor information to determine appropriate cycle times. Other devices such as microwaves, coffeemakers, and toasters are quickly joining the collection.

In addition, specialized equipments have been designed in response to the growing interest in assistive environments. AT&T's Kids Communicator resembles a hamster exercise ball and is equipped with a wireless videophone and remote maneuverability to monitor the environment from any location. A large collection of companies including Friendly Robotics, Husqvarna, Technical Solutions, and University of Florida's Lawn Nibbler have developed robotic lawnmowers to ease the burden of this time-consuming task, and indoor robot vacuum cleaners including Roomba and vacuums from Electrolux, Dyson, and Hitachi are gaining in popularity and usability. Researchers at MIT's Media Lab are investigating new specialized devices, such as an oven mitt that can tell if food has been warmed all the way through. A breakthrough development from companies such as Philips is an interactive tablecloth that provides cable-free power to all chargeable objects placed on the table's surface. An environment that can combine the features of these devices with information gathering and remote control power of previous research will realize many of the intended goals of smart environment designers.

13.2.5 Networking Standards

A smart environment will be able to control all of its various networked devices (see Fig. 13.3) such as computers, sensors, cameras, and appliances, from anywhere

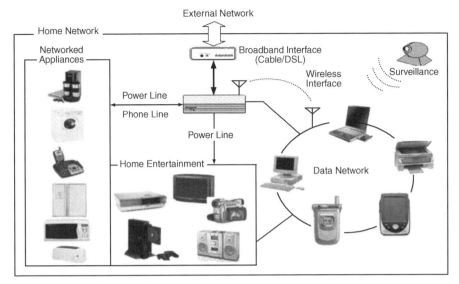

Figure 13.3 Networked devices in smart environments.

and at anytime through the Internet. For example, when the inhabitant is away, she can still be in contact with her different environments to monitor their status and/or access her personal database. From that perspective, all the hardware and software for enabling the smart environments should be based on open standards. Moreover, they should be easy to install, configure, and operate in order to be user-friendly to the nonprofessional inhabitants or consumers. IEEE 802.11- and IEEE 802.15- based wireless LANs, and Bluetooth using spread-spectrum techniques under 2.4GHz or 5GHz unlicensed ISM (Industrial, Science and Medical) wireless spectrum and Home RF (radiofrequency) technology have been applied to wireless networking infrastructures for smart environments. Alongside, Ethernet (IEEE 802.3), phoneline networking alliance (PNA), and X10 powerline networking have emerged as smart-environment-wired networking technologies on the market. These technologies have advantages and disadvantages. For example, X10 power-line networking has the widest availability; however, it has a much lower speed than do other PNA and wireless standards. Performance comparison, coexistence capability, and interoperability of these technologies have begun in the academic and industry research realms while implementing prototypes of smart environments using the above mentioned standards.

13.2.6 Predictive Decisionmaking Capabilities

The features of a smart environment described up to this point provide the potential for fulfilling the goal of a smart environment, that is, improving the experience of inhabitants of the environment. However, control of these capabilities is mostly in the hands of the users. Only through explicit remote manipulation or careful programming can these devices, sensors, and controllers adjust the environment to fit the needs of the inhabitants. Full automation and adaptation rely on the software itself to learn, or acquire information that allows the software to improve its performance with experience.

Specific features of the more recent smart environments that meet these criteria incorporate predictive and automatic decisionmaking capabilities into the control paradigm. Contexts (mobility, activity, etc.) of inhabitants as well as of the environment can be predicted with good accuracy based on observed activities and known features. Models can also be built using inhabitant patterns to customize the environment for future interactions. For example, an intelligent car can collect information about the driver including typical times and routes to go to work, theater, restaurant, and store preferences, and frequently used gas stations. Combining this information with data collected by the inhabitant's home and office as well as Internet-gathered specifics on movie times, restaurant menus and locations, and sales at various stores, the car can make recommendations based on the learned model of activity patterns and preferences.

Similarly, building a model of device performance can allow the environment to optimize its behaviors and performance. For example, a smart kitchen may learn that the coffeemaker requires 10 min to completely brewing a full pot of coffee, and will start it up 10 min before it expects the inhabitants to want their first cup. Smart lightbulbs may warn when they are about to expire, letting the factory automatically deliver replacements before the need is critical.

As a complement to predictive capabilities, a smart environment will be able to make decisions on how to automate its own behaviors to meet the specified goals. Device settings and timings of events are now under the control of the environment. Such a smart environment will also have to elect between alternate methods of achieving a goal, such as turning on lights in each room entered by an inhabitant or anticipating where the inhabitant is heading and illuminating just enough of the environment to direct the individual to their goal. In fact, this learning and prediction aspect of smart environments will be the focus of the rest of this chapter.

13.3 THE MavHome SMART HOME

The MavHome at the University of Texas at Arlington represents an environment that acts as an intelligent agent, perceiving the state of the home through sensors and acting on the environment through device controllers. The goal is to maximize inhabitants' comfort and minimize the home's operating cost. To achieve this goal, the house must be able to reason about, learn, predict, and adapt to its inhabitants.

In MavHome, the desired smart home capabilities are organized into an agent-based software architecture that seamlessly connects the components while allowing improvements to be made to any of the supporting technologies. Figure 13.4 describes the architecture of a MavHome agent that separates the technologies and functions into four cooperating layers. The *decision* layer selects actions for

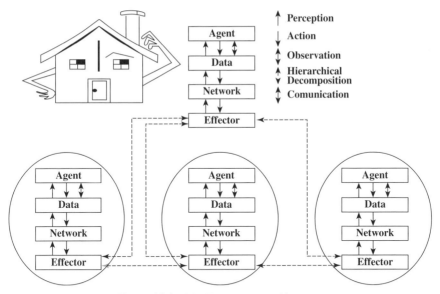

Figure 13.4 MavHome agent architecture.

the agent to execute. The *information* layer collects information and generates inferences useful for making decisions. The *communication* layer is responsible for routing and sharing information between agents. The *physical* layer contains the environment hardware, including devices, transducers, and network equipment. The MavHome software components are connected using a distributed interprocess communication interface.

Because controlling an entire house is a large-scale complex learning and reasoning problem, it is decomposed into reconfigurable tasks. Thus, the physical layer for one agent may represent another agent somewhere in the hierarchy, which is capable of executing the task selected by the requesting agent.

Perception is a bottom–up process. Sensors monitor the environment (e.g., lawn moisture level) and, if necessary, transmit the information to another agent through the communication layer. The database records the information in the information layer, updates its learned concepts and predictions accordingly, and alerts the decision layer of the presence of new data. During action execution, information flows top–down. The decision layer selects an action (e.g., run the sprinklers) and relates the decision to the information layer. After updating the database, the communication layer routes the action to the appropriate effector to execute. If the effector is actually another agent, the agent receives the command through its effector as perceived information and must decide on the best method of executing the desired action. Specialized interface agents allow interaction with users, robots, and external resources such as the Internet. Agents can communicate with each other using the hierarchical flow shown in Figure 13.4. In the remaining discussions, a smart home will generically represent a smart environment.

13.4   AUTOMATION THROUGH LEARNING AND PREDICTION

In order to maximize comfort, minimize cost, and adapt to the inhabitants, a smart home must rely on sophisticated tools for intelligence building such as learning, prediction, and making automated decisions. We will demonstrate that learning and prediction indeed play an important role in determining the inhabitant's next action and anticipating mobility patterns within the home. MavHome uses these predictions in order to automate selected repetitive tasks for the inhabitant. The home will need to make this prediction solely on the basis of past mobility patterns and previously observed inhabitant interaction with various devices (e.g., motion detectors, sensors, device controllers, video monitors), as well as the current state of the inhabitant and/or the house. The captured information can be used to build sophisticated models that aid in efficient prediction algorithms. The number of prediction errors must be minimal, and the algorithms must be able to deliver predictions with minimal delays for computation. Prediction is then handed over to a decisionmaking algorithm that selects actions for the house to meet its desired goals. The underlying concepts of MavHome prediction schemes lie in the text compression, online parsing, and information theory. Well-investigated text compression methods [9,35] have established that good compression algorithms are also good learners and hence good predictors. According to information theory [9], a predictor with an order (size of history used) that grows at a rate approximating the entropy rate of the source is an optimal predictor. In the following, we summarize our novel paradigm for inhabitant's mobility and activity predictions.

13.4.1   Inhabitant Location Prediction

By definition, a "smart" environment is *context-aware* in the sense that by combining inputs from multiple sensing devices, it should be able to deduce the inhabitant's intent or attributes without explicit manual input. Location is perhaps the most common example of context. Hence, it is crucial for a smart environment to track an inhabitant's mobility accurately by determining and predicting the person's location. The prediction also helps in optimal allocation of resources and activation of effectors in location-aware applications [12,25]. We first proposed [2] a model-independent algorithm for location prediction in wireless cellular networks, which we later adopted for indoor location tracking and predicting inhabitants' future locations in smart homes [16,29]. This approach is based on symbolic representation of location information that is specified not in absolute terms, but relative to the topology of the corresponding access infrastructure (e.g., sensor ids or zones through which the inhabitant passes), thus making our approach universal or technology/model-independent. At a conceptual level, prediction involves some form of statistical inference, where some sample of the inhabitant's past movement history (profile) is used to provide intelligent estimates of that individual's future location, thereby reducing the location uncertainty associated with this prediction [12,28].

Hypothesizing that the inhabitant's mobility has repetitive patterns that can be learned, and assuming the inhabitant's mobility process to be stochastically random, we proved the following result [2,3]. It is impossible to optimally track mobility with less information exchange between the system (in this case smart environment) and the device (detecting the inhabitant's mobility) than the entropy rate (in bits per second) of the stochastic mobility process. Specifically, given all past observations of the inhabitant's position and the best possible predictors of future position, some uncertainty in the position will *always* exist unless the device and the system exchange location information. The actual *method* by which this exchange takes place is irrelevant to this bound. All that matters is that the exchange exceeds the entropy rate of the mobility process. Therefore, a key issue in establishing bounds is to characterize the mobility process (and therefore its entropy rate) in an adaptive manner. To this end, on the basis of the information theoretic framework, an optimal online adaptive location management algorithm, called *LeZi-update*, was proposed [2,3] for cellular communication networks. Rather than assuming a standard mobility model of the node, LeZi-update learns node movement history stored in a Lempel-Ziv (LZ) type of compressed dictionary [35], builds a *universal* mobility model by minimizing entropy, and predicts future locations with a high degree of accuracy. In other words, LeZi-update offers a model-independent solution to manage uncertainty related to node mobility. This framework is quite general and applicable to other contexts such as activity prediction [17], resource provisioning [12,28], and anomaly detection.

Figure 13.5a depicts a typical floorplan layout of MavHome together with the placement of motion (in-building) sensors along the inhabitant's routes, by partitioning MavHome's coverage area into sensor *zones* or sectors. When the

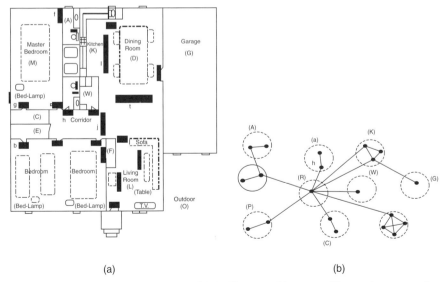

(a)   (b)

Figure 13.5 (a) Typical floorplan of MavHome architecture; (b) graph representing connectivity of sensor zones.

system (environment) needs to contact the inhabitant, it will initiate a location pre-diction scheme. In order to control the location uncertainty, the system also relies on the location information as sampled by the sensors, which in turn helps reduce the search space for subsequent prediction. As shown in Figure 13.5b, the floorplan can be represented as a connected graph $G = (V, E)$ where the node set $V = \{a, b, c, \ldots\}$ denotes the zones (sensor ids) and the edge set E denotes the neighborhood adjacency between a pair of zones. While moving from one zone to another, the inhabitant crosses an array of sensors along a route. For example, the movement from the corridor (R) to the dining room (D) in the floorplan can be expressed by the collection of sensors $\{j,l\}$ or $\{j,k\}$.

The LeZi-update framework uses a *symbolic space* to represent the sensing zone of the smart environment as an alphabetic symbol and thus captures the inhabitant's movement history as a string of symbols. Thus, while the geographic location data are often useful in obtaining precise location coordinates, the symbolic information removes the burden of frequent coordinate translation and is capable of achieving universality across different networks [25,28]. (The blessing of symbolic represen-tation also helps us hierarchically abstract the indoor connectivity infrastructure into different levels of granularity.) Tacit in this formulation is that every node has some movement patterns that can be learned in an online fashion. Essentially, we assume that node itineraries are inherently *compressible*, and this allows appli-cation of *universal data compression* algorithms [35], which make very basic and broad assumptions, and yet minimize the source entropy for stationary ergodic stochastic processes [27].

In LeZi-update, the symbols (sensor ids) are processed in chunks and the entire sequence of symbols withheld until the last update is reported in a compressed (encoded) form. For example, referring to the abstract representation of mobility route in Figure 13.6a, let `ajlloojhhaajlloojaajlloojaajll...` be the inhabitant's movement history at any instant. This string of symbols can be parsed as distinct substrings (or phrases) `a, j, l, lo, o, jh, h, aa, jl, loo, ja, aj, ll, oo, jaa, jll`, and so on. As shown in Figure 13.6b, such a symbolwise

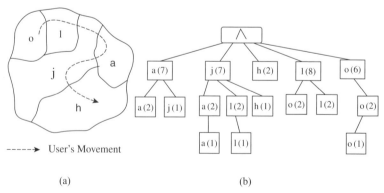

(a)   (b)

Figure 13.6 (a) Symbolic representation of mobility; (b) trie holding zones and their frequencies.

TABLE 13.1   **Phrases and Their Frequencies at Context** j1, j, **and** Λ

Jl	J	Λ		
$l\|jl(1)$	$a\|j(1)$	$a(4)$	$aa(2)$	$aj(1)$
$\Lambda\|jl(1)$	$aa\|j(1)$	$j(2)$	$ja(1)$	$jaa(1)$
	$l\|j(1)$	$jl(1)$	$jh(1)$	$l(4)$
	$ll\|j(1)$	$lo(1)$	$loo(1)$	$ll(2)$
	$h\|j(1)$	$o(4)$	$oo(2)$	$h(2)$
	$\Lambda\|j(2)$	$\Lambda(1)$		

context model, based on variable- to fixed-length coding, can be efficiently stored in a dictionary implemented by a *trie*. Essentially, the mobile acts as an *encoder* while the system acts as a *decoder* and the frequency of every symbol is incremented for *every prefix of every suffix* of each phrase. By accumulating larger and larger contexts, one can affect a paradigm shift from traditional position update to *route update*. For stationary ergodic sources with n symbols, this framework achieves asymptotic optimality, with improved update cost bounded by $\Omega(\log n - \log \log n)$, where $\log n$ denotes logarithm base 2.

One major objective of the LeZi-update scheme is to endow the prediction process, by which the system finds nodes whose position is uncertain, with sufficient information regarding the node mobility profile. Each node in the trie preserves the relevant frequencies provided by the update mechanism in the current context. Thus, considering jll as the latest update message, the usable contexts are its prefixes, namely, $j1$, j, and Λ (null symbol). All predictable routes (parsed phrases) with frequencies in this context are listed in Table 13.1. Following the blending technique of *prediction by partial match* (PPM) [7], the probability computation starts from the leaf nodes (highest level) of the trie and *escapes* to the lower levels until the root is reached. According to the principle of *insufficient reasoning* [27], every phrase probability is distributed among individual symbols (zones) according to their relative occurrence in a particular phrase. The total residence probability of every zone (symbol) is computed by adding all the probabilities that it has accumulated from all possible phrases at this context. The optimal prediction order is now determined by polling the zones in decreasing order of these residence probabilities.

So overall, the application of information-theoretic methods to location prediction has allowed quantification of minimum information exchanges to maintain accurate location information, provided an online method by which to characterize mobility, and in addition, endowed an optimal prediction sequence [12]. Through learning, this approach allows us to build a higher-order mobility model rather than assuming a finite model, and thus minimizes entropy and leads to optimal performance.

While the basic LeZi-update algorithm was used to predict only the current location from past movement patterns, this approach has also been extended [29] to predict the likely future routes (or trajectories) of inhabitants in smart homes and also for heterogeneous environments [24]. The route prediction exploits the *asymptotic equipartition property* in information theory [9], which states that for a random process X with entropy $H(X)$, the number of observed unique paths of length n is $2^{H(X)}$ with probability 1. In other words, for reasonably large n, most of the probability mass is concentrated in only a small subset (called the *typical set*) of routes, which encompasses the inhabitant's most likely routes and captures the *average nature of long-length sequences*. Accordingly, the algorithm simply predicts a relatively small set of likely paths (one of which the user will almost surely take next). A smart home environment can then act on this information by activating resources (e.g., by turning on the lights in corridors that constitute one or more of these routes) in a minimal and efficient manner rather than turning on all lights in the house. Experiments demonstrate that our predictive framework can save up to 70% (electrical) energy in a typical smart home environment [29]. The accuracy of prediction is up to 86%, and only 11% of routes constitute the typical set.

13.4.2 Inhabitant Action Prediction

A smart home inhabitant typically interacts with various devices as part of her or his routine activities. These interactions may be considered as a sequence of events, with some inherent pattern of recurrence. Again, this repeatability leads us to the conclusion that the sequence can be modeled as a stationary stochastic process in terms of mobility. Inhabitant action prediction consists of first mining the data to identify sequences of actions that are sufficiently regular and repeatable to generate predictions, and then using a sequence matching approach to predict the next action in one of these sequences.

To mine the data, a window can be moved in a single pass through the history of inhabitant actions, looking for sequences within the window that merit attention. Each sequence is evaluated using the *minimum-description-length* principle [27], which favors sequences that minimize the description length of the sequence once it is compressed by replacing each instance of the discovered pattern with a pointer to the pattern definition. A regularity factor (daily, weekly, or monthly) helps compress the data and thus increases the value of a pattern. Action sequences are first filtered by the mined sequences. If a sequence is considered significant by the mining algorithm, then predictions can be made for events within the sequence window. Using this algorithm as a filter for two alternative prediction algorithms, the resulting accuracy increases on average by 50%. This filter ensures that MavHome will not erroneously seek to automate anomalous and highly variable activities [19,20].

As described above, the action prediction algorithm parses the input string (history of interactions) into substrings representing phrases. Because of the prefix property used by the algorithm, parsed substrings can be efficiently maintained in a trie along with frequency information. To perform prediction, the algorithm calculates the probability of each symbol (action) occurring in the parsed sequence, and predicts the action with the highest probability. To achieve optimal predictability, the predictor must use a mixture of all possible order models (phrase sizes) when determining the probability estimate. To accomplish this, techniques from the PPM family of predictors are incorporated, which generate weighted Markov models of different orders. This blending strategy assigns greater weight to higher-order models, in keeping with the advisability of making the most informed decision.

In experiments run on sample smart home data, predictive accuracy of this approach converged on 100% for perfectly repeatable data with no variation, and converged on 86% accuracy for data containing variations and anomalies [17].

13.4.3 Automated Decisionmaking

As mentioned earlier, the goal of MavHome is to enable the home to automate basic functions in order to maximize the inhabitants' comfort and minimize the operating cost of the home. We assume that comfort is a function of the number of manual interactions with the home and the operating cost of energy usage.

Because the goal is a combination of these two factors, blind automation of all inhabitant actions is seldom the desired solution. For example, an inhabitant might turn on the hallway light in the morning before opening the blinds in the living room. MavHome could, on the other hand, open the blinds in the living room before the inhabitant leaves the bedroom, thus alleviating the need for the hallway lights. Similarly, turning down the air conditioning after leaving the house and turning it back up before returning would be more energy-efficient than turning the air conditioning to maximum after arriving home in order to cool it as quickly as possible [29].

To achieve its goal, MavHome uses reinforcement learning to acquire an optimal decision policy. In this framework, the agent learns autonomously from potentially delayed rewards rather than from a teacher, reducing the requirement for the home's inhabitant to supervise or program the system. To learn a strategy, the agent explores the effects of its actions over time and uses this experience to form control policies that optimize the expected future reward.

MavHome learns a policy based on a state space, $S = \{s_i\}$, consisting of the states of the devices in the home, the predictions of the next event, and expected energy utilization over the next time unit. A reward function, r, takes into account the amount of required user interaction, the energy consumption of the house, and other parameters that quantify the performance of the home. This reward function can be tuned to the particular preferences of the inhabitants, thus providing a simple means to customize the home's performance. Q-learning is used [34] to approximate an optimal action strategy by estimating the predicted value, $Q(s_t, a_t)$, of executing action a_t in state s_t at time t. After each action, the utility is updated as $Q(s_t, a_t) \leftarrow \alpha \left[r_{t+1} + \gamma \max_{a \in A} Q(s_{t+1}, a) - Q(s_t, a_t) \right]$. After learning, the optimal action, a_t, can be determined as: $a_t = \arg \max_{a \in A} Q(s_t, a)$.

13.5 MavHome IMPLEMENTATION

In the MavHome smart home project at the University of Texas at Arlington, student activity data are collected continuously according to their interactions with

Figure 13.7 Web camera views of MavHome environment (left) and ResiSim visualization (right).

devices in the environment. Off-the-shelf X10 controllers automate most devices and thus inhabitant's actions. Arrays of sensors track their mobility.

Using the ResiSim 3D simulator, a graphical model has been constructed of the intelligent environment. The model allows a visitor at a remote location to monitor or change the status of devices in MavHome, as shown in Figures 13.7 and 13.8. Images in the left column of Figure 13.7 show Web cameras placed throughout the

Figure 13.8 ResiSim update after desk lamp (lower left) is turned on.

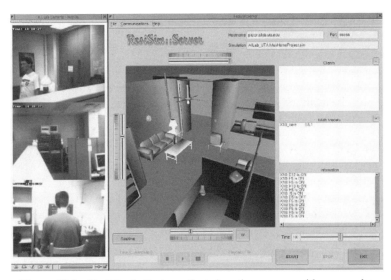

Figure 13.9   ResiSim indicates activated motion sensors with green orbs.

environment, and the simulator visualization is shown on the right. The "information" window in the lower right indicates that devices have recently been manipulated, either manually or by MavHome. Figure 13.8 shows that the light in the entryway (upper left) is illuminated once Darin enters the environment and the lamp on Ryan's desk (lower left) turns on to assist him with work. The updated status of the lamp is shown by the yellow circle in the ResiSim model (right). The model will indicate the status of sensors as well — the orbs in Figure 13.9 indicate that there are two areas of activity captured by motion sensors.

A live demonstration of MavHome was conducted in the fall of 2004. During the previous weeks, activity data were collected for one of the project participants ("MavHome Bob"). Actions included turning on lights en route to his desk in the morning, watching a live news feed on the computer, taking a coffee and TV break, and turning off devices on the way out at the end of the day. Despite the presence of approximately 50 people during the live demonstration (who were setting off motion sensors throughout the environment), MavHome correctly predicted and automated each activity. Figure 13.10 reflects the movements of MavHome Bob as he moves through the environment, and lights are illuminated reflecting his typical activities.

Figure 13.10   Bob's movements in MavHome. Bob's position is indicated by a dashed box.

13.6 PRACTICAL CONSIDERATIONS

So, how easily can the features of a smart home be integrated into new or existing homes? The software described in the MavHome implementation consists of commercial X10 controllers, a computer, a variety of sensors, and a wireless network. A simple implementation can be integrated into an existing house for under a thousand dollars, in many cases. If robots or customized devices are introduced, the cost increases.

A computer interface to a smart home must be very simple. Manual control of devices can override home decisions, and alternative interfaces including voice control are offered. Other than starting or resetting the software, no interaction with the computer is required. In our experiments, the software adapted to user activities in a couple of weeks, but the training time will vary according to the complexity of inhabitant's actions and the number of people in the home. Although minimal expertise is required, various types of interaction are possible depending on the needs of the inhabitant. The user can certainly vary the threshold at which activities are automated, although this is not necessary because manual resetting of actions selected by the house constitute negative reward and eventually the house will not automate those particular commands. The inhabitant can also request that the home simply make suggestions for automation; selection of rules for automation will be made by the inhabitant on a case-by-case basis.

Introducing intelligent control into a house can result in a number of privacy and safety issues. Safety constraints must be placed on each device to ensure that the house will not select an action that endangers its inhabitants. The house may not be allowed, for example, to select a temperature setting below 50°F or above 90°F. The entire automation can be quickly disabled with one mouse click or voice command — each device can operate with or without computer control. The inhabitant also needs to specify the type of data that can be collected, and which data, if any, can be disseminated for learning across multiple households or cities.

Similarly, smart homes typically benefit from collecting information about the health, typical patterns, and other features of their inhabitants. This leads to a number of privacy and security issues. Data should be collected only on features allowed by the inhabitants, and shared with other sites only as volunteered. New smart homes in neighboring locations could, for example, benefit from patterns learned in an older home, but care must be taken to share information without violating the privacy of home inhabitants.

13.7 CONCLUSIONS

This chapter demonstrated the effectiveness of a paradigm based on learning and prediction in a smart home environment. Efficient prediction algorithms provide information useful for future locations and activities, automating activities, optimizing design and control methods for devices and tasks within the environment, and identifying anomalies. These technologies reduce the workload of maintaining a home, reducing energy utilization, and providing special benefits for elderly and people with disabilities. In the future, these abilities will be generalized to a conglomeration of environments, including smart offices, smart roads, smart hospitals, smart automobiles, and smart airports, through which a user may pass through in daily life. Another research challenge is how to characterize mobility and activity profiles of multiple inhabitants (e.g., living in the same home) in the same dictionary and predict or trigger events to meet the common goals of the house under conflicting requirements of individual inhabitants.

ACKNOWLEDGMENTS

This research was supported by the U.S. National Science Foundation grants under award numbers IIS-0121297 and IIS-0326505.

REFERENCES

1. G. D. Abowd, Classroom 2000: An experiment with the instrumentation of a living educational environment, *IBM Syst. J.* (special issue on human–computer interaction: a focus on pervasive computing) 38(4), pp. 508–530, 1999.

2. A. Bhattacharya and S. K. Das, LeZi-Update: An information-theoretic approach to track mobile users in PCS networks, *Proc. ACM Int. Conf. Mobile Comput. Networking* (*MobiCom*), Aug. 1999, pp. 1–12.

3. A. Bhattacharya and S. K. Das, LeZi-Update: An information-theoretic approach for personal mobility tracking in PCS networks, *Wireless Networks* **8**(2–3):121–135 (March–May 2002).

4. A. Bobick, S. Intille, J. Davis, F. Baird, C. Pinhanez, L. Campbell, Y. Ivanov, A. Schutte, and A. Wilson, The KidsRoom: A perceptually-based interactive and immersive story environment, *Presence* **8**(4):369–393 (Aug. 1999).

5. B. Brumitt, J. Kumm, B. Meyers, and S. Shafer, Ubiquitous computing and the role of geometry, *IEEE Pers. Commun.* **7**(5):41–43 (Aug. 2000).

6. Cisco, `http://www.cisco.com/warp/public//3/uk/ihome`.

7. J. G. Cleary and I. H. Witten, Data compression using adaptive coding and partial string matching, *IEEE Trans. Commun.* **32**(4):396–402 (April 1984).

8. D. J. Cook and S. K. Das, *Smart Environments: Technology, Protocols, and Applications*, Wiley, 2005, Chapter 1.

9. T. M. Cover and J. A. Thomas, *Elements of Information Theory*, Wiley, 1991.

10. S. K. Das, D. J. Cook, A. Bhattacharya, E. O. Heierman, and T.-Y. Lin, The role of prediction algorithms in the MavHome smart home architecture, *IEEE Wireless Commun.* **9**(6):77–84 (Dec. 2002).

11. S. K. Das and D. J. Cook, Agent based health monitoring in smart homes, *Proc. Int. Conf. Smart Homes and Health Telematics* (*ICOST*), Singapore, Sept. 2004 (keynote talk).

12. S. K. Das and C. Rose, Coping with uncertainty in wireless mobile networks, *Proc. IEEE Personal, Indoor and Mobile Radio Communications, Conf.*, Barcelona, Spain, Sept. 2004 (invited paper).

13. Edinvar, http://www.stakes.fi/tidecong/732bonne.html.

14. A. Fox, B. Johanson, P. Hanrahan, and T. Winograd, Integrating information appliances into an interactive space, *IEEE Comput. Graph. Appl.* **20**(3):54–65 (2000).

15. Gloucester, http://www.dementua-voice.org.uk/Projects_Gloucester-Project.html.

16. K. Gopalratnam and D. J. Cook, Online sequential prediction via incremental parsing: The active LeZi algorithm, *IEEE Intelligent Syst.* (in press).

17. K. Gopalratnam and D. J. Cook, Active LeZi: An incremental parsing algorithm for sequential prediction, *Int. J. Artificial Intelligence Tools* **14**(1–2) (2004).

18. S. Helal, B. Winkler, C. Lee, Y. Kaddoura, L. Ran, C. Giralo, S. Kuchibholta, and W. Mann, Enabling location-aware pervasive computing applications for the elderly, *Proc. IEEE Int. Conf. Pervasive Computing and Communications* (*PerCom'03*), March 2003, pp. 531–538.

19. E. Heierman, M. Youngblood, and D. J. Cook, Mining temporal sequences to discover interesting patterns, *Proc. KDD Workshop on Mining Temporal and Sequential Data*, 2004.

20. E. Heierman and D. J. Cook, Improving home automation by discovering regularly occurring device usage patterns, *Proc. Int. Conf. Data Mining*, 2003.

21. Intel, http://www.intel.com/research/prohealth.

22. C. Kidd, R. J. Orr, G. D. Abowd, D. Atkeson, I. Essa, B. MacIntyre, E. D. Mynatt, T. E. Starner, and W. Newstetters, The aware home: A living laboratory for ubiquitous computing, *Proc. 2nd Int. Workshop on Cooperative Buildings*, 1999.

23. C. Le Gal, J. Martin, A. Lux, and J. L. Crowley, Smart office: Design of an intelligent environment, *IEEE Intelligent Syst.* **16**(4) (July–Aug. 2001).

24. A. Misra, A. Roy, and S. K. Das, An information theoretic framework for optimal location tracking in multi-system 4G wireless networks, *Proc. IEEE InfoCom*, March 2004.

25. A. Misra and S. K. Das, Location estimation (determination and prediction) techniques in smart environments, in D. J. Cook and S. K. Das, eds., *Smart Environments*, Wiley, 2005, Chapter 8, pp. 193–228.

26. M. Mozer, The neural network house: An environment that adapts to its inhabitants, *Proc. AAAI Spring Symp. Intelligent Environments*, 1998.

27. J. Rissanen, *Stochastic Complexity in Statistical Inquiry*, World Scientific Publishers, 1989.

28. A. Roy, S. K. Das, and A. Misra, Exploiting information theory for adaptive mobility and resource management in future wireless cellular networks, *IEEE Wireless Commun.* **11**(4):59–64 (Aug. 2004).

29. A. Roy, S. K. Das Bhaumik, A. Bhattacharya, K. Basu, D. J. Cook, and S. K. Das, Location aware resource management in smart homes, *Proc. IEEE Int. Conf. Pervasive Computing and Communications* (*PerCom'03*), March 2003, pp. 481–488.

30. Smart sofa, http://www.dsg.cs.tcd.ie/?category_id=350.

31. M. B. Srivastava, R. Muntz, and M. Potkonjak, Smart kindergarten: Sensor-based wireless networks for smart developmental problem-solving environments, *Proc. 7th ACM Int. Conf. Mobile Computing and Networking* (*MobiCom'01*), 2001.

32. P. Steurer and M. B. Srivastava, System design of smart table, *Proc. 1st IEEE Int. Conf. Pervasive Computing and Communications (PerCom'03)*, March 2003.

33. M. C. Torrance, Advances in human-computer interaction: The intelligent room, *Working Notes of the CHI 95 Research Symp.*, 1995.

34. C. J. Watkins, *Learning from Delayed Rewards*, PhD thesis, Cambridge Univ., 1989.

35. J. Ziv and A. Lempel, Compression of individual sequences via variable rate coding, *IEEE Trans. Inform. Theory* **24**(5):530–536 (Sept. 1978).

CHAPTER 14

Enforcing Security in Mobile Networks: Challenges and Solutions

ROBERT H. DENG

School of Information System, Singapore Management University

FENG BAO, YING QIU, and JIANYING ZHOU

Institute for Infocomm Research, Singapore

14.1 INTRODUCTION

The vision of pervasive computing is about the creation and management of largely invisible digital or smart spaces where many devices interact seamlessly in wired as well as wireless environments, and where information can be managed and accessed quickly, securely, efficiently, and effortlessly anywhere, anytime. A key technology in realizing the vision of pervasive computing is mobile networking.

In mobile networking, communication and computing activities are not disrupted while the user roams from one subnet to another. Instead, all the needed reconnection occurs seamlessly. In today's Internet, the Internet Protocol (IP) routing depends on a well-ordered hierarchy. Routers deliver packets from source to destination according to the subnet prefix derived from the destination IP address by masking off some of the low-order bits. Thus, an IP address typically carries with it information that specifies the IP node's point of attachment to the Internet. The IP routing hierarchy depends on nodes that remain fixed to a subnet and on subnets that don't move between larger networks. If a node is unplugged from one subnet and reconnected to another, it will lose its old IP address and get a new one. This is not a problem if the user is willing to log off and on whenever his/her computer changes point of attachment to the Internet. In fact, this phenomenon has been referred to as the "road warrior" scenario, not mobile networking [1]. From a user's perspective, truly mobility translates to the ability to be reachable

Mobile, Wireless, and Sensor Networks: Technology, Applications, and Future Directions
Edited by Rajeev Shorey, Akkihebbal L. Ananda, Mun Choon Chan, and Wei Tsang Ooi
Copyright © 2006 John Wiley & Sons, Inc.

in spite of movement across networks, as well as an ability to maintain existing communication during such movements. Confident access to the Internet anytime, anywhere will help free us from the ties that bind us to our desktops. Consider how cellular phones have given people new freedom in carrying out their work. Taking along an entire computing environment in one's pocket has the potential not just to extend that flexibility but to fundamentally change the existing work ethic. Having the Internet available to us as we move will give us the tools to build new computing environments wherever we go.

However, risks are inherent in any mobile networking technology. Some of these risks are similar to those of fixed networks, some are exacerbated by the underlying wireless connectivity, and some are entirely new. If mobile networks are to succeed in the commercial world, their security aspect naturally assumes paramount importance. There is a need to devise security solutions to prevent attacks that jeopardize the secure network operation.

Two related security services, mutual entity authentication and network access control, are particularly prominent in mobile and wireless environments:

Mutual Entity Authentication. A network wants to ensure that it is communicating with a genuine mobile node; otherwise, there is a danger that a spurious node will be able to fraudulently gain a level of service without ever intending to pay for the service. Authentication of the network to the mobile node is also necessary in order to prevent a type of man-in-the-middle attack as described by Mishra and Arbaugh [2].

Access Control. Only authorized mobile nodes can obtain access to the network. The airwave of the underlying wireless access network is openly exposed to intruders, making it the logical equivalent of placing an Ethernet port in the parking lot.

The mobility solution consists of support for roaming, which provides "always on" global reachability, and support for redirection of traffic, which provides existing session continuity. Both roaming and traffic redirection introduces new avenues for a hacker to launch various attacks, in particular the so-called redirect attacks, which redirect the user's traffic to locations chosen by attackers [1]. The lack of security infrastructure implies that there is no central authority, which can be referenced when it comes to making trust decisions about other parties in the network and that accountability cannot be easily implemented. The transient relationships in mobile networking do not help in building trust based on direct reciprocity and give additional incentives for nodes to cheat.

The IETF's Mobile IP (MIP) specifications support mobile networking by allowing a mobile node to be addressed by two IP addresses, a home address and a "care of" address. The former is an IP address assigned to the mobile node within its subnet prefix on its home subnet, and the latter is a temporary address acquired by the mobile node while visiting a foreign subnet. The dual-address mechanism in MIP allows packets to be routed to the mobile node regardless of its current point of attachment, and the movement of the mobile node away from its home subnet is transparent to transport and higher-layer protocols. MIP version 4 (MIPv4) was specified by Perkins [3], and the most recent specification for MIP version 6 (MIPv6) was published by the IETF Mobile IP Working Group [4]. How to prevent redirect attacks in both MIPv4 and MIPv6 have proved to be technically very difficult. We feel that understanding the security issues of MIP in general, and the redirect attacks in particular, will enable the reader gain a good appreciation of the security challenges in mobile networking.

The remainder of the chapter is devoted to MIP security. In order to keep the presentation compact and without loss of generality, we will focus our discussion on MIPv6. In Section 14.2, we briefly overview the operation in MIPv6 and detail the types of redirect attacks. In Section 14.3, we introduce some basic concepts and terminologies in cryptography that will be used often in the following sections. Section 14.4 reviews in detail three techniques for securing binding update: the return routability (RR) protocol [4,6], the cryptographically generated addresses (CGA) protocol [7,8], and the home agent proxy (HAP) [9] protocol. In Section 14.5, we compare the three protocols in terms of security, performance, and scalability. Finally, we conclude the chapter by pointing out possible future research directions.

14.2   OPERATION AND REDIRECT ATTACKS IN MOBILE IPV6

14.2.1   Mobile IPv6 Operation

In MIPv6 [4], every mobile node has a home address (HoA), an IP address assigned to a mobile node within its home subnet. A mobile node is always addressable by its home address, whether it is currently attached to its home subnet or is away from home. While a mobile node is at home, packets addressed to its home address are routed using the normal IPv6 routing mechanisms in the same way as if the node were never mobile. Since the subnet prefix of a mobile node's home address is the subnet prefix of its home subnet, packets addressed to it will be routed to its home subnet.

While a mobile node is away from home and attached to some foreign subnet (see Fig 14.1), it is also addressable by one or more care-of addresses ($CoAs$), in addition to its home address. A care-of address is an IP address associated with a mobile node while visiting a particular foreign link. The subnet prefix of the mobile node's care-of address is the subnet prefix on the foreign subnet being visited by the node. A mobile node typically acquires its CoA through stateless [10] or stateful (e. g., DHCPv6 [11]) address autoconfiguration. While on the foreign subnet, the mobile node registers its CoA with its home agent by sending a *home binding update* message to the agent:

$$BU_{HA} = \{CoA, HAA, HoA | LT, \ldots\}$$

where CoA and HAA (the IP address of the home agent) are the source and destination addresses of the message. The home binding update message creates an

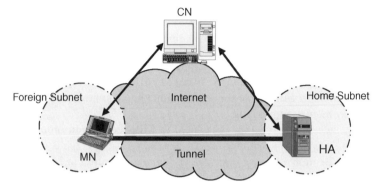

Figure 14.1 Basic operation in MIPv6.

association between HoA and CoA for the mobile node with lifetime LT at the home agent. The home agent thereafter uses proxy *neighbor discovery* to intercept any IPv6 packets addressed to the mobile node's HoA on the home subnet, and tunnels each intercepted packet to the mobile node's CoA [4]. To tunnel intercepted packets, the home agent encapsulates the packets using IPv6 encapsulation, with the outer IPv6 header addressed to the mobile node's CoA.

A mobile node may at any time initiate route optimization operation with a correspondent node by sending a correspondent binding update message to the correspondent node:

$$BU_{CN} = \{CoA, \ CNA, \ HoA \mid LT, \ \ldots\}$$

where CNA is the IP address of the correspondent node and is used as the destination address in this message. The correspondent binding update message allows the correspondent node to dynamically learn and cache the mobile node's current CoA. When sending a packet to the mobile node, the correspondent node checks its cached bindings for an entry for the packet's destination address. If a cached binding for this destination address is found, the node uses an IPv6 routing header [12] to route the packet to the mobile node by way of the CoA indicated in this binding. If, instead, the correspondent node has no cached binding for this destination address, the node sends the packet normally (i.e., to the mobile node's home address with no routing header), and the packet is subsequently intercepted and tunneled to the mobile node by its home agent as described above. Therefore, route optimization allows a correspondent node to communicate directly with the mobile node, avoiding delivery of traffic via the mobile node's home agent.

14.2.2 Redirect Attacks

During the rest of the chapter we focus on redirect attacks and their countermeasures in MIPv6. Security issues such as network access control, traffic confidentiality, and traffic integrity are beyond the scope of MIPv6 and therefore are not

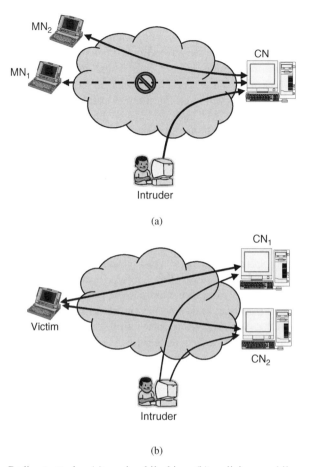

Figure 14.2   Redirect attacks: (a) session hijacking; (b) malicious mobile node flooding.

discussed here. It is apparent that the binding update operation, if implemented as described in Section 14.2.1, would introduce serious new security vulnerabilities. Unauthenticated binding updates are subject to the so-called redirect attacks, specifically, malicious acts to redirect traffic from correspondent nodes to locations chosen by an intruder through forging, replaying, and modification of binding update messages. We classify redirect attacks into two categories, session hijacking and malicious mobile node flooding, as depicted in Figure 14.2.

1. *Session Hijacking.* In the session hijacking redirect attack shown in Figure 14.2a, assume that a mobile node MN_1 is communicating with a correspondent node *CN*. An intruder sends a forged binding update message (or replays an old binding update message) to *CN*, claiming that MN_1 has moved to a new care-of address belonging to a node MN_2. If *CN* accepts the fake binding update, it will redirect to MN_2 all packets that are intended for MN_1. This

attack allows the intruder to hijack ongoing connections between MN_1 and CN or start new connections with CN pretending to be MN_1. This is an "outsider" attack since the intruder tries to redirect other nodes' traffic. Such an attack may result in information leakage, impersonation of the mobile node MN_1, or flooding of MN_2. This attack is serious because MN_1, MN_2, CN and the intruder can be any nodes anywhere on the Internet. All the intruder needs to know is the IP addresses of MN_1 and CN. Since there is no structural difference between a mobile node home address and a stationary IP address, the attack works as well against stationary Internet nodes as against mobile nodes. The threat of this attack caused IEFT to halt the MIPv6 process until a solution for authenticating binding updates was found. It is believed that deployment of the binding update protocol without security could result in a breakdown of the entire Internet [6].

2. *Malicious Mobile Node Flooding.* In the malicious mobile node flooding attack depicted in Figure 14.2b, an intruder, namely, a malicious mobile node, sends valid binding update messages to its correspondent nodes CN_1 and CN_2, claiming that it has moved to the victim's location. Here the victim can be either a node or a network. For example, the intruder could initiate requests to videostreaming servers, and flood the victim's node or network by redirecting traffic from the video servers to the victim. This is an "insider" attack since the malicious mobile node is a legitimate mobile node and its actions are "legal" binding update operations. The consequence of this attack is also grave because it can be used by any node to flood any victim node. It provides hackers with a convenient and yet powerful tool to launch DoS or DDoS attacks.

We note that, instead of targeting correspondent nodes, these attacks apply equally to home agents of mobile nodes. Specifically, by sending forged or malicious binding update messages to a mobile node's home agent, an intruder can redirect traffic intended to the mobile node to a location of its choice.

14.3 CRYPTOGRAPHIC PRIMITIVES

Before discussing countermeasures to redirect attacks, we review the following cryptographic primitives to be used throughout the chapter:

One-Way Hash Function. A hash function takes a variable-length input string and converts it to a fixed-length output string, called a *hash value*. A one-way hash function, denoted as $h()$, is a hash function that works in one direction; it is easy to compute a hash value $h(m)$ from a preimage m; however, it is computationally infeasible to find a preimage that hashes to a particular hash value. Examples of widely used one-way hash functions are MD5 [13] and SHA [14].

Keyed Pseudorandom Function. A keyed pseudorandom function, denoted as $prf(k,m)$, accepts a secret key k and a message m, and generates a pseudorandom output that is computationally infeasible to distinguish from a true random sequence for anyone who does not know the secret key k. This function is often implemented using a keyed one-way hash function [15] and is used for both generation of message authentication codes and derivation of cryptographic keys.

Digital Signature Scheme. A digital signature scheme is a cryptographic tool for generating nonrepudiation evidence, authenticating the integrity of a signed message as well as its origin. In a digital signature scheme, an entity X has a public key P_X and private key S_X. To digitally sign on a message m, X uses a signature generation function to compute a signature $s = \sigma(S_X, m)$ on m. Any entity wishing to verify the authenticity of m first obtains an authentic copy of X's public key P_X and then checks the validity of the signature s using a verification function $v(P_X, m, s)$, which gives a yes/no output depending on the validity of the signature. Examples of well-known digital signature schemes are RSA [16] and DSS [17].

Public Key Certificate. A public key certificate is a data structure, digitally signed by a certification authority (CA), used to identify an entity (e.g., a user, an IP node, a server, a router) and to associate the entity with a public key. The public key certificate of an entity X is denoted as $Cert_X = \{X, P_X, VI, SIG_{CA}\}$, where P_X is X's public key, VI is the valid interval of the certificate, and SIG_{CA} is CA's signature on $\{X, P_X, VI\}$. This certificate attests that entity X is the one associated with the public key P_X.

14.4 PROTOCOLS FOR AUTHENTICATING BINDING UPDATE MESSAGES

Obviously, countermeasures to redirect attacks presented in Section 14.2.2 are to authenticate binding update messages. IETF [2] assumes that mobile nodes and home agents know each other, and thus have a preestablished security association between them. A *security association* is a data record shared by two entities that includes mutually agreed cryptographic algorithms and parameters (e.g., secret keys). The MIPv6 specification [4] stipulates that the IPsec's encapsulating security payload (ESP) [18] be used to set up a secure tunnel between a mobile node and its home agent. The secure tunnel protects home binding updates sent from the mobile node to its home agent as well as all other messages exchanged between the two entities. Therefore, authenticating home binding update messages, that is, binding update messages from a mobile node to its home agent, is straightforward.

In the following text we focus on protecting correspondent binding update messages, that is, binding update messages from a mobile node to its correspondent nodes. This has been a challenging problem and has received considerable attention in the mobile IP research community. It is expected that MIPv6 will be used on a

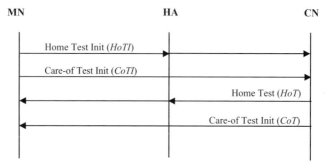

Figure 14.3 Return routability protocol.

global basis between nodes belonging to different administrative domains; therefore, it is only practical and realistic to assume that there is no preestablished security association between a mobile node and a random correspondent node. We present three representative protocols for authenticating correspondent binding update messages:the return routability (RR) protocol, the cryptographically generated addresses (CGA) protocol, and the home agent proxy (HAP) protocol. For each protocol, we first describe its operations and then discuss its security and performance.

14.4.1 Return Routability Protocol

Protocol Operation In IETF's Return Routability (RR) protocol [4], a correspondent node CN keeps a secret key k_{CN} and generates a nonce at regular intervals, say, every few minutes. CN uses the same key k_{CN} and nonce with all the mobile nodes with which it is in communication, so that it does not need to generate and store a new nonce when a new mobile node contacts it. Each nonce is identified by a nonce index. When a new nonce is generated, it must be associated with a new nonce index; for example, j. CN keeps both the current value of N_j and a small set of previous nonce values, N_{j-1}, N_{j-2}, and so on. Older values are discarded, and messages using them will be rejected as replays. Message exchanges in the RR protocol are shown in Figure 14.3, where the $HoTI$ (home test init) and $CoTI$ (care-of test init) messages are sent to CN by a mobile node MN simultaneously. The HoT (home test) and CoT (care-of test) are replies from CN. All RR protocol messages are sent as the IPv6 "mobility header" in IPv6 packets. In the representation of a protocol message, we will use the first two fields to denote source IP address and destination IP address, respectively. We will use CNA to denote the IP address of the correspondent node CN.

When MN wants to perform route optimization, it sends

$$HoTI = \{HoA, CNA, r_H\}$$

and

$$CoTI = \{CoA, CNA, r_C\}$$

to CN, where r_H and r_C are random values used to match responses with requests. $HoTI$ tells MN's home address HoA to CN. It is reverse tunneled through the home agent HA, while $CoTI$ informs MN's care-of address CoA and is sent directly to CN.

When CN receives $HoTI$, it takes the source IP address of $HoTI$ as input and generates a *home cookie*

$$C_H = prf(k_{CN}, HoA \,|\, N_j \,|\, 0)$$

and replies to MN with

$$HoT = \{CNA, HoA, r_H, C_H, j\},$$

where | denotes concatenation and the final 0 inside the pseudorandom function is a single zero octet, used to distinguish home and care-of cookies from each other. The index j is carried along to allow CN later efficiently finding the nonce value N_j that it used in creating the cookie C_H. Similarly, when CN receives $CoTI$, it takes the source IP address of $CoTI$ as input and generates a *care-of cookie*

$$C_C = prf(k_{CN}, CoA \,|\, N_i \,|\, 1)$$

and sends

$$CoT = \{CNA, CoA, r_C, C_C, i\}$$

to MN, where the final 1 inside the pseudorandom function is a single octet 0×01. Note that HoT is sent via MN's home agent HA while CoT is delivered directly to MN.

When MN receives both HoT and CoT, it hashes together the two cookies to form a session key

$$k_{Bu} = h(C_H | C_C),$$

which is then used to authenticate the correspondent binding update message to CN

$$BU_{CN} = \{CoA, CNA, HoA, Seq\#, i, j, MAC_{BU}\},$$

where $Seq\#$ is a sequence number used to detect replay attack and

$$MAC_{BU} = prf(k_{BU}, CoA \,|\, CNA \,|\, HoA \,|\, Seq\# \,|\, i \,|\, j)$$

is a message authentication code (MAC) protected by the session key k_{BU}. MAC_{BU} is used to ensure that BU_{CN} was sent by the same node that received both HoT and CoT. The message BU_{CN} contains j and i, so that CN knows which nonce values N_j and N_i to use to first recompute C_H and C_C and then the session key k_{BU}. Note that CN is stateless until it receives BU_{CN} and verifies MAC_{BU}. If MAC_{BU} is verified positive, CN may reply with a binding acknowledgement message

$$BA = \{CNA, CoA, HoA, Seq\#, MAC_{BA}\},$$

where $Seq\#$ is copied from the BU_{CN} message and

$$MAC_{BA} = prf(k_{BU}, CNA \,|\, CoA \,|\, HoA \,|\, Seq\#)$$

Entry for MN: HoA, CoA, Seq#	k_{CN}, N_j, N_{j-1}, N_{j-2}
Entries for other mobile nodes	

Figure 14.4   A binding cache implementation at CN in the RR protocol.

is a MAC generated using k_{BU} to authenticate the BA message. CN then creates a binding cache entry for the mobile node MN. The binding cache entry binds HoA with CoA, which allows future packets to MN be sent to CoA directly. Bindings established with correspondent nodes using keys created by way of the RR protocol is limited to a maximum of 420 s [4].

An example implementation of the binding cache at CN is shown in Figure 14.4, where HoA is used as an index for searching the binding cache for the destination address of a packet being sent and the sequence number $Seq\#$ is used by CN to sequence binding updates and by MN to match a return binding acknowledgment with a binding update. Each binding update sent by MN must use a $Seq\#$ greater than (modulo 2^{16}) the one sent in the previous binding update by the same HoA. There is no requirement, however, that the sequence number value strictly increase by 1 with each new binding update sent or received [4]. Note that the session key k_{BU} is not kept in the cache entry. When CN receives a binding update message, based on the nonce indexes i and j in the message, it recomputes the session key using k_{CN} and the list of the most recent nonce values, say, $\{N_j, N_{j-1}, N_{j-2}\}$, and then verifies BU_{CN} using the newly computed session key.

The mobile node MN maintains a binding update list for each binding update message sent by it, for which the lifetime has not yet expired. A binding update list for a correspondent node CN consists of CN's IP address, MN's home address HoA and care-of address CoA, the remaining lifetime of the binding, the maximum value of the sequence number sent in previous binding updates to CN, and the session key k_{BU}.

Discussion   In the RR protocol, the two cookie exchanges verify that a mobile node MN is alive at its addresses, that is, is at least able to transmit and receive traffic at both its home address HoA and care-of address CoA, respectively. The eventual binding update is cryptographically protected with the session key k_{BU} obtained by hashing the concatenation of the two cookies C_H and C_C. Obviously, the RR protocol protects binding updates against intruders who are unable to monitor the $HA-CN$ path and the $MN-CN$ path simultaneously.

The IETF MIPv6 documents [4,5] stated that the motivation for designing the RR protocol was to have sufficient support for mobile IP, without creating major new security problems. It was not the goal of the Mobile IP Working Group to protect against attacks that were already possible before the introduction of IP mobility. The protocol does not defend against an intruder who can monitor the $CN-HA$ path. The argument was that such intruders would in any case be able to mount an active attack against MN when it is at its home location.

However, the design principle of the RR protocol, specifically, defending against an intruder who can monitor the $CN-MN$ path but not the $CN-HA$ path, is flawed since it violates the well known "weakest link" principle in security. After all, one has no reason to assume that an intruder will monitor one link and not the other, especially when the intruder knows that monitoring a given link is particularly effective to expedite its attack. Although intruders are in fact able to mount active attacks when a node is at home in the static IPv6, we demonstrate below that it is much easier to launch redirect attacks in MIPv6 than in the static IPv6.

First, let's consider the session hijacking attack shown in Figure 14.2a. In the case of the static IPv6 without mobility (which is equivalent to the mobile node MN at its home subnet in MIPv6), to succeed in the attack, the intruder must be constantly present on the $CN-HA$ path. In order to redirect CN's traffic intended for MN to a malicious node, the intruder most likely has to gain control of a router or a switch along the $CN-HA$ path. Furthermore, after taking over the session from MN, if the malicious node wants to continue the session with CN while pretending to be MN, the malicious node and the router need to collaborate throughout the session. For example, the router tunnels CN's traffic to the malicious node and vice versa.

In the case of MIPv6, the effort committed to break the RR protocol to launch a session hijacking attack could be considerably lesser. Assume that MN_1 and CN in Figure 14.2a are having an ongoing communication session and the intruder wants to redirect CN's traffic to his collaborator MN_2. The intruder monitors the $CN-HA$ path (i.e., anywhere from MN_1's home network to CN's network) to obtain HoT, extracts the home cookie C_H, and sends it to MN_2. On receiving C_H, MN_2 sends a $CoTI$ to CN, and CN will reply with a care-of cookie C_C. MN_2 simply hashes the two cookies to obtain a valid session key, and uses the key to send a binding update message to CN on behalf of MN_1. The binding update will be accepted by CN, which will in turn direct its traffic to MN_2.

Another related attack is when a mobile node MN rapidly moves from one care-of address CoA to another CoA'. Since MN runs the RR protocol whenever it moves to a new location, an intruder can intercept the care-of cookie in the current RR session and the home cookie in the next RR session, hash the two cookies, and send a binding update message with the CoA in the current session to the correspondent node. The correspondent node will send its traffic back to CoA. Hence, MN, which has moved to CoA', will not receive data from the correspondent node. Note that in this attack the intruder does not have to intercept the two cookies at the "same time."

The RR protocol is also subject to a "traffic permutation" attack. Consider Figure 14.5, where a correspondent node provides online services to many mobile clients. An intruder can simply eavesdrop on the RR protocol messages to collect cookies on the border between the correspondent node and the Internet. The intruder then hashes random pairs of cookies to form session keys, and sends binding update messages to the correspondent node. Such a forged binding update message will be accepted by the correspondent node with probability $\frac{1}{4}$. This will cause

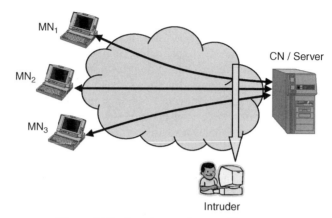

Figure 14.5   Intruder attacks an online server.

redirection of traffic to randomly selected mobile clients and eventually bring down the services of the correspondent node.

The attacks outlined in the preceding three paragraphs are due to the decoupling of HoA and CoA in RR messages. In the original RR protocol, the home cookie $C_H = prf(k_{CN}, HoA|N_j | 0)$ and the care-of cookie $C_C = prf(k_{CN}, CoA|N_i|1)$ are delivered without any stated relationship. Any pair of home cookie C_H and care-of cookie C_C can generate a valid session key k_{BU} as long as the indexes, i and j, are still valid.

However, these attacks can be prevented by modifying the RR protocol to include both CoA and HoA in the generation of home cookie and care-of cookie, respectively. In the following improved RR protocol, HoA and CoA are bound together. A mobile node MN sends $HoTI = \{HoA, CNA, CoA, r_H\}$ and $CoTI = \{CoA, CNA, HoA, r_C\}$ to a correspondent node CN, which replies with the home cookie $C_H = prf(k_{CN}, HoA|N_j|CoA| 0)$ and the care-of cookie $C_C = prf(k_{CN}, CoA|N_i|HoA| 1)$.

Next, consider the malicious mobile node flooding attack shown in Figure 14.2b. In the static IPv6 without mobility, perhaps the best example of flooding attack is the DDoS attack, in which a multitude of compromised systems attack a single target. There are many ways to launch a malicious mobile node flooding attack against a victim (which can be either a node or a network) in MIPv6. For example, the malicious node starts some traffic-intensive sessions with correspondent nodes and moves to the victim's network or the border between the victim network and the outside world. It then runs the RR protocol to redirect traffic from the correspondent nodes to the victim's network by sending them binding update messages. The malicious mobile node does not need any special software or networking skill to launch this attack.

Finally, we point out that the IETF MIPv6 specification limits the lifetime of an RR-authorized binding to a maximum of 420 s [4]. This will have performance implications. Imagine having a time-sensitive session between a mobile node and a correspondent node where the mobile node must perform the RR protocol every 420 s or less. Quality of communication will suffer if the RR protocol cannot be executed in a timely manner because of congestion or malfunction of the home agent, home subnet, or the $CN-HA$ path.

14.4.2 Cryptographically Generated Addresses Protocol

An IPv6 address consists of 128 bits and is divided into two portions: a subnet prefix and an interface identifier. The home addresses of all the mobile nodes associated with a home link share the same home link subnet prefix and are differentiated by their unique interface identifiers. The CGA protocol [7,8] generates an IPv6 home address for a mobile node where the interface identifier portion is created from a one-way hash of the mobile node's public key. The mobile node uses the corresponding private key to sign correspondent binding up messages.

Protocol Operation Each mobile node MN has a public/private key pair P_{MN} and S_{MN} in a digital signature scheme. MN's home address is given by $HoA = \{HL|II\}$, where HL is the n-bit home link subnet prefix and II is the $(128-n)$-bit interface identifier. The II field is obtained by taking the leftmost $(128-n)$ bits of the hash function output $h(P_{MN})$. A binding update message from MN to a correspondent node CN is given by

$$BU = \{CN,\ CoA,\ HoA,\ Seq\#,\ P_{MN},\ 128-n,\ SIG_{MN}\}$$

where

$$SIG_{MN} = \sigma(S_{MN},\ CoA|CN|HoA|Seq\#|P_{MN}|128-n)$$

is MN's digital signature generated using its private key S_{MN}. On receiving the BU, the correspondent node CN computes $h(P_{MN})$, compares the leftmost $(128-n)$ bits of $h(P_{MN})$ with the rightmost $(128-n)$-bit II in HoA, and verifies the signature using the public key P_{MN}. If the hash value matches the value of II and if the signature verification is positive, CN accepts the binding update message.

Discussion The hash function $h()$ here acts as a "one to one" mapping from a public key value to an interface identifier; it binds a public key value with an interface identifier. Since it is computationally difficult to either find the private key or forge a digital signature given the public key, a match of $h(P_{MN})$ with II in HoA as well as positive verification of the signature on BU proves that BU was generated by the mobile node whose interface identifier portion is II and who knows the private key S_{MN}. This is the only assurance a correspondent node gets from BU. As a consequence, the protocol is able to provide good protection against the session hijacking attack provided the number of bits in II, $(128-n)$, is large enough. If $(128-n)$ is small, an intruder can randomly generate pairs of public and private keys, hash the public keys, and look for a match to a target node's II. Once a match is found, the intruder is able to impersonate the target node and forge-binding updates. The computational complexity of this brute-force attack is on the order of $o(2^{(128-n)})$. A clever method [7] effectively removes the $(128-n)$-bit limit on the hash length by artificially increasing both the cost of generating a new CGA address and the cost of a brute-force attack while keeping the cost of CGA-based authentication constant. The interested reader can refer to the paper by Aura [8] for technical details.

On the other hand, since this protocol does not provide any proof of the authorization of *MN* to use the particular *HoA*, it is not able to protect against the malicious mobile node flooding attacks. Actually, an intruder can simply generate a public/private key pair, hashs the public key to form a home address, sign a binding update message that contains a victim's address as *CoA*, and send it to a correspondent node. The correspondent node will accept the binding update and start sending traffic to flood the victim node.

Compared with the RR protocol, the CGA protocol is computationally intensive since every binding update message requires the mobile node to generate a digital signature and the correspondent node to perform a verification of digital signature.

14.4.3 Home Agent Proxy Protocol

The HAP protocol [9] employs public key cryptosystems in order to provide strong security and good scalability. There are two important design considerations in protocols using public key cryptosystems. The first is performance, since public key cryptosystem operations are computationally intensive. Portable devices with constraint computational power, such as PDAs and cellular phones, are predicted to account for a majority or at least a substantial fraction of the population of mobile devices. It is crucial to keep the number of public key cryptosystem operations in mobile devices to the absolute minimum. The second consideration is the mechanism used to securely bind a subject's name with its public key since they have significant impact on the entire system architecture and operation. Such a binding is typically achieved using public key certificates issued by a trusted certification authority, or *CA* for short. A public key certificate at the minimum consists of a subject's name, its public key, valid time interval, and *CA*'s digital signature on the data items mentioned above. In the MIPv6 environment, a mobile node could be issued a public key certificate with its home address as the subject's name. However, having public key certificates with IP addresses as subjects' names is not recommended practice for several reasons:

1. IP addresses are often obtained by DNS (Directory Name Service) lookup and DNS does not provide a secure way of mapping names to IP addresses.
2. IP addresses are subject to renumbering, both when service providers change and when configurations change, so they may not be as persistent as other subject names (e.g., domain names) [19].

3. IP addresses are leased to an interface for a fixed length of time. When an IP address's lease time expires, the association of the address with the interface becomes invalid and the address may be reassigned to another interface elsewhere in the Internet. There might be various reasons for keeping IP addresses' lease time short, such as for privacy protection. For devices that function as client devices, Narten and Draves [20] recommend changing their IP addresses periodically to prevent eavesdroppers and other information collectors from correlating the clients' seemingly unrelated activities over an extended period of time.

Therefore, it is very difficult in practice for CAs to keep track of correct associations between IP addresses and all devices' interfaces in a consistent and timely manner, not to mention issuing and revoking public key certificates for them. Subnet prefixes for home links, however, are much more tractable and manageable because (1) a home subnet prefix is normally much more persistent than a mobile node's home address, (2) the number of home links is significantly smaller than the number of mobile nodes, and (3) subnet prefixes are managed by system administration staff who can much more efficiently keep track prefix changes than of which IP address is associated with which individual mobile node. Motivated by these observations, the HAP protocol is designed to possess the following features:

1. It performs one-way authenticated key exchange between MN and CN, where MN authenticates itself to CN and the exchanged session key is used to secure binding update messages from MN to CN.
2. It employs public key cryptosystems and is secure against any powerful adversary who is able to launch both passive (e.g., eavesdropping at multiple points) and active (e.g., man-in-the-middle) attacks.
3. It is easy to manage and scalable. Instead of issuing public key certificates containing home addresses as subject names for individual mobile nodes, we issue public key certificates containing home subnet prefixes as subject names for home links.
4. No public key cryptographic operations are performed at mobile nodes. MIPv6 assumes that home agents are trusted by mobile nodes as well as correspondent nodes and that communications between mobile nodes and their home agents are protected with preestablished security associations; home agents function as trusted security proxies for mobile nodes in the protocol. They verify the legitimacy of mobile nodes' home addresses, facilitate authentication of mobile nodes to correspondent nodes, and establish shared secret session keys for them.

System Setup   A home subnet is associated with a public/private key pair P_H and S_H in a digital signature scheme. The private key S_H is kept by a home agent HA in the home link, probably inside in a tamperproof hardware cryptographic processing device. The home subnet obtains a public key certificate

$$Cert_H = \{HS, P_H, VI, SIG_{CA}\}$$

from a certification authority CA, where HS is the home subnet prefix, VI is the valid duration of the certificate, and SIG_{CA} is CA's signature on HS, P_H, and VI. The protocol also uses the Diffie–Hellman key exchange algorithm to arrive at a mutual secret value between parties of the protocol. Let p and g be the public Diffie–Hellman parameters, where p is a large prime and g is a generator of the multiplicative group $Z_p{}^*$. To keep the notation compact, we will write g^x mod p simply as g^x. Since generation of large primes in real time can be very time-consuming, we assume that the values of p and g are agreed on beforehand by all the parties concerned or are embedded in $Cert_H$.

Protocol Operation   As in the RR protocol, all the protocol messages in HAP are carried within IPv6 "mobility header," which allows protocol messages to be piggybacked on any existing IPv6 packets. The protocol messages exchanged among a mobile node MN, its home agent HA, and its correspondent node CN are shown in Figure 14.6. In the protocol, the existence of and operations performed by HA are transparent to both MN and CN. As far as MN is concerned, it sends message REQ to and receives REP from CN. Similarly, from CN's perspective, it receives $COOKIE0$, $EXCH0$, and $CONFIRM$ from and sends $COOKIE1$ and $EXCH1$ to MN.

The use of cookies during the key exchange is a weak form of protection against an intruder who generates a series of request packets, each with a different spoofed source IP address and sends them to a protocol party. For each request, the protocol party will first validate cookies before performing computationally expensive public key cryptographic operations. For details on cookie generation and validation, please refer to the report by Karn and Simpson [21].

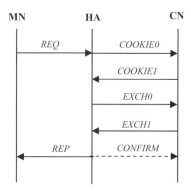

Figure 14.6   Message exchange in the proposed protocol.

As before, the first two fields in a protocol message are the source IP address and destination IP address, respectively. When MN wants to start route optimization operation with CN, it sends a route optimization request

$$REQ = \{HoA, CNA, n_0\}$$

to CN via reserve tunneling, where n_0 is a nonce value used to match the reply message REP and CNA is the IP address of the correspondent node CN. Message REQ is sent to MN's home subnet via the IPsec-protected secure tunnel. IPsec provides replay protection only when dynamic security association establishment is used. This may not always be possible, and manual keying might be preferred in certain circumstances. For this reason, we have included n_0 to counter message replay. On arriving at the home link, REQ is intercepted by HA using IPv6 "neighbor discovery" [4,22]. HA will not forward REQ to CN; instead, it creates a cookie C_0 and sends

$$COOKIE0 = \{HoA, CNA, C_0\}$$

to CN. In reply, CN creates a nonce n_1 and a cookie C_1, and sends

$$COOKIE1 = \{CNA, HoA, C_0, C_1, n_1\}$$

to MN. Note that the destination address in $COOKIE1$ is MN's home address HoA. As a result, this message is delivered to MN's home subnet and intercepted by HA using IPv6 neighbor discovery. After receiving $COOKIE1$, HA checks on the validity of C_0, generates a nonce n_2 and a Diffie–Hellman secret value $x < p$, computes its Diffie–Hellman public value g^x and its signature

$$SIG_H = \sigma(S_H, HoA|CNA|g^x|n_1|n_2|TS)$$

using home link's private key S_H, where TS is a timestamp. This timestamp does not have to be checked by the recipient during the message exchange. It will be used to trace back the culprit in the event that a malicious mobile node flooding attack occurred. This point will be further clarified later. Finally, HA replies to CN with

$$EXCH0 = \{HoA, CNA, C_0, C_1, n_1, n_2, g^x, TS, SIG_H, Cert_H\},$$

where $Cert_H = \{HS, P_H, VI, SIG_{CA}\}$ is the public key certificate of the home subnet as defined before. Note that the values of n_1 and n_2 are included in the signature SIG_H in order to counter replay of old signatures and to resist chosen message attacks to the signature scheme, respectively.

When CN receives $EXCH0$, it validates the cookies, the home link's public key certificate $Cert_H$, the signature, and, perhaps more importantly, checks for equality of the home subnet prefix strings embedded in both $Cert_H$ and HoA. If all the validations and checking are positive, CN can be confident that the home address HoA of MN is authorized by its home subnet and that the Diffie–Hellman public vaule g^x is freshly generated by MN's home subnet. CN next generates its Diffie–Hellman secret value $y < p$. It then computes its Diffie–Hellman public value g^y, the Diffie–Hellman key $k_{DH} = (g^x)^y$, a session key

$$k_{BU} = prf(k_{DH}, n_1 | n_2)$$

and a MAC

$$MAC_1 = prf(k_{BU}, g^y | EXCH0)$$

and sends

$$EXCH1 = \{CNA, HoA, C_0, C_1, g^y, MAC_1\}$$

to MN. Again, this message is intercepted by HA, which first validates the cookies, calculates the Diffie–Hellman key $k_{DH} = (g^y)^x$ and the session key $k_{BU} = prf(k_{DH}, n_1|n_2)$. HA then computes

$$MAC_2 = prf(k_{BU}, EXCH1)$$

and sends

$$CONFIRM = \{HoA, CNA, MAC_2\}$$

to CN. The validity of MAC_2 is checked by CN, and if it is valid, CN creates a cache entry for HoA and the session key k_{BU}, which will be used for authenticating binding update messages from MN.

On positive verification of MAC_1, HA also sends

$$REP = \{CNA, HoA, n_0, k_{BU}\}$$

to MN through the secure IPsec ESP–protected tunnel. After receiving REP, MN checks that n_0 is the same as the one it sent out in REQ. If so, MN proceeds to send CN binding update messages protected using k_{BU} as in the RR protocol. It should be noted that the $CONFIRM$ message serves to confirm the key to CN and hence is optional.

Discussion A misconception among some people is that public key cryptography–based security solutions over the Internet necessitate the existence of a global *public key infrastructure* (PKI). A living counterexample of this is the extensive deployment of the *secure socket layer* (SSL) [19]. In the SSL protocol, a SSL-enabled Web server is authenticated to SSL-aware browsers, proving its identity at each SSL connection. This proof of identity is conducted through the use of a public/private key pair by the server where the public key is validated with a X.509 public key certificate issued by a CA. Under the SSL architecture, Web server authentication can be the only validation performed, which may be all that is needed in some circumstances. This is applicable for those applications where

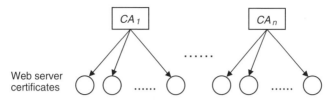

Figure 14.7 The fragmented PKI model used in SSL.

the user needs to be assured of the identity of the target Web server, such as when placing an order from an online merchant. The SSL trust model is built on a so-called fragmented PKI as shown in Figure 14.7, in which multiple independent CAs issue public key certificates directly to Web servers. CA's public keys are embedded in popular Web browsers. At the time of writing, 89 CA public keys are embedded in Microsoft's Internet Explorer version 6 and 58 CA public keys are embedded in Netscape's Browser version 7.1. In the fragmented PKI, trust as to the validity of CA public keys rests with the developers of browser software as well as with the integrity of the software.

The trust model and the design principle of the HAP protocol follow those of the SSL. In the HAP, a CN is the equivalent of a Web browser and a HA is the equivalent of a Web server. CAs issue public key certificates directly to HAs. The HAP performs a strong one-way authentication of MN|HoA to CN and provides CN with the confidence that it shares a secret session key with MN. Here we would like to point out that the most important message is $EXCH0$. Recall that after receiving $EXCH0$, CN checks on the equality of the home subnet prefix contained in both $Cert_H$ and HoA. This check is critical to detect a man-in-the-middle attack. The signature $SIG_H = S_H(HoA|CN|Alg^x|n_1|n_2|TS)$ serves two purposes: (1) it certifies that the Diffie–Hellman value g^x was originated by MN's home agent HA on behalf of MN and (2) it testifies that HoA is under HA's (or equivalently the home link's) jurisdiction and is a legitimate home address for its mobile node MN. This authenticates MN's HoA to CN.

Since a successful completion of the protocol allows CN to authenticate MN's HoA and also allows the two nodes to set up a secret session key for securing binding updates, the protocol prevents the session hijacking attack shown in Figure 14.2a. This protocol, as are any other protocols, is not able to completely prevent malicious mobile node flooding attacks. However, if a correspondent node were accused of having bombarded a network service or site, it could present the signature $SIG_H = \sigma(S_H, HoA|CN|Alg^x|n_1|n_2|TS)$ and point its fingers at the home agent HA. HA can subsequently nail down the mobile node MN that had a home address HoA and perform a binding update at the time specified by TS.

In the HAP protocol, mobile nodes are not required to perform any public key cryptographic operations, but correspondent nodes are. Public key cryptographic operations may not be a great concern if a correspondent node is a server machine. However, a correspondent node can also be a mobile node with limited computational power and battery life. In this case, public key operations supposedly

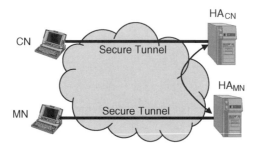

Figure 14.8   Scenario where *CN* is a mobile node.

performed by the correspondent node can be offloaded to its home agent. This scenario is depicted in Figure 14.8, where HA_{MN} and HA_{CN} are the home agents of *MN* and *CN*, respectively. Since it is assumed in MIPv6 that a mobile node has a pre-established security association with its home agent, it is logical in our protocol to have HA_{MN} and HA_{CN} perform public key cryptographic operations on behalf of *MN* and *CN*, respectively. Also, because of the symmetric arrangement of the entities, it is possible to perform a mutual authenticated key exchange between *MN* and *CN* and establish session keys to secure binding update messages in both directions.

Finally, since the HAP protocol uses strong cryptosystems, the secret session key k_{BU} established from the protocol could be used for a long period of time. This is in contrast to the RR protocol, where the protocol must be executed at least every 420 s even if the mobile node stays at the same foreign location.

14.5   CONCLUDING REMARKS AND FUTURE DIRECTION

Mobile IP allows mobile nodes to have seamless communications when they change their points of attachment in the Internet and is poised to take off in a big way in the not-too-distant future. However, introduction of mobility into IP also brought with it new security issues and attacks, including redirect attacks, which perhaps warrant the most attention.

In this chapter, we first classified redirect attacks into two types: (1) *session hijacking attacks*, where an intruder hijacks an existing session between a mobile node and a correspondent node and redirects the correspondent node's traffic to a malicious location; and (2) *malicious mobile node flooding attacks*, in which a mischievous mobile node sets up communication sessions with correspondent nodes, and then redirects traffic from the correspondent nodes to flood a victim node or network.

Next we presented and analyzed three very different protocols that are designed to secure correspondent binding updates in order to prevent redirect attack. The major advantage of RR protocol and the CGA protocol is that they do not assume the existence of an Internet wide public key infrastructure (PKI). However, they provide only limited security protection against redirect attacks.

The HAP protocol makes use of a digital signature scheme and the Diffie–Hellman key exchange algorithm, where public key certificates are not issued for each and every mobile node, but are issued for home subnets according to home subnet prefixes. Such an approach makes certificate issuing, tracking, and revocation much more practical and manageable. In HAP, a home agent functions as security proxy for its mobile nodes and testifies as to the legitimacy of a mobile node's home address to a correspondent node during protocol execution. Recognizing that most mobile nodes are constrained in processing power and battery life and that home agents can be easily equipped with increasingly low-cost yet powerful cryptographic processing hardware accelerators, the protocol was designed to off load all the expensive public key cryptosystem operations from mobile nodes to their home agents. The underlying assumption in HAP protocol is the existence of fragmented certification authorities or fragmented PKI in the Internet. This is a practical and workable assumption. In fact, the HAP follows the same trust model underlying the tremendously successful SSL protocol.

In the long term, we believe that secure and truly global-scale operation of MIPv6 demands the existence of the Internetwide PKI — the set of infrastructural services that support the wide-scale use of public-key-based digital signatures and encryption. Although there are several PKI standard efforts and many commercial PKI offerings on the market, deployment of Internetwide PKI is believed to a complex and groundbreaking undertaking. Ren et al. [23] suggest a PKI for mobile IPv6 that adopts a three-layer hierarchical trust management framework. However, further research is required to study the feasibility of the framework from security, management, operational, and performance perspectives.

Another issue beginning to receive increasing attention in the mobile networking community is location privacy. In MIPv6, mobile nodes roam from one network to another. Location-aware applications and services may take advantage of such location information to provide better services to users; however, the same location information can also be used to track users' movements and utilized against users' interests by malicious individuals or organizations. How to strike a balance between user-friendly services and user privacy is still an open research problem.

REFERENCES

1. C. E. Perkins, Mobile networking through Mobile IP, http://www.computer.org/internet/v2n1/perkins.htm.

2. A. Mishra and W. A. Arbaugh, *An Initial Security Analysis of the IEEE 802.1X Standard*, Technical Report CS-TR-4328, UMIACS-TR-2002-10, Univ. Maryland, Feb. 2002.

3. C. Perkins, *IP Mobility Support*, IETF RFC 2002, Oct. 1996.

4. D. Johnson, C. Perkins, and J. Arkko, *Mobility Support in IPv6*, IETF RFC 3775, June 2004.

5. Mankin et al., Threat models introduced by mobile Ipv6 and requirements for security in mobile Ipv6, IETF draft-ietf-mipv6-scrty-reqts-02.txt, May 2001.

6. T. Aura, Mobile IPv6 security, *Proc. 10th Int. Workshop on Security Protocols*, LNCS 2467, Cambridge, UK, April 2002.

7. G. O'Shea and M. Roe, Child-proof authentication for MIPv6 (CAM), *Comput. Commun. Rev.* (April 2001).

8. T. Aura, Cryptographically generated addresses (CGA), *Proc. 6th Information Security Conf.*, LNCS 2851, Bristol, UK, 2003.

9. R. Deng, J. Zhou, and F. Bao, Defending against redirect attacks in mobile IP, *Proc. 9th ACM Conf. Computer and Communications Security*, Nov. 2002, pp. 59–67.

10. S. Thomas and T. Narten, *IPv6 Stateless Address Autoconfiguration*, IETF RFC 2462, Dec. 1998.

11. J. Bound et al., *Dynamic Host Configuration Protocol for IPv6 (DHCPv6)*, IETF RFC 3315, July 2003.

12. S. Deering and R. Hinden, *Internet Protocol, Version 6 (IPv6) Specifications*, IETF RFC 2460, December 1998.

13. R. Rivest, *The MD5 Message Digest Algorithms*, IETF RFC 1321, April 1992.

14. NIST, *Secure Hash Standard*, NIST FIPS PUB 180, May 1993.

15. H. Krawczyk, M. Bellare, and R. Canetti, *HMAC: Keyed-Hashing for Messaging Authentication*, IETF RFC 2104, Feb. 1997.

16. R. Rivest, A. Shamir, and L. Adleman, A method for obtaining digital signatures and public-key cryptosystems, *Commun. ACM.* **21**: 120–126 (Feb. 1978).

17. NIST, *Digital Signature Standard*, NIST FIPS PUB 186, May 1994.

18. S. Kent and R. Atkinson, *IP Encapsulating Security Payload (ESP)*, IETF RFC 2406, Nov. 1998.

19. E. Rescorla, *SSL and TLS: Designing and Building Secure Systems*, Addison-Wesley, 2001.

20. T. Narten and R. Draves, *Privacy Extensions for Stateless Address Autoconfiguration in IPv6*, IETF RFC 3041, Jan. 2001.

21. P. Karn and W. Simpson, *Photuris: Session-Key Management Protocol*, IETF RFC 2522, 1999.

22. T. Narten, E. Nordmark, and W. Simpson, *Neighbor Discovery for IP Version 6 (IPv6)*, IETF RFC 2461, Dec. 1998.

23. K. Ren, W. Lou, K. Zeng, F. Bao, J. Zhou, and R. H. Deng, Routing optimization security in mobile IPv6, *Computer Networks Journal* (accepted for publication).

CHAPTER 15

On-Demand Business: Network Challenges in a Global Pervasive Ecosystem

CRAIG FELLENSTEIN, JOSHY JOSEPH, DONGWOOK LIM, and J. CANDICE D'ORSAY

IBM Global Services, Network Services (NS) Organization, Brookfield, Connecticut

15.1 INTRODUCTION

This chapter presents highlights of and resulting conclusions concerning on-demand business network challenges in a global pervasive ecosystem.

To set the stage for our discussions, let's consider the fact that the technology areas of mobile, wireless, and sensor networks have realized an exponential growth worldwide since the late 1990s; the same with pervasive devices since the mid-1990s. Let's also consider the fact that wireless LANs in many countries have become a common household practice. These types of *wireless networks* are seemingly becoming the only solution to getting broadband (e.g., DSL, cable, satellite, and utility companies) network access into urban, suburban, semirural, and rural areas of the world.

This notion of managed broadband access is also being planned and further considered in many industries, such as the automotive industry (e.g., Honda, Hyundai, Fords). Automotive companies provide network access systems such as operational in-dash Internet access, conversational navigation systems, email and messaging systems, and life/safety monitoring systems — to name just a few. Virtually all vertical industries, worldwide, are affected by the discussions presented in this chapter.

Few broadband managed services providers today exist with strategies to cover all broadband markets and vertical industry segments. The solutions in this chapter will cross all industries, and the companies in these industries. A company called

Mobile, Wireless, and Sensor Networks: Technology, Applications, and Future Directions
Edited by Rajeev Shorey, Akkihebbal L. Ananda, Mun Choon Chan, and Wei Tsang Ooi
Copyright © 2006 John Wiley & Sons, Inc.

Vision Media Technologies, Inc.[1] is involved in enabling American Tribal Nations, and also has strategic plans for deployment of many of these advanced types of devices and network services solutions; their view of commerce is to provide vehicles for economic growth and transformation. Advancing technologically savvy countries such as India, China, and South Vietnam are also realizing how such innovative networks can significantly contribute toward any country's economic growth and visionary progress.

The subjects discussed in this chapter apply to virtually all types of cultural endeavors, whether it is cross-industry solutions, cross-government endeavors, worldwide Tribal Nation solutions, and/or public-sector solutions. Internet business and personal consumer solution activities continue to amaze and beat everyone's expectations, worldwide, even though the e-commerce phenomenon has purportedly passed. This continued Internet flurry of innovation involves advanced forms of media and content, and innovative new types of network services delivery mechanisms, which are what all that end users see and experience; furthermore, these same advanced content media are often required by many applications, which we reference in various types of examples throughout this chapter.

The serving of this content is also a major incentive for the economies of many countries. It is capable of affecting both economical growth models and acting as stimuli for enhancing additional critical skills and capabilities within the country. Vietnam, India, and China have all considered offshore capabilities in one form or another: Some are deeply involved in many initiatives surrounding this topic. This is due (in part) to the cost of operation being lower than that in other countries currently engaged in the businesses surrounding "content hosting." This is a fundamental shift in global economies, most notably in independent approaches to entering the global commerce marketplace.

The world is now witnessing the emergence of public Internet access points, as WiFi (wireless fidelity) hotspots: examples are Lufthansa Airlines, Starbucks Coffee, and Borders Book and Music stores. These enterprises have all instituted wireless Internet environments for their consumer base to utilize on a daily basis for business or personal use.

This massively distributed Internet enables wireless Internet service providers and other business enterprises to deliver advanced managed services. Examples of these services are wireless wide area networks and broadband wireless network services. These services can be utilized as carrier services to these same consumers of commercial wireless services, to allow them to subscribe to further services, such as email, messaging, advanced media, dating, and many other types of third-generation (3G) applications. These services can be accessed and utilized from anywhere, at anytime, on any type of pervasive computing device.

In addition, *sensor networks* are envisioned to continue to emerge as key instrumentation for very creative applications, across a wide variety of human endeavors. Applications of sensor networks range from biomedicine to battlefield monitoring,

---

[1]For more information on Vision Media Technologies, Inc., please refer to their Website, http://www.vmtl.com.

and many other cultural and habitat venues. Advanced areas of focus such as tornado activity monitoring, seismic prediction and eruption activity monitoring, and advanced forms of security tracking movements in rooms where physical security is paramount. These are only a few innovative application areas — there are many more opportunistic areas of exploration already under way. In fact, the subject of sensor networks even finds itsself implanted into popular science fiction (at least it used to be) books. One very good example of this is the book entitled *Prey* by Michael Creighton [1]. This book by Creighton describes nanobot technology and sensor networks in a somewhat astounding fashion.

We believe that the theme and focus of the subjects discussed in this chapter are very timely, especially as this material applies to developing on-demand business strategies until 2007–2010 or so. This chapter provides full treatment of wireless mobile, and sensor networks — contrasted to on-demand business solution areas. This chapter delivers information that is easy to understand and focuses on issues ranging from advanced forms of on-demand business service architectures, on demand business operating environments, a variety of networking protocols in a pervasive computing ecosystem, business modeling, and security matters, and finally concludes with an analysis of economic, culture and market trends.

15.2 ON-DEMAND BUSINESS

This discussion introduces on-demand business,[2] which is (simply speaking) a state of operations that the world is evolving to that reaches across all global industry sectors. We will explore this evolution in much greater detail, as we address a few very interesting topics of on-demand business. We will also discuss some on-demand business solutions integration challenges of mobile, wireless, and sensor networks.

Let's take a look at the meaning of the term "on-demand business" (see Fig. 15.1). On-demand business is all about integration of services, in ways unlike

An on demand business is an enterprise whose business processes—integrated end-to-end across the company and with key partners, suppliers and customers—can respond with speed to any customer demand, market opportunity or external threat.

Figure 15.1   Illustration of what it means to become an on-demand business.

[2]See IBM, On demand business; available on the Web at http://www-306.ibm.com/
e-business/ondemand/us/index.html.

the past world of distributed computing. On-demand business is (in many ways) the realizable result of an evolution of many networked technologies, in conjunction with some key business process transformation activities.

Given this clear explanation of what it means to be an on-demand business, let's now dig a bit deeper into what it takes to deliver this kind of environment. First, an on-demand business operating environment must be established.

15.2.1   On-Demand Business Operating Environment

The *on-demand business operating environment* defines a set of integration and infrastructure management capabilities that customers utilize, in a modular and incremental fashion, to become an on-demand business. The on-demand business operating environment provides the linkages between business needs and IT abilities, by allowing

- Businesses to stay focused on the core business needs
- People, processes, and information to become fully integrated
- Information technology (IT) to quickly sense and respond to changes in business requirements
- Infrastructure designs to absolutely match the business designs
- Standards-based, modular, built-for-change applications to be more network-aware, and the networks to become more application-aware

What does all of this really mean? It means that large enterprises and small businesses, alike, can now more easily manage capacity, security, and availability — on demand. These businesses can (for the first time) respond faster and more effectively to planned and unplanned situations in the marketplaces.

Companies can now integrate and manage more effectively large distributed environments, including wireless, direct subscriber line (DSL), cable, satellite, and utility company solution approaches. These combinatorial approaches to solution development have and will continue to unveil incredible innovations and technologies across many fields of communication. The on-demand business operating environment adopts itself well in all these advanced types of communications systems solutions.

Reducing costs while increasing revenue streams across all lines of business is the key. There is worldwide pressure on the services provider industry to reduce capital and operational expenditures, while increasing revenues and profits [e.g., some global telcos (telecommunications carriers)]. So, consider this: Why is it that you can plug a credit-card-sized device into your laptop, and become a wireless end user in about 60 s; yet, it takes 32 people about 8 months to install a "billing system" that costs about 21 million U.S. dollars in systems integration costs?

Becoming an on-demand business includes reducing costs and reapplying technology in areas of the business that increase efficiency and are a part of a planned on-demand business transformation roadmap. Collaboration and integration with

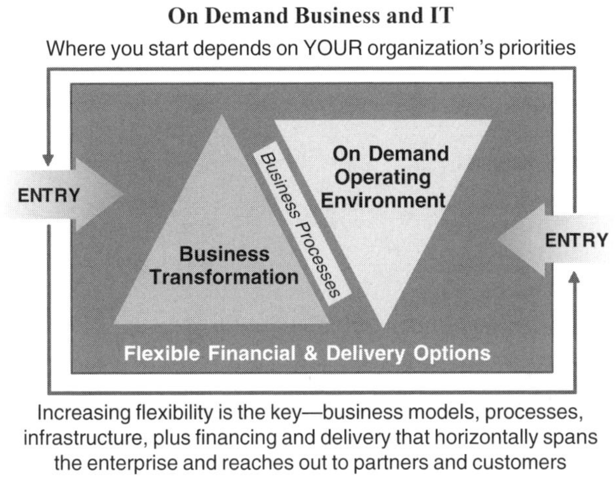
On Demand Business and IT
Where you start depends on YOUR organization's priorities
ENTRY
Business Processes
On Demand Operating Environment
ENTRY
Business Transformation
Flexible Financial & Delivery Options
Increasing flexibility is the key—business models, processes, infrastructure, plus financing and delivery that horizontally spans the enterprise and reaches out to partners and customers

Figure 15.2 Flowchart showings entries on which a business can begin the transformation roadmap of adopting on-demand business models.

best-of-breed partners becomes the goal. The world is focusing on this now, which is a paradigm long overdue, and is quite simple to do through on-demand business solutions and services.

Position yourself for future needs; look 3–5 years ahead. Accept the fact that the term *on-demand business operations* is not an example of short-lived technical jargon (or academic experiment) terminology; in fact, it is a reality brought upon the world as a result of faster networks and more sophisticated forms of utility computing (e.g., self-healing systems, self-provisioning systems, sensor network solutions, metered billing). This is a transformation that is identifying with this new evolution. It involves a keen focus toward on-demand business operations. This is not a revolution — it is simply an evolution. All worldwide industries are faced with this on-demand business transformation in one way or another. How these enterprises enter this transformation is entirely up to their priorities. The entry paradigm, shown in Figure 15.2, is priority-based.

Throughout history, we can see a pattern. It is a pattern of evolution that started with traditional concepts in computing, then it was abruptly advanced by virtue of the Internet, and now we are cycling back around some very traditional ideas, yet far more complex and powerful than we once new them. For example, grid computing [2] is distributed computing with something far more powerful and complex than distributed computing concepts of the past. Sensor networks offer many new types of global solutions, and some impacts will be noted to the global networks as a result of these implementations. Web services since the early 1990s have evolved from being nonexistent to prolific and abound with complexities.

The only thing for sure is that on-demand business, advanced solutions, and change itself, are far more pervasive than what first meets the eye. Let's take a look at this computing paradigm transformation in Figure 15.3.

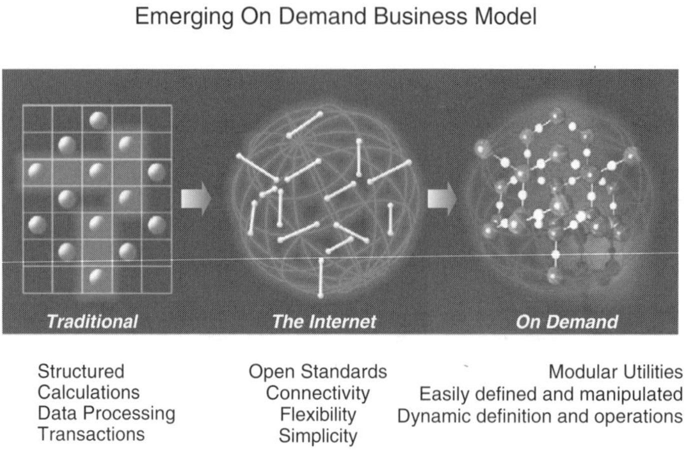

Emerging On Demand Business Model

Traditional	**The Internet**	**On Demand**
Structured	Open Standards	Modular Utilities
Calculations	Connectivity	Easily defined and manipulated
Data Processing	Flexibility	Dynamic definition and operations
Transactions	Simplicity	

Figure 15.3 Illustration of various stages of transformation toward on-demand business models.

The IBM on-demand business operating environment is the end-to-end enabling IT infrastructure and capabilities that allow an on-demand business to execute advanced IT operations — tightly aligned with the business strategy. These operations align with the business strategy and enable the on-demand business to:

- Become more responsive to markets
- Focus on core competencies of the business
- Benefit from a plurality of variable cost structures
- Be resilient to external threats

An on-demand business operating environment helps any business easily manage its IT operations as one cooperative entity, as well as effectively deal with the opportunities and the disruptions that influence its growth and prosperity. An on-demand business operating environment helps businesses (both large and small) to readily take advantage of opportunities that realize financial benefits.

An on-demand business operating environment unlocks the value within existing IT infrastructures, to be applied to solving business problems. It is an integrated platform, based on open standards, to enable rapid deployment and integration of business applications and processes. Combined with an environment that allows true virtualization and automation of the infrastructure, it enables delivery of IT capabilities — on demand.

Typical evolutions to conducting on demand business also require an approach to acquiring capabilities that are always addressing key requirements. This is accomplished by following a roadmap consisting of modular incremental steps to becoming an on-demand business. Building upon existing capabilities, on-demand solution offerings support any company's evolution to becoming an on-demand business. The on-demand business operating environment embodies two fundamental

On Demand Business Operating Environment

The on demand business Operating Environment enables horizontal business process integration, and a service provider interface that leverages common application containers, an enterprise services bus, and a wide variety of infrastructure services.

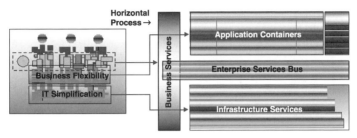

Figure 15.4 Schematic representation of various on-demand business operating environment characteristics.

concepts: business flexibility and IT simplification. Figure 15.4 illustrates this distinction in more detail.

The first concept is to increase *business flexibility* through capabilities designed to expedite implementation of integration initiatives. The ability to connect people, processes, and information in a way that allows the organization to become more flexible and responsive to the dynamics of its markets, customers, and competitors is critical. This becomes increasingly important as the value net is extended, in order to tightly integrate partners, suppliers, and customers into the business processes.

The second concept is *IT simplification*, the creation of an infrastructure that's easier to provision, deploy, and manage. This is accomplished through the creation of a single, consolidated, logical view of, and access to, all available resources in a network. Many organizations have become comfortable with the practice of over-provisioning; that is, buying excess capacity in order to handle the occasional spikes that almost every system experiences. Eliminating the practice of overprovisioning networks by moving to an infrastructure that accommodates dynamic resource provisioning is very important. This virtualized infrastructure will significantly reduce an organization's capital and operational expenditures. Figure 15.5 delineates areas of consideration, regarding these types of environment. The following discussion describes the on-demand business operating environment architecture.

The on-demand business operating environment is based on the concepts of a *service-oriented architecture* (SOA), which views every application or resource as a service implementing a specific, identifiable set of (business) functions. In addition to the business functions, services in an on-demand environment may also implement management interfaces to participate in the broader configuration, operation, and monitoring of the environment. The conceptual model of a

On Demand Business Operating Environment Architecture

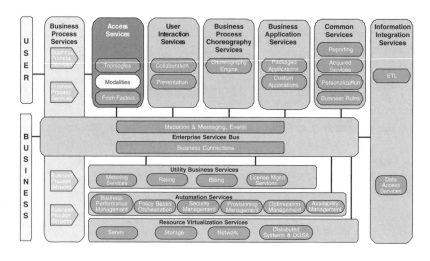

Figure 15.5 Schematic representation illustrating various levels of architecture for the on-demand business operating environment.

service-oriented architecture applies to the virtualization of both business functions and physical infrastructure. This interface-based integration is achieved through open standards and Web services.

Services communicate with each other by exchanging structured information — messages or documents. Their capabilities are defined by interfaces declaring messages that they can produce or consume, policy annotations declaring quality of service required or provided, and choreography annotations declaring behavioral constraints that must be respected in service interactions. The actual implementation is hidden from the requester of a service; thus service-oriented architectures are a convenient way to achieve application integration by allowing new and existing applications to be quickly combined into new contexts. Existing applications are "adapted" to service declarations through the service interface, and transform messages into operation on the existing application.

As illustrated in Figure 15.5, the interactions between services flow through the *enterprise service bus* (ESB), which provides a set of infrastructure capabilities, implemented by middleware technologies that enable service orientation. ESB supports service-, message-, and event-based interactions in a heterogeneous environment with appropriate service levels and manageability. However, we should note that all interactions don't require network communication and XML messages.

What is *multimodal*?

- It denotes the ability to combine multiple human–machine interaction modes. Personal preferences, social situations, and device and network capabilities determine the selected mode.
  - ► Examples of input include keyboard, touch- or tapscreen, handwriting, and voice recognition. Similarly, the output can be varied as well, including visual or text-to-speech.

Figure 15.6 Definition and examples of multimodality.

Note in Figure 15.5 the designation of "modalities." This is an important capability that we will explore, and as you will see, these communication modalities introduce several of the current challenges in the global pervasive networks. *Modalities* introduce complex machine communications capabilities to effectively deal with a plurality of human–machine interactions. Examples of these interactions are widely practiced today from voice communications to data communications, intersecting a wide variety of pervasive devices and personal configurations.

Examples of different *modalities* include keyboards, touch- or tapscreens, handwriting and voice recognition, and audio/videostreaming (see Fig. 15.6). This input and output is then varied to fit the device and consumer communications need; that is, end-to-end communications may originate on a laptop device and end up on a cell phone (e.g., SMS messaging). Incorporating visual displays, text-to-speech communications, speech-to-text communications, touchscreens, and other forms of device and networking–human interactions accomplishes this multimodal, pervasive computing environment.

Multiple modalities enable the capability to combine multiple human–machine interaction modes; that is, personal preferences are always considered, along with social and cultural situations. With such varying usage scenarios, the pervasive device and the on-demand network capabilities dynamically determine the selected mode of communications transport and delivery.

What does all of this discussion really mean when you consider the global network challenges of these types of pervasive environments? This ultimately means that the networks, from all global networking service providers, must continually embrace new and more effective capabilities to acquire more "knowledge" of the applications and devices within their network domains. Interestingly enough — according to *Webster's Dictionary*, the word "intelligence" is partly defined as "the capacity to acquire and apply knowledge." This is fairly straightforward to understand; however, this continues to be a challenging and somewhat evolutionary process across all industries that are successfully delivering these types of on-demand business global networks.

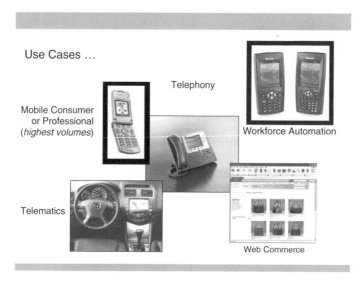

Figure 15.7 Various types of multimodal access.

One notable accomplishment is that a "services integration" tier has seemingly upstaged the *enterprise services bus* (ESB). The ESB provides a set of infrastructure capabilities, implemented by middleware technologies that enable service orientation. The ESB supports service, message- and event-based interactions in a heterogeneous environment with appropriate service levels and manageability.

This services integration convergence will continue to better define itself over time (through a series of global, industrial on-demand business initiatives) and hence become a more functional and enriched enterprise services bus across industries. This services integration convergence activity will continue to emphasize and reinforce the importance of open standards across all industries and enterprises. Equally important to consider is this is not simply a business issue. As shown in Figure 15.7, it is the general consumer public driving many of the most challenging demands on the networks. To a large degree, the specific computing device dictates many of the services integration convergence points.

One fascinating benefit of being involved in this industry today is that those individuals with a strong desire to learn will observe technologies converging. For instance, my cell phone is now an Internet browser, my laptop is now a cell phone, and my belt calculator is now a computer and a cell phone. My glasses now contain a very small Internet screen that I can easily view — technological innovation is converging throughout several dimensions of the networks.

When and where these brilliant new human–machine capabilities ever end is not yet known. By the year 2010, will my car be able speak to me when it notices my eyes beginning to indicate signs of fatigue as I drive? If so, will my car's male or female voice then stimulate me by asking if I wish to play a "name that song" game, or if I wish to have my email read to me so that I can verbally respond to selected emails, as I drive across the country? Today, it automatically navigates me

Why Multimodal?

- Pervasive computing is about anytime/anyplace access to applications, information, and services
- Enablers
  - ▸ Wireless
  - ▸ Miniaturization
  - ▸ Standards
- As the interface to technology becomes more pervasive, the number of users increases

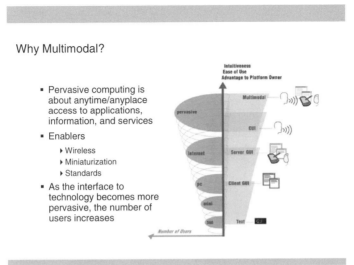

Figure 15.8   Timeline of the pervasive trends since the days of the IBM 360.

around road situations and finds my way home when I am lost, and does so at my verbal command of "home." It finds my way to the restaurant I need to be at, even when I am in a strange city and have no paper (hardcopy) maps.

The least common denominator here is simple — the networks and their transport of data. These innovative solutions will always continue to be challenging for these types of on-demand solutions. Additionally, this least common denominator will always be at the core of this on-demand business "evolution." In fact, as shown in Figure 15.8 — as the technology interfaces become simpler and more pervasive — the end user populations will increase in size. Is this a phenomenon or a natural part of this on-demand business evolution? We suggest that it is the latter. Either way, we must pay close attention to the networks and transport of data. Networks are often overlooked, only to suddenly realize that they are either over- or underprovisioned. Simpler interfaces only mean more complex situations to manage within the infrastructure.

As perhaps implied by Figure 15.8 telecommunications carriers (the telcos) prosper when their consumers use their cell phones prior to or instead of going to their laptop or desktop computers to communicate a message. Telcos from around the world are making great strides in this area. Telcos such as SK Telecom or Korea Telecom in Korea, Reliance Infocomm and Reliance India Mobile, China Telecom, or NTT DoCoMo in Japan are only a few of the world-class players in this multimodal space.

A key message here is that the cell phones could very likely become the computers of the future. Mobility is important and is finally taking hold. The shift of technology today is totally consumer-led, and consumers are demanding device-independent mobility. As the capabilities of the cell phones increase, and are adopted by the masses, the ability and need to better manage and store personal

What are popular examples of Multimodal Portals?

Figure 15.9 Two examples of popular multimodal portals: personal digital assistants (PDAs) and cell phones.

content will increase. Consumers are also beginning to demand interoperability of communications technology with home entertainment systems, although for the most part this is still in its infancy. Mass adoption in this area probably will not fully happen until technology is truly plug-and-play. Open standards play a significant role here, and are the basis for much of our current progress in this area. Mobility will encompass home, work, automobiles, and cell phones — and consumers will demand interoperability and seamless transport of content, simply because they can.

As we have all probably either seen or experienced, computers and PDA and cell phone (see Fig. 15.9) technologies have been intersecting since 2002 or so. However, for a moment, let's view this as a new agenda; think of it now as a personal mobile portal for enterprise operations and consumer-based, managed broadband services. What is *broadband*, you ask? It consists of several categories of methods for attaining Internet or telecommunications connectivity, including DSL, cable, wireless, satellite, and utility company access methods of the future.

This new agenda still remains a tremendous opportunity for carriers and service providers to help us all better manage our daily lives both personally and professionally — with the devices we chose to utilize. This now helps address both professional growth and personal lifestyle needs, affecting many aspects involving culture and way of life.

Although these examples may not be new to some, they remain a challenge in the pervasive networks of the world. As we continue to strive and meet with each of these challenges, we continue to enhance our daily lives, thanks to these brilliant ecosystem approaches, on-demand business operating environments, and our global pervasive networks. Furthering this example, combine this kind of network demand with the fact that anyone can now make long-distance calls for virtually free, using voice and Internet protocol applications (e.g., Skype[3]). This type of functionality is typically referred to as *VoIP* (Voice over Internet Protocol). The challenges will not get simpler in the future.

Different defaults are always based on consumer devices or business devices, and separate market segmentations will always exist. For example, a homemaker is generally speaking as an interested consumer of domestic types of service, whereas a home office or mobile telecommunications worker may seek different types of service. Oddly enough, both of these individuals are often the same person, just working through different parts of a busy day. In either case, social groups or personas can take advantages of these multiple approaches, and in some cases the services will best support those activities involved in the home whereas in other cases specialized services will sustain those typical needed across the global business communities. Finally, let us not forget that we could not do this in 1995, when only a few had this type of futuristic vision.

Content changes and transformations, based on multimodal usage patterns, are the key capability needed in these types pervasive domains. Why should enterprises invest in this? Call center and other data services are now available to customers not only via a more pervasive device but also with more convenient speech-enabled interfaces than simply a graphical user interface (GUI).

Why should carriers invest in this? Their business model morphs from simply being a voice/data channel, and now augments personal and professional persons into a portal (e.g., MSN or Yahoo!). While Microsoft is trying to make carriers just a pipe, other industries are viewing many forms of carriers as services providers, as well as a pipe.

On-demand business strategies are resonating around the world in this regard. We are noting AT&T Wireless, Bell Mobility, DoCoMo, China Telecom, France Telecom, KDDI, SingTel, Nextel, Orange, Sprint PCS, Swisscom, Reliance Infocomm, and T-Mobile — with a multitude of business solutions development, key briefings, and rich services and sales strategies. These types of portal activities are landing significant interests throughout the world. IBM continues to help many of our customers and business partners transform their enterprises, in order to become an on demand business.

Let's now explore yet another concept, which plays a large role in the services integration arena. This is referred to across the industry as the "manager of managers" effort. This involves a single enterprise and the need to manage multiple managers of functionalities and services. Figure 15.10 shows the complexities involved in this type of activity, from a systematic viewpoint.

As illustrated in Figure 15.10, the fundamental premise in on-demand business operations involves a significant amount of collaboration and integration of networking partners to deliver what would appear to be a single on-demand business solution. This approach implies that all partners are able to adhere to and provide on-demand business functionalities.

---

[3]For more information on "Skype," a free VoIP telecommunications Internet phone tool, please refer to the Website `http://www.skype.com`.

Managing the Managers of On Demand Business

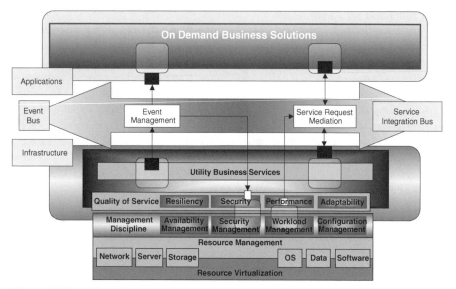

Figure 15.10 Schematic representation of various levels of operation within the on-demand business operating environment. At this level, collaboration and integration between networked partners is a primary goal.

Let's now take a deeper look at some of the elements of pervasive computing that play a tremendous role in on-demand business solutions delivery to the consumer. In the following discussion, we will explore a world-class example of pervasive services.

15.2.2 End Systems and Pervasive Computing Ecosystem

Mobile devices, as we discussed earlier in examples of multimodal portals and cell phones, become more sophisticated with different embedded technologies due to the evolution of the service content and requirements of the advanced functions. It was the "user experience" and "user perception" that gave rise to new services and business models. Starting from text-based SMS, mobile users began to demand more rich contents service, urging service providers to create more new mobile services "on demand."

For example, in Korea, popular single music albums are first released through mobile services, generating a new business model for both service providers and music publishers. Although there are controversial cases of the copyright issues around MP3 music download service, mobile service has already become a new channel of music service. By downloading music files, whether MP3 or service provider's proprietary formats, the revenue could be shared between service providers and music publishers, while airtime charge is solely kept by the service providers.

New mobile services taking advantage of the advanced technology of the human–machine interface (HMI) are introduced to the customers. Music search services, by capturing the rhythm from the radio through the mobile phone, is a new service model more recently launched by KTF, one of the service providers in Korea.

Extending a typical HMI of the number keypads on the mobile phone to barcode serial number can find a new business model[4] in the mobile advertisement in the form of mobile games. You can now feed your cyber pet in your mobile phone by inputting a real cat food product barcode number, which then can be converted to cat food in the "cyber pet caring game" (also known as *Damagochi*), or even to a whole new game item in a mobile adventure game. Moreover, enhanced HMI such as a camera on the mobile phone can scan the barcode stamps instead of punching in all the numbers on the keypad.

End-to-end systems in the pervasive computing environment are the fundamental requirement from the service providers (as well as device manufacturers) to sense and respond to the customer's rapidly changing demands, and respond by launching new services. Service providers want to extend their service platform to device environment seamlessly, and device manufacturers want to influence and involve themselves in the service provider access domains.

Embedded software constitutes the building blocks of the mobile platform in the client domain. As Java contents become more popular and the Java platform is accepted as a de facto standard service platform by service providers, there is a continuous race toward "platform leadership" [3] among the players in the mobile environment. Service providers need to have a well-defined service provider ecosystem across several service access domains, to intelligently respond to the customer's demand. This race seems to only be getting fiercer as we mature in this portion of the ecosystem.

The Java virtual machine (JVM) is a software platform that resides in mobile devices as runtime environment to run various applications. JVM positions on top of the interface know as the *hardware abstraction layer* (HAL) or *porting layer* to communicate through the operating system to the hardware chipset of the mobile devices, which then controls the native functions (memory, screen, etc.) of the hardware while running Java applications on top. IBM, as world is biggest supporter for Java, provides total product-level embedded solutions including JVM to device manufacturers with independent Java licensing. Through the "porting partner program," IBM partners with the leading embedded technology solution providers in the world to provide IBM Java technology to mobile device manufacturers.

15.3 SERVICE-ORIENTED ARCHITECTURE

15.3.1 Principles of Service Orientation

Service-oriented architecture (SOA) introduces a loosely coupled architecture concept with a set of abstractions over a component. These couplings are granular

---

[4] Korea Patent 0376762, registration number 2000-0066102, registration date March 6, 2003.

Services Offerings
and Services Oriented Architectures

- Capability-based match-
  making between service
  providers and
  consumers/requesters

- Requires rich meta-info about
  interaction end points—what
  requesters want, and what
  providers can deliver

- The focus is on the interface,
  behavior, and policy
  definitions, such as: WSDL,
  BPEL, and WS-Policy

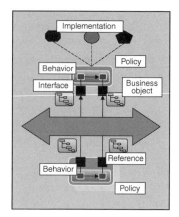

Figure 15.11   Schematic representation of the concept of services offerings and a simple example of architecture in a services solution.

enough for consumption by clients and are accessible over the network with well-defined policies as dictated by the components.

In SOA, the resources are made available to other participants in the network as independent services that are accessed in a standardized way. This provides for more flexible coupling of resources than in traditional systems architectures. This concept of service offering is illustrated in Figure 15.11. We submit that SOA is more of an architectural philosophy than an architectural (single) blueprint.

Figure 15.11 illustrates the principles of an SOA and the behaviors that are realized by the application of this philosophy. Let's now take a look at how this relates when applied across various service provider domains.

15.3.2   Service Access Domains

We have discussed the manager of managers. We have discussed the enterprise services bus. We have commented on the need for sophisticated middleware, including mechanisms of workflow management between services providers and a plurality of independent systems. We will now take a conceptual look at what we mean when we consider the subject of "services access domains."

Services, in this context, will now be related to the services providers. Figure 15.12 depicts the notion of a "hub and spoke" infrastructure environment, illustrating how service providers (SP*x*) on the left, work through and across a hub-and-spoke approach in order to communicate with independent systems, for the sole reasons of delivering back into the SP some form of functionalities. This illustrates the process of managing the many managers involved in a services solution.

Also note in Figure 15.12 the small, stacked cubes labeled 1–6 and A–E. These suggest a programmatic language set of interfaces or utilities [e.g., eXtensible

Service Provider Interface

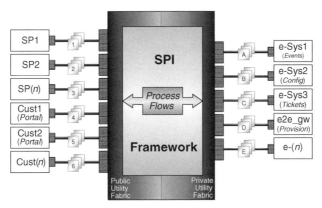

Figure 15.12   Schematic representation of the concept of a service provider interface "bus" framework.

Markup Language (XML)] that function as a conduit between domains. Note the need for programmatic, reusable "utilities," as noted in the private and public utility fabric. These utilities sense and respond to the *application awareness* of the intelligent on-demand business network. These concepts mirror some of the concepts in utility computing. All this said, these topics are beyond the scope of this chapter, and could warrant a whole separate book on this topic.

This strategic *utility fabric* is very interesting in terms of service integration capabilities between multiple on-demand business enterprises. In Figure 15.12, note that both sides of the fabric [service providers (SP1-*n*) and the systems (e-Sys1-3)] can now communicate with each other, regardless of the networking protocols. The service providers, the on-demand networks, and the on-demand business applications are essentially becoming more aware (i.e., *application aware*) of each other through this public–private fabric. Also involved in this service provider interface solution design point is sophisticated middleware, with an awareness of the workflow requirements of the on-demand business applications.

So, we speak of services. What kinds of "services" are we really considering at this stage? There are a wide range of services, which we will illustrate later in this chapter. There are also many utilities still to be created; yet many utilities already exist today. Some of these services can best be described as small utility services, such as a "metered billing," "bandwidth on demand," or "event correlation." The notion of any utility is simple to apply — it is simply what a typical utility company provides. Utilities, like the lights in a building, accrue cost only according to the time they are turned on. Likewise, metered billing provides for access and utilization of advanced services, while paying for those services only as they were consumed. *Bandwidth on demand* is simply the act of provisioning in times of unexpected (or expected) needs, additional network transport capabilities, and additional application server capabilities — all of this can be provided, yet only as needed. Think of these types of utility services as a wide variety of small agents, enabling services through a plurality software solution component approach. This type of software services bundling, in actuality, implements software as a service (SaaS).

The global industries are just becoming aware of the compelling factors in the SaaS portion of the ecosystem; however, it is important to understand that underlying all of this discussion is a particular framework implemented in many instances around the world. This framework is based on open standards.

SaaS providers focus on obtaining additional cost benefits from their service delivery models, where retention of customers is their key focus. They are able do this by considering the service requirements from the beginning of any service development. This can be achieved by building highly scalable stateless architectures to minimize costs and maximize utilization. SaaS helps customers quickly realize on-demand business application environments. As an example, access a business function delivered as a variable-price network service, and being able to do this without being concerned about the means necessary to enable that business function. Providers of hosting services then naturally provide added value by becoming the trusted and reliable environment for SaaS providers.

SaaS is leveraging the current and emerging global standards for Web services. This consequently allows customers to more easily produce and utilize on-demand business functionalities — delivered as a software service. New global standards have been driven by a wide variety of customer requirements. These standards allow for the creation of new on-demand business services that are not possible by utilizing yesterday's technologies and development approaches. Service providers need to focus on working with their customers in order to learn what is needed in their product and service offerings. The importance of this collaboration and convergence point is so that they, too, can establish themselves in this emerging global space as an on-demand business service provider.

It is highly likely that during the near future, more and more companies will rent software as they require it, or obtain software that is delivered through SaaS solutions, rather than purchase software. For example, a company may choose to rent sensors and the software required to leverage the advantages brought forward by these sensors across a wider network. This introduces a new and improved means for integration in on-demand business operating environments. The risk and qualities of services rendered now becomes a services orientation question that is targeted at the independent provider of the software service. Let's explore this notion of risk mitigation in the following discussion.

Figure 15.13 illustration provides a service orientation approach toward becoming an on-demand business in the ecosystem that we have been exploring. Service-level agreements (SLAs) and quality of service (QoS) all become very significant goals that any service provider must be able to guarantee and maintain for any on-demand business. Let's further explore the meaning of "service orientation" in this context.

There are many underlying assumptions surrounding the networks. Again, the networks are the least common denominator when considering these types of

Service orientation and on-demand business:

An on-demand business operating environment enables the end user to incorporate existing and new utilities and to build higher-level services—and this depends on networks.

- Low latency and high throughput are performance-critical, and have always been the quest of networking services

- New on-demand business environments allow end users to build their own virtual organizations, facilitating the need for creating problem solving organizations

- Premium broadband managed services, such as bandwidth on demand, metered billing, automated provisioning, and more become a "Software as a service" (SaaS) focus

- Networking services are required in every on-demand business solution; the most obvious areas include provisioning benefits, event correlation, problem management, SLA management, and QoS

> *On-demand business grids facilitate the ability for any individuals, worldwide, to join virtual organizations that render advanced problem solving services*

Figure 15.13   Some key factors in utility computing.

on-demand business services, especially when one considers how these services are actually transported and delivered. In the following discussion we will unveil some specifics regarding these service domains, and introduce the aspects of operational and business support systems, which are very critical in the pervasive ecosystem.

15.3.3 Service Domains

The areas of service domains clearly have complexities across several dimensions, including utilities and a complex array of service delivery functions. Operational aspects influence how services are delivered and maintained according to a predetermined SLA. All SLAs are significant to both consumers and the services providers — to consumers because of the obvious need for the specific information according to some level of high availability, and to service providers because they are held responsible for making this information available — when and where it is requested.

Across the industry, especially with telcos and cable companies, there is a strong need to reduce operational expenditures (OpEx) and capital expenditures (CapEx). One primary focus helping to advance this endeavor is focusing attention on the OSS/BSS layers of an enterprise. OSS/BSS (as shown in Fig. 15.14) is an acronyon that has been adopted by the global industries, that stands for the operational support systems and the business support systems across industrial and enterprise

Services Provider Ecosystem and OSS

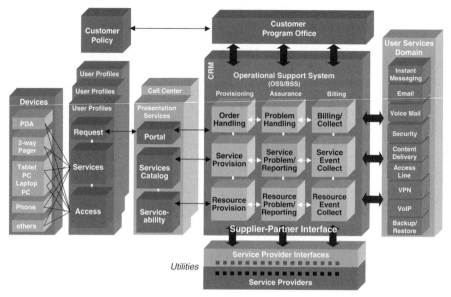

Figure 15.14 This illustration shows concepts of a service provider ecosystem, with a focus towards the operational support systems (OSS).

solutions. The most notable effort in this area is among the telcos, namely, the TeleManagement Forum.[5] The TMF has dedicated an aggressive strategy to bring together global telecommunications firms that share this common goal of reducing OpEx and CapEx.

Figure 15.14 shows the various dimensions of the OSS/BSS model, in a services provider ecosystem. The OSS/BSS networking challenges are immense, considering the global industries and their business enterprise arenas. This challenge is then compounded by each enterprise's strategy toward providing these types of services, and their chosen infrastructure for delivery. If you again look at Figure 15.14 (from left to right), you will note that starting with the consumers of services (far left) using a variety of pervasive devices, the consumers can request and receive content created (far right) — sent through the carrier or enterprise (center). This can be, in fact, an ecosystem of pervasive on-demand business services being provided to a culturally wide, global consumer base. Note the utility interfaces in the lower section of this diagram, perhaps indicating the presence of many on demand business relationships — collaborating through some form of a utility fabric as we have already previously discussed.

Several service provider companies have very innovative solutions addressing this overall type of pervasive collaboration ecosystem. IBM has, in fact, helped

---

[5]For more information on the TMF, please refer to http://www.tmforum.org.

On-demand business managed network services
Focus on rich, new operational and business support systems

- In the context of a service-oriented grid computing applications, the focus is not on the various protocol levels but rather the interfaces to the network transport services

- Network transport services must embrace on-demand business with open standards solution patterns, while enabling a wide range of new operational support systems and business support services (OSS/BSS) market development initiatives

- Communication protocols are becoming more integrated, working with solution patterns and open standards

You can plug in a credit-card-size adapter and be on a network in 10 s, but it takes 6 months and millions of dollars to install a "billing" system

Figure 15.15 List describing the importance of operational and business support systems.

many customers achieve this advanced state of operations, while we are also helping them become an on-demand business. This is a transformation activity not unknown to many enterprises throughout the global markets. The biggest question is where are you in this transformation endeavor? Can you see it, and have you already started on your own transformation journey? Or, do you need a partner to help, especially one that is well on its way in this journey?

In order to achieve a pervasive collaboration ecosystem we need on-demand business managed network services as illustrated in Figure 15.15.

The OSS/BSS areas are key focal points for inspection, when considering the fact that cost must be removed, while attempting to enhance customer satisfaction levels.

15.4 GENERAL ISSUES

The general issues are complex when it comes to services integration across any pervasive ecosystem. In fact, this will probably always be the case. We have transformed our cultures, our global theater of business, and even our personal lives through the introduction of this pervasive ecosystem. Think about how many people you encounter each week who are using these devices. Also, how many of these devices leverage some of the concepts that we have discussed in this chapter? Let us start by looking at some of the more conspicuous issues that warrant further consideration.

15.4.1 Network Layering and Standardization Considerations

Several very key considerations are imposed on the local, metro, regional, and global networks. Figure 15.16 describes some of these very important considerations within the network layers where many of the appliances we have been discussing communicate data, voice, and information.

At first it might not be obvious exactly why this really matters. In fact, worldwide, this particular layer of the network structure matters a lot, especially when introducing sensor network devices.

Today, we have on the horizon Internet enabled toasters that have a weather emblem burned onto a piece of toast — as an indicator of that day's weather. We have toothbrushes that record video imagery of your teeth while you are brushing them, and then transmit this imagery to the dentist so that she or he can review with you any concerns in the imagery. We have walls with very tiny sensor network wireless devices embedded in paint, transmitting a variety of information regarding the space and movements within that facility. We have sensor devices being injected into tornados to study unknown weather phenomena. Many of these advanced solutions rely on the layer 4–7 network structure for communications transport.

Layer 4–7 switching refers to the content-aware intelligent network switching of Internet traffic. The layer 4–7 switch knows links and important network information about the computing communications session. Layer 4–7 also knows about application-level specifics, such as what type of user or device is requesting the content to be sent to the pervasive device. This could be a laptop computer, a handheld device, a frequent e-commerce shopper, or a first-time Website visitor — to name only a few of the end-user persona characteristics. Layer 4–7 also deals

Network Layer 4–7
New appliances introduce network challenges.

- Network layer 4–7 middle boxes, such as firewalls, intrusion detection, SSL accelerators, traffic shaping appliances, and load balancing devices introduce networking challenges in some grids

- Increasing importance of security, mobility, gigabit appreciation for layers 4–7

- Vision of network layers 4–7 supports explosive growth of IP networks since the 1990s

> *A service provider interface will have to discover and signal "middle boxes" in order to access QoS and services behavior of choice*

Figure 15.16   Displayed list questioning the layer 4–7 area, introducing many new types of technological challenges.

with the type of content that the user is requesting; for instance, an executable script, static content, cached content, dynamic content, streaming Webcast videos, audiostreams, videostreams, shopping cart comparisons are all examples of content — the list is lengthy.

All of this information passing through these layers requires some innovative and sophisticated switching schemes, involving application switches that offer traffic management, load balancing, application redirection, bandwidth management, sensors and actuators, and mesh devices, along with high-performance security services to server farms, data centers, and other vendor networks. Layers 4–7 present both a complex and complicated environment.

The massive amounts of information routing through layers 4–7 of any pervasive ecosystem today do indeed challenge the networks. However, standardization of the networks, and standardization of the products and languages operating within the networks will in effect accelerate the promotion of cost-effective services integration to occur. Standardization is helping to strengthen and advance the industry players that are aware of the layer 4–7 challenges. For these aggressive enterprise contenders, they will quickly realize (or continue to benefit from) profitable networks and regulatory reform. This has, and will continue to act as, long-term growth stimuli for the globally competitive telecommunications industries — the on-demand business service providers — to engage in their own on-demand business transformation journey.

Telecommunications standards are usually specifications of system requirements, features, interfaces, or protocols in both the network and product layers. Standards in these layers are generally agreed on and/or developed through a "standardization" process. Standards are critical for a number of on-demand business reasons, including the following:

- Standards are evolving after years of careful analysis of the repeating patterns in a realistic environment.
- Standards provide the most essential framework and underpinnings for the adoption of new technologies and on-demand business service.
- An on-demand business operates in multivendor networks and must demand that others maintain some form of global standards compliance — this is necessary in order to try and support the wider desire for a realizable decrease in CapEx.
- Standards are a foundation and long-term growth stimuli for regulatory reform and are a stimulus for competitive telecommunication environments in many parts of the world to become more efficient and services-oriented.

In contrast to the interconnection of networks that we have discussed, standards are also the primary basis for operational cost savings within networks. Global service providers increasingly depend on standards compliance, even by their own chosen services/equipment providers. This allows them to better architect, and deliver, more profitable multivendor on-demand business networks while at the same time reducing CapEx and OpEx and increasing profits.

Wireless Networks and Security
2600 Hacker organization discloses wireless security holes

"WEP: Not for me"
2600 Magazine (Winter 2003-2004)

☑ Security of wireless router "admin" functions exposed

"McWireless Exposed"
2600 Magazine (Summer-Fall 2003)

☑ IBM 4Q03 Ethical Hack for a USA wireless wholesaler

"We stopped at a rather long light and one SSID said 'Linksys.' I remembered the default setup, so I checked 'join.' DHCP gave me an IP, I browsed to 192.168.1.1, a dialog popped up and I typed 'admin' as the password, and 2 seconds later ..." by 0x20Cowboy {Hacker at-large}

Figure 15.17   Snapshots identifying the defensive *"hacking"* reality of wireless environments.

15.4.2   Wireless Security Considerations

The following discussion brings to the forefront the security challenges surrounding the wireless communications area. Security can provide for (or break) the most innovative wireless solution and therefore must be closely attended to in the early design phases. It is not enough to assume that you are always secured in the wireless networks, especially while transmitting in public wireless hotspots.

Although the wireless environments today are providing much more effective defensive mechanisms and more highly secured approaches to access, this is not the time for complacency. Listen closely to the services providers in these areas; they are very keen on this topic, and diligently working to close security gaps.

Figure 15.17 describes a couple of very interesting articles regarding this subject of wireless security. The topic of security goes beyond the scope of this chapter. The topic does however, have enough merit to introduce and point out a couple of very challenging network situations that occurred in late 2004 and early 2005.

The overall point of Figure 15.17 is for you, the reader, to be cognizant of any and all networking security issues, especially if you plan on deploying wireless footprints. Document strategies, plans, and findings, while paying close attention to any identifiable security risks and then, in federation with each of the other providers (where necessary), agree to accept and manage these risks in appropriate ways.

Despite issues with securing wireless networks, the number of wireless network applications has continued to grow. Most recently there have been striking advances in the product tracking category, from placing chips on warehouse packaging to tiny circuits sewn into clothing product labels. Figure 15.18 lists some key security

WiFi security
Network security is paramount for wireless communication

Currently, all 802.11a, b, and g devices support WEP (wired
  equivalent privacy) encryption, which has had flaws and exploits
  well documented

The ultimate goal is 802.11i, a robust set of security improvements.
  We are on the road to 802.11i

The Wi-Fi Alliance has required WPA (wireless fidelity protected access),
  which fixes all of WEP's problems. This is a subset of 802.11i, and
  allows full backward compatibility for most 802.11a and b devices
  made prior to 2003.

*Standards do not forego the need for close
inspections of wireless, or any security risks*

Figure 15.18 Identification of the "802.xx" protocols of interest in wireless communications.

advances in the wireless communications standards areas, which are important to
many types of wireless solutions.

Sensor networks and actuators warrant important discussion at this stage of the
chapter. These devices and advanced wireless concepts introduce yet another
dimension to the challenges presented to the networks of an on-demand business
ecosystem.

15.4.3 Sensor Networks

Ad hoc wireless networks have existed in our lives since the 1990s; yet, the key to
this topic is the more recent proactive presence of these types of network sensors,
their actuators, and their seamless service integration and communications into a
highly dynamic and configurable network services environment. This involves
many seamless service integration design points, a wide variety of open standards,
and layer 4–7 (L4–L7) communications of different classes utilized by the wide
variety of computing devices — soon, all of this will be well engrained as an integral part of our daily routines. This is very likely to be the status quo for many of us
in the near future.

The world is on the verge of yet another unbelievable, evolutionary technological step. Consider smart dust; "smart dust" particles are sensor network devices
that are very tiny wireless microelectromechanical sensors (MEMSs). These smart
dust particle devices can detect everything from light to vibrations. They can be
mixed into paint and applied to walls, enabling them to function as "smart" walls.
MEMS could even be released in the atmosphere, if there were a justifiable cause.

Sensor networks
Network security is implemented at L2 and L3, but L4 is less common

On-demand business networks and sensor devices can be pervasive,
but layer 4 will still require attention ...

■ Security is at the link level L2 (e.g., WEP or FrameRelay)

■ Security is at the network level L3 (e.g., IPsec)

■ L4 does not present a "one size fits all" security environment

■ Mesh networks, and respective devices warrant attention

*Network security standards are less
focused in the L4 areas—standards exist
at L2 and L3*

Figure 15.19   Some of the challenges facing sensor network technology.

As a result of more recent R&D breakthroughs in silicon chip technology, and new fabrication techniques, these smart dust "motes" will eventually be the size of a grain of sand (or smaller). Each particle will be able to contain tiny sensors, microscopic computing circuits, bidirectional wireless network communications technologies, and (of course) a remote power supply. These motes will gather large amounts of data, run complex computations, and then communicate that information to another tiny device particle, using two-way-band radio between motes — at distances approaching more than 1000 ft.

That said, considering sensor networks and what makes all this effort worthwhile is a growing feeling among researchers that these technologies may eventually have a huge impact on society. This also helps explain why the U.S. Defense Advanced Research Projects Agency (DARPA) began funding aspects of this kind of work at the University of California, Berkeley, back in 1998. However, building sensor networks with security and energy efficiency is still not a mature technology (see Figure 15.19).

Mesh networks [4] are regularly distributed networks that generally allow transmission only to a node's nearest neighbors. The nodes in these networks are generally identical, so that mesh nets are also referred to as *peer-to-peer networks*. Mesh nets can be good models for large-scale networks of wireless sensors that are distributed over a geographic region (as within a city). Typical applications of this today are found in personnel or vehicle security surveillance systems, city-wide emergency response applications, and traffic management systems.

In these types of environments, the regular mesh structure always reflects the communications topology of the networks; the actual geographic distribution of the mesh network nodes does not necessarily have to be a regular mesh. Since there are generally multiple routing paths between nodes, these mesh nets are robust enough to withstand failure situations of individual nodes or links.

Advantages of mesh networks could be described as "self-forming" and "self-healing" functionalities. In "self-forming," each mesh node finds the other node to form and optimize the network automatically, which each computing algorithm of the nodes then calculates. Even though all nodes may be identical and have the same computing and transmission functionalities, certain mesh nodes can be designated as "mesh group leaders," which then take on additional leadership functions to form a mesh network. If a group leader is suddenly disabled, another node will then inherit and take over these group leadership duties, healing the disabled network in a real-time fashion; this is "self-healing."

An advantage of mesh networks is that even though all nodes may be identical and have the same computing and transmission functionalities, certain mesh nodes can be designated as "mesh group leaders," which then take on additional leadership functions. If a group leader is suddenly disabled, another node will then inherit and take over these group leadership duties.

All the advances that we have just explored will continue to grow at incredible rates, throughout the world. It is interesting to observe how Korean and Japanese societies have integrated so many of these advanced technologies into their daily lives and also to note how countries such as India and Vietnam have transformed their technological capabilities since the mid-1990s or so. Transformations in the governments of these countries also somewhat reflect this technological progress as new applications of these devices and services are developed in these countries and are delivered on a global scale. As we continue to evolve as an Internet-enabled global culture, sensor networks, mesh networks, actuators, and other devices and services will all continue to play extremely significant roles.

15.5 CONCLUSION

In this chapter we have discussed issues ranging from advanced forms of on-demand business services architectures, to on-demand business itself, on-demand operating environments, a variety of networking protocols for a pervasive computing ecosystem, automated business processes, security, and analysis of economic, cultural, and market trends. As the global connectivity and application of pervasive devices grow, the network challenges in a globally pervasive ecosystem become more complex — as do the networking services.

We require more than ever a wide variety of consistent efforts to overcome these artificial boundaries caused by the disconnected global ecosystem. Applications of sensor networks, mesh networks, interconnected service providers, and intelligent networks will all help provide solutions ranging from biomedicine, to battlefield monitoring, and many other cultural and habitat venues. We conclude our chapter with the thought that the applications of senor networks, along with mesh and grid computing networks, are significant future trends and directions.

We predict a globally pervasive ecosystem that will become a routine way of life in each of our daily activities, even more so than today. This addresses a global economy that, together, overcomes the worldwide "digital divide" [8].

The transformation to an on-demand business includes reducing costs and reapplying technology in areas of the business that increase efficiency and are a part of a planned on-demand business transformation roadmap. Collaboration and integration with best-of-breed partners becomes the goal. We know that on-demand business needs an intelligent network with quick response attributes to meet the global challenges.

The ecosystem we describe, including a dramatic technology evolution, open standards, and service-based integration, have now become a critical catalyst for the future interconnected systems, networks, mobile devices, and sensors of our new global economy.

ACKNOWLEDGMENTS

We would like to thank Ms. Ashley Gillespie (Newtown, Connecticut) and Ms. Jessica Hu (New Milford, Connecticut) for providing outstanding editorial development support for this chapter.

REFERENCES

1. M. Creighton, *Prey*, November 2002; available on the Web at http://www.amazon.com/exec/obidos/tg/detail/-/0066214122/104-62832973430359?v=glance.

2. J. Joseph and C. Fellenstein, *Grid Computing*, December 2003; available on the Web at http://www.amazon.com/exec/obidos/search-handle-url/index=books&field-author=Joseph%2C%20Joshy/002-3731803-5787221.

3. A. Gawer and M. Cusumano, *Platform Leadership*, Harvard Business School Press, Boston, April 2002.

4. Mesh networks; available on the Web at http://www.meshnetworks.com/.

5. Project MESA; available on the Web at http://www.projectmesa.org.

6. J. Joseph, M. Ernest, and C. Fellenstein, Evolution of grid computing architecture and grid adoption models, *IBM Systems Journal*, Vol. 43, No. 4, 2004; available on the Web at http://www.research.ibm.com/journal/sj/434/joseph.html.

7. C. Fellenstein, *On Demand Computing: Technologies and Strategies*, August 2004; available on the Web at http://www.amazon.com/exec/obidos/tg/detail/-/0131440241/qid=1100708937/sr=1-3/ref=sr_1_3/104-62832973430359?v=glance&s=books.

8. *Digital Divide*; available on the Web at http://www.pbs.org/digitaldivide/.

INDEX

Access control, 360
Access points (APs)
  building, 34
  campus WLAN studies, 17–19, 32–40
  client-side tools, 15–16
  loss, 14–15
  malfunctioning, 6, 25
  management information bases (MIBs),
    11
  mapping, 16, 24
  network sniffing, 13
  nonacademic WLANs, 20–21
  personal digital assistants (PDAs), 18
  quality of service (QoS), implications of,
    47, 53
  single-queue comparisons, 58, 63
  SNMP community, 16, 25
  syslog messages, 6–9, 16, 25
  usage studies, 24, 30
  VoIP, single queue, 61
  wireless sniffing, 14–15
Acknowledgment (ACK) packet
  MANETs, 77
  VoIP, 49, 56, 61, 63–64, 66
Acknowledgment spoofing, 299
ACM SIGCOMM, 20
ACS log, 34–36
Ad hoc networks, security issues, 284.
    See also Mobile ad hoc networks
    (MANETs)
Ad hoc on-demand distance vector (AODV)
    protocol, multihop wireless
    networks, 107–109, 125, 127,
    129–131, 134–135

Adaptation algorithms, 47
Adaptive House (University of Colorado-
    Boulder), 338
ADDTS request, 55
Adjusted best-point heuristic, 235
Admission control
  contention-based, 48, 54–55, 70
  quality of service (QoS) provisioning,
    47–48, 58
Admission control mandatory (ACM)
    subfield, 54
AES encryption, 286, 290, 303
Aggregate-commit-prove, 300
Aggregation, wireless sensor networks,
    299–302
Aging problem, storage management,
    271–272
AIFS, quality of service (QoS), 46–47
All-pair shortest-path algorithm (APSP), 232
All-sensor field intensity, 231
Altered routing information, 299
Always-on devices, 24
Ambient noise, 120
Annealing, simulated, 247
Announcement traffic indication message
    (ATIM), MANETs, 97–98
Anomaly detection, 297
Anonymization, 33
Antenna, *see specific types of antennae*
  functions of, 15, 197
  MANETs, 83
  power control design and, 77
Apple Airport, 31
Art gallery problem (AGP), 224, 238

Mobile, Wireless, and Sensor Networks: Technology, Applications, and Future Directions
Edited by Rajeev Shorey, Akkihebbal L. Ananda, Mun Choon Chan, and Wei Tsang Ooi
Copyright © 2006 John Wiley & Sons, Inc.

Assistive environments, 342

Association
campus WLAN studies, 19
syslog messages, 9
wireless-side measurement studies, 21

Asymptotic equipartition property, 349–350

Atheros, 14

Attacks. *See also* Hackers
denial-of-service, 223, 286–287, 364
event fabrication, 301
insider, 364
link layer, 288
malicious mobile node flooding, 363–364,
370, 375, 378
man-in-the-middle, 360, 373
redirect, 360, 362–364, 369, 378
sinkhole, 299
Sybil, 288, 299
traffic permutation, 369–370
on wireless sensor networks, 287–289

Attenuation, field gathering wireless sensor
networks, 212

Attractive forces, wireless sensor networks,
241

Audio/video (AV) streaming, quality
considerations, 45

Augmented reality applications, storage
management, 260, 279

Authentication
campus WLAN usage studies
implications of, 18, 31, 36
LEAP, 32
home binding, 361–362, 366, 368,
372–373, 376
logs, 11–12, 17, 30, 32
mobile target tracking, 192
mutual entity, 360
protocols, 223
security issues, 287
wireless sensor networks, 289–291

Backoff algorithms, 46
Backoff process, 48–49
Bandwidth
allocation, MANETs, 80
coherence, 99
on demand, 397–398
sensor networks, 176–177, 180
Barrier coverage, 224

Basestation
placement of, 2
security issues, 290, 292, 299
wireless sensor networks, 222

Battery/batteries
capacity, 109
level, 78
power, wireless sensor networks, 221, 261
replacement, 292

Battlefield monitoring/surveillance, 191,
197, 223

Bayes' rule, 150, 152

Beacon(s)
intervals, MANETs, 96–97
wireless sensor networks, 285

Belief state, mobile target tracking
182–183

Bellman–Ford shortest path algorithm, 82,
115, 208

Best-point heuristic, 235

Bidding protocol (BIDP), 243–244, 248

Billing, metered, 397

Binary hypothesis testing, wireless sensor
networks, 145–146, 149–152, 154

Binary phase shift keying (BPSK), 120

Binary sensing model, 225–226

Binary sensors, 174, 177–180

Bit error rate (BER), 92, 119–120, 126–129,
136

Blanket coverage, 224

Bluetooth, 343

Bootstrapping protocol, 292, 296

Bottlenecks, sources of, 91, 109, 213, 216,
278

Boundary conditions, 233, 240

Breadth-first search (BFS), 236

Broadband
defined, 392
managed services, 381–382
sensor networks, 191–194

Broadcast ID, multihop wireless networks,
129, 131–134

Broadcast incremental power (BIP)
algorithm, 83

Broadcast storm problem, 80

Building-access point relation, 34

Building maps, 16, 24

Business flexibility, 387

Busy-tone channel, MANETs, 87–88

Campus waypoint, 18
Campus wireless network usage
  methodologies
    analysis, 33–34
    anonymization, 33
    authentication logs, 32
    trace collection, 32–33
  network environment, 31–32
  overview of, 29–30, 42–43
  related work, 30–31
  results
    ACS log, 34–36
    roaming patterns, 36–40
    trace data, 40–42
Campus WLANs, 16–20
Candidate routes, 3
Capital expenditures (CapEx), 399–400,
    403
Carrier sense multiple access with collision
    avoidance (CSMA/CA), 45, 48, 75,
    83
CBC-MAC, 290
Cellular networks, 2
Cellular phones, 105, 372, 389–392
Cellular systems, power control design, 84,
    86, 91
Center of stimulus (CoS), 301
Certification authority (CA), 365, 374
Channel
  access
    controlled, 51
    DCF, 49
    EDCA, 51–52
    quality of service (QoS), 47
  conditions, VoIP, 58
  error rates, multihop wireless networks,
    122
  load, quality of service (QoS), 47
  utilization measurement, 48
Cisco
  Aironet 350, 6–7, 31
  AP1200 access points, 31
  AP350 access points, 31
  Internetworking Operating System (IOS),
    6, 8
  LEAP (Lightweight Extensible
    Authentication Protocol), 18, 31–32
  probe response studies, 22
  Secure ACS, 32

Clear-to-send (CTS) packet
  MANETs, 75, 81, 83, 87–90, 93–94,
    96–97
  multihop wireless networks, 127
Client-side tools
  campus WLAN studies, 18
  characteristics of, 24
  WLAN measurement, 15–16
Cluster-based collaborative storage (CBCS)
  components of, 265–270
  security issues, 286
Cluster-based communication model, 296
Cluster head (CH)
  characteristics of, 91
  cluster-based collaborative storage, 266
  election phase, cluster-based collaborative
    storage, 266, 286
  intrusion detection, 298
  key management, 293, 296–297
  security issues, 302
Clustering
  field gathering wireless sensor network,
    211–213, 216
  MANETs, 90–91
  wireless sensor networks, 146, 253
Cluster rotation, 266–267
CMMBCR algorithm, 109
Coherence bandwidth, 99
Coherence time (T_c), 90
Collaboration model, wireless sensor
    network, 216
Collaborative signal and information
    processing (CSIP), 222
Collaborative storage
  benefits of, 264
  coordinated sensor management, 266–267
  design space, 265
  experimental evaluation, 267–271
  multiresolution-based storage, 271–273
  protocols, 265–266
Collectors, wireless sensor networks, 198,
    205–208, 212
Collision(s)
  avoidance information, 87, 89, 99
  MANETs, 76, 95
  mobile target tracking, 194
  VoIP, 68, 70
COM+, 320
Combinatorial optimization, 252

Communication graph, 228
Communication model, 226
Compression, 55. *See also* Data compression
Concurrent transmissions, 76, 86
Conditional probability, 246
Confidentiality, 286–287, 290
Configuration grid, 244–245
Connected environments, 341
Connected network, 225
Connected topology, mobile target tracking, 187
Connectivity
    global network, 251
    graph, field gathering wireless sensor networks, 215
    influential factors, 77
    MANETs, 82
    sensor networks, 143
    wireless sensor networks, 228–229, 295
Connectivity-based clusters, 266
Connectivity set (CS), MANETs, 89
Constant-bit-rate (CBR), 56
Contention window (CW), 49
Context-aware smart environment, 346
Continuous density function, 204–205
Controller nodes, 295
Control packet overhead, mobile target tracking, 194
Convex hulls, binary sensor network tracking, 177–178
Convoy tree, mobile target tracking
    characterized, 185
    construction of, 186
    expansion and pruning, 186
    reconfiguration, 186–187
Cookies, 367–369, 374–375
Cooperative tracking, binary detection sensor networks, 179–180
Coordinated collaborative storage (CCS), 267–269
Coordinated local storage (CLS), 267–270
Coordinated sensor management, 266–267
Coordinators, MANETs, 98
CoralReef
    analysis package, 34
    toolset, 40
CORBA, 309, 320
Cordless telephones, 95
Corrupt data packets, 222
Cougar, 271

Covering coding, 188
Critical transmission range (CTR) assignment, 229
Crossing point, 249
Cross-layering, 74
Cryptographically generation address (CGA) protocol, 361, 371–372, 378
Cumulative energy, 106
Current interference, MANETs, 84
CWmin, 46–47
Cyber pet caring game, 395

Damagochi, 395
Dartmouth College, campus WLAN usage studies, 18–19, 30–31
Data
    access patterns, 259
    acquisition, 283
    analysis software, testing of, 25
    collection
        methodologies, 5–6
        sensor networks, 259–262, 283
        Simple Network Management Protocol (SNMP), 11
        software, testing of, 25
        wireless sensor networks, 283
    compression
        storage management, 267
        universal, 348
        wireless sensor networks, 205–208, 214
    corruption, detection of, 263
    exchange phase, cluster-based collaborative storage, 266
    freshness, 289
    gathering, wireless sensor networks, 252–253
    generation, rate of, 258–259
    manipulation costs, 258
    processing, wireless sensor networks, 159
    retention, 262
    transmission, wireless sensor networks, 159–160
Data-centric storage (DCS), 271, 274–275, 277
DCF interframe space (DIFS), 48–49, 52, 56
DCOM, 309, 320
Deauthenticate events, syslog messages, 9
Deauthentication, campus WLAN usage studies, 31

Decentralized algorithms, 216
Decision rules, 147, 151–154, 168–170
Decryption, 294, 299
Degree of coverage, sensor networks, 224–225, 229, 240
Delaunay triangulation, 237
Delay
   AODV, 135
   MANETs, 95
   propagation, 90
   VoIP, 60–61
   wireless sensor networks, 122–123
Denial-of-service (DoS) attacks, 223, 286–287, 364
Deployment probability distribution, 204
DES, 290
Destination node, 198
Device control, 340–341
DHCP servers, access to, 16, 40
DHCPv6, 361
Diffie–Hellman key exchange algorithm, 374–376, 379
Digital signature, 286, 365, 373
Dijkstra algorithm, 115, 232, 234
Directed diffusion, 271
Directional antennas (DAs), 2, 93–95, 99, 209, 299
Directional network allocation vector (DNAV), 94
Directory Name Service (DNS), 372
Direct subscriber line (DSL), 383–384
Disassociation, 9–10, 19
Disk space, 13, 25
Distance mapping, mobile target tracking (MTT), 181–182
Distributed coordination function (DCF)
   characterized, 1
   interframe space, 48–49, 52, 56
   legacy, 48–49, 58
   quality of service (QoS)
      channel access, 52
      comparative performance evaluation, 56, 63
      defined, 45
      dual-queue scheme, 51
      legacy, 53
      modification of, 46
Distributed data compression, 206
Distributed hash table (DHT) system, 274

Distributed index for features in sensor networks (DIFS)
   architecture, 278–279
   high-level event, 278
   overview of, 277–278
   simple quad tree approach, 278
Distributed self-spreading algorithm (DSSA), 241–242, 247
Distributed tracking schemes, mobile target tracking (MTT)
   group management for track initiation, 182–184, 191
   maintenance, 182, 184–185
   tracking tree management, 185–187
Diurnal usage patterns, campus WLAN studies, 17–19
Doppler spread, 90
Drilldown constraints, in storage management, 273
Dual busy-tone multiple access (DBTMA), MANETs, 88
Dual-queue scheme
   implications of, 49–50
   legacy DCF, 48
   MDQ, 45–46
   VoIP, 68
Dynamic adjustment scheme, 47
Dynamic convoy tree-based collaboration (DCTC), 185–186
Dynamic source routing (DSR)
   MANETs, 79, 91
   multihop wireless networks, 107, 109, 125

Eavesdropping, 222, 289, 373
E-commerce, 382
Edinvar Assisted Interactive Dwelling House, 338
Effective bit energy-to-noise spectral density ratio, 92
Effective reliable throughput, 121
Embedded software, 395
Enclosure graph, 82
Encryption
   algorithms, 223
   implications of, 13
   mobile target tracking, 192
   wireless sensor networks, 286, 289–291, 294–295, 299

End-to-end
   communication, 389
   enabling IT infrastucture, 386
   packet delivery, MANETs, 78, 80
   reliability, 106
   retransmission (EER), multihop wireless
      networks
      assigning link costs, 115–118
      characterized, 107–108
      on-demand routing protocols, 130,
         132–134
      optimal minimum-energy paths,
         110–115
      performance evaluation, 118, 121,
         123–125
   systems, 395
Energy-aware (EA) routing, 105, 118, 122,
   135
Energy conservation, 252–253
Energy-constrained nodes, 249
Energy consumption
   MANETs, 78–79
   storage management, 261
   wireless sensor networks, 158–161, 223,
      250, 297
Energy cost analysis, 109–115
Energy-efficient
   communication, 2
   networks, MANETs, 78, 92–93
   protocol stack, 82
   routing, multihop wireless networks, 114
Enhanced distributed channel access
      functions (EDCAFs), 53
Enhanced distributed channel access
      (EDCA)
   characteristics of, 2
   comparisons with
      default access vs. PIFS access, 66–68,
         70
      MDQ, 63–66
   quality of service (QoS), see Enhanced
      distributed channel access (EDCA)
      quality of service
   parameter set, 53
Enhanced distributed channel access
      (EDCA) quality of service (QoS)
   characterized, 45–46
   comparative performance evaluation,
      VoIP
      access, default vs. PIFS, 66–68, 70

   characterized, 56–57
   MDQ, 63–66
   defined, 51
   802.11e provisions, 51–54
   transmission opportunity (TXOP), 51,
      54
Enterprise service bus (ESB), 388, 390
Entropy coding, 207–208, 214
Environmental conditions, 259
Environmental hazard monitoring, 197, 221,
   224, 289
Error-free delivery, 3
Error rate, multihop wireless networks,
   109–110. *See also* Bit error rate
   (BER); Link error rate
Ethernet, 12–14, 325, 360
EtherPeek, 18, 32
Euclidean distance, 225, 240
Euler–Lagrange equation, 233
Event correlation, 397
Event fabrication attacks, 301
Execution phase, 245
Expected average ability, 230
Exposure paths
   implications of, 251–253
   maximal
      best-case-coverage, 230, 234–235
      breach path, worst-case coverage, 230,
         235–237
      support path, best-case coverage, 230,
         237
   minimal, worst-case coverage, 230–234
Extensible Markup Language (XML), 319,
   326–329, 331, 388, 396–397

Fabrication technologies, 173
Fading, multihop wireless networks, 106
Farfield propagation, field gathering wireless
      sensor network, 210
Fast-backoff scheme, 47
Fault tolerance
   sensor collaboration, 175, 180
   wireless sensor networks, 194
FDMA, 95
Feedback channel, MANETs, 84
FFS, 263
Fidelity, 31, 98, 138, 173, 382,
   404–405
Field gathering wireless sensor network
   characteristics of, 142

data compression, 205–208, 210, 214
defined, 198
fluid flow model, 201
implications of, 197–199
lifetime limits, energy constraints
  data compression, 205–208
  defined, 198–199, 216
  mathematical framework, 200–204
  model and assumptions, 199–200
  network layout and, 204–205
  throughput tradeoff, 216
open problems, 215–216
throughput limits
  data compression, 214
  defined, 198, 208–210
  many-to-one communication results,
    209–210, 212–214
  model and assumptions, 210–211
  one-on-one communication results,
    211–212
  practical algorithms, 214–215
Field intensity, 231
File-sharing, 19
Filesystem limits, 13
Firewalls, 25
Firmware
  changes, 25
  802.11 card, 15
  upgrades, 6
First-in/first-out (FIFO) transmission queue,
  48, 58
First-order derivatives, 233
Fixed power level, 75
Fixed-transmission power, 128
Flash memory, storage management,
  261–264
Flat architecture, 216
Flood-fill algorithm, 245
Flooding, 223
Flow
  augmentation, 78, 202
  fluid, 201
  redirection, 202
  traffic, 252
  TCP, 56
  unidirectional, 56
Fluid flow model, 201
Forward-error correction codes,
  106
Forward progress rate, 77

Frequency
  channels, CDMA-based MANETs, 96
  reuse distance, 91
FTP/TCP, unidirectional flow, 56
Functional lifetime, 201

Gain, MANETs, 85, 93
Gateways, MANETs, 91
Gaussian distribution, 246
Gaussian noise, 146
Generic loss, 14
Geographic adaptive fidelity (GAF)
  MANETs, 98
  multihop wireless networks, 138
Geographically-based clusters, 266
Geographic hash table (GHT) system,
    storage management:
  canonical methods, 275
  characteristics of, 274–276, 279
  distributed index for features in sensor
    networks (DIFS), 277–279
  interaction with greedy perimeter stateless
    routing (GPSR), 276
  perimeter refresh protocol, 276–277
  structured replication, 277
Geometric fidelity, 173
Geometric random graph (GRG), sensor
    network coverage, 227–228
Georgia Tech
  Aware Home, 338
  wireless LANs, 17
Global networks
  on-demand, 389
  topology, mobile target tracking, 185
Global positioning system (GPS), 16, 24, 82
Global signaling channel, 83
GLOBUS, 309
Gloucester Smart Home, 338
Goal resolution phase, 245
Goal selection phase, 244–245
Gossiping, 223
Graph EMbedding (GEM) for sensor
    networks, 277
Graphical user interface (GUI), 393
Graphs, sensor networks
  characterized, 224–225
  graph-theoretic perspective
    geometric random graph (GRG),
      227–228
    graph connectivity, 228–229

Greedy perimeter stateless routing (GPSR), 276–277, 299
Grid
  deployment, mobile target tracking, 192–193
  network, wireless sensor network, 251
  topology, multihop wireless networks, 136–137
Group(s), mobile target tracking (MTT)
  formation, 183
  management strategies, 183–184
G.711, defined, 55

Hackers, 360, 364, 404
Handheld computers, 32
Handheld devices, 93
Handshake packets, 75, 87
Hardware abstraction layer (HAL), 395
Hash value, 364
HCF controlled channel access (HCCA), 51
Health monitoring, 283
Hierarchical architecture, 216
Hierarchical sensor networks, 191–194
Hierarchical summarization, 272
High-mobile users, campus WLAN usage studies, 20, 31
Histograms, quad trees, 278
Home
  binding update, 361–362
  location, 19
  node, geographic hash table (GHT) system, 276–277
Home agent proxy (HAP), 361–362, 372–379
Homeland defense, 283
HomeRf wireless networking system, 95
Hop-by-hop retransmission (HHR), multihop wireless networks
  assigning link costs, 115–116
  characterized, 107–108, 137
  on-demand routing protocols, 132–134
  optimal minimum-energy paths, 114–115
  performance evaluation, 118, 121–123
Hops, *see specific types of hops*
  energy-efficient reliable packet delivery, 105–106
  number of, 3
  transmission range and, 76–77
Hostile environment, 191

Hostnames, 16
Hotspots, 1, 5, 46, 277, 382, 404
HTTP, 17, 40, 42, 325, 328
Hub and spoke infrastructure, 396
Human-machine interface (HMI), 395
Human-subject research, 13, 24
Hybrid coordination function (HCF):
  defined, 45, 50
  IEEE 802.11e standards, quality of service (QoS), 50–51

IBM, 386, 393, 395, 400
ICQ, 17
Idle listening, 192, 194
Idle power consumption, 286
IEEE 802.11 standards:
  access points (APs), 6, 8, 53
  authentication, 12
  deauthentication, 12
  infrastructure network, 6, 8–9
  MAC layer, 14, 21–22, 45, 49, 268
  management information base (MIB), 10
  network interface cards (NICs), 10, 14–15, 22
  PHY/MAC layer, 14, 21–22, 45, 49
  power control, 74–76, 95
  power saving modes (PSM), 96–99
  QoS provisioning, 45–70
  smart environments, 343
  syslog message, 6, 9
  VoIP, 58
  wireless sniffing, 14
IEEE 802.11b standard, 99
IEEE 802.11e
  admission control, contention-based, 48, 54–55
  enhanced distributed channel access (EDCA), 48, 51–54
  implications of, 2, 50–51
  quality of service (QoS), 46
IEEE 802.15.4 standard, 304
IEEE 802.2, 55
IETF Mobile IP Working Group, 361, 365–366
Imprecise detections algorithms (IDA), 238–239, 247
Incremental self-deployment algorithm (ISDA), 244–245, 248
Index node selection, 279

Industrial, Scientific, and Medical (ISM):
radio band, 2.4-GHz, 95
wireless sensor network, 343
Industrial sensing, 141
Information-driven sensor network tracking,
175–177
Information processing, sensor networks,
277
Information technology (IT)
functions of, 384
simplification, 387
Information Technology Services Division
(ITS), 30, 32–33
Initialization phase, 244
In-network processing, 303
"Insider" attacks, 364
Inspection, MANETs, 84
Institutional review boards, functions of,
13, 24
Insufficient reasoning, 349
Integer linear programming algorithm
(ILPA), 244, 246, 248
Integrity, security issues, 287
Intelligent control, 354
"Intelligent" routing protocol, 78
Intel Proactive Health Project, 338
Intended receiver, MANETs, 87
Interface identifier, 371
Interference
field gathering wireless sensor networks,
210–211, 215
MANETs, 76–77, 86–87, 89
margin, MANETs, 87, 89
Interference-aware protocol, MANETs,
84–91, 99
Interframe space (IFS), 46
Intersil Prism, 14
Intrusion detection system (IDS), 197, 284,
297–298
Inventory management, 197
IP address, 16, 360, 372–373
IP protocols, campus wireless networks,
40–41
IP telephony, see Voice over IP
(VoIP)
IRC, 17
ISM (Industrial, Science, and Medical)
wireless networking, 343
Isotropic antenna, 94

Jamming, 287–288
Java virtual machine (JVM), 395
Jitter, VoIP, 68–70

k
connectivity, 225
-edge connectivity, 229
neighbors, 239–240
node-connected, 225
sensors, 253
Kerberos authentication logs, 17
Keyed pseudorandom function, 365
Key exchange algorithm, 374–379
Key management, security issues, 287,
292–297
Key predistribution scheme (KPS), 289,
295–296
Kids Communicator (AT&T), 342

Laptops, 19, 93, 105, 384, 389
Large attenuation, 212
Large-scale sensor networks, 249
Large-scale wireless sniffing, 24
Latency
cooperative mobile target tracking, 180
wireless sensor networks, 223, 252
Layer 4–7 switching, 402–403, 405
LEACH, 268, 299
Leader-based mobile target tracking,
182–184
Legacy DCF, 48–49, 58
Lempel–Ziv (LZ), 347
Lexicographic max-min node lifetime
problem, 204
LeZi-update scheme, MavHome, 347–349
Likelihood ratio, 150, 177
Linear programming, wireless sensor
networks lifetime, 201–205
Link error rate
implications of, 2
multihop wireless networks
characteristics of, 107, 112, 114, 117
estimation of, 126–129, 135
Link layer attacks, 288
Link-layer quality indicator (LQI), 47–48
Link loss rates, 106
Link scheduling, 109
Load-aware routing, field gathering wireless
sensor networks, 213

Load balancing, storage management, 270
Loading, MANETs, 96
Lobe interference, MANETs, 94, 99
Local area network, 37
Localized encryption and authentication
  protocol (LEAP), 18, 31–32, 296
Local storage (LS), 267–270
Login, remote, 12
Log-structured file system, 263–264
Long-distance hops, 106, 112
Long-range hops, 107
Loss recovery, multihop wireless networks,
  117–118
Low attenuation, 212
Lucent NIC, 22

Malicious mobile node flooding attacks,
  363–364, 370, 375, 378
MALITDA, 338–339
Management information base (MIB), 10
"Manager of managers," 393
Man-in-the-middle attack, 360, 373
Many-to-many communication, field
  gathering wireless sensor networks,
  209–210, 212–214
Mapping
  access points (APs), 16, 24
  campus wireless roaming studies, 39–40
  graph connectivity, 229
Markov chain, applications of, 17–18
Markov predictors, campus WLAN studies,
  20
M-ary FSK (frequency shift keying), 120
Matchbox file system, 258, 263–264
Mathematical framework
  communication model, 143, 226
  coverage model, 226–227
  sensing model, 225–226
MavHome (University of Texas—Arlington):
  architecture, 344–345
  automated
    decisionmaking, 351
    inhabitant action prediction, 350–351
    inhabitant location prediction, 346–350
  characteristics of, 310, 338, 344–345
  funding resource, 339
  goal of, 351
  implementation, 351–354
  live demonstration of, 353

  trie, 349
  zones, 347–348
Maximum permissible range, 116
Maximum segment size (MSS), 56
Maximum weight matching, 95
Measurement tools
  authentication logs, 11–12
  client-side tools, 15–16
  network sniffing, 12–13
  SNMP, 10–11
  syslog, 6–10
  wireless sniffing, 14–15
Medical equipment, 95
Medium access control (MAC)
  campus WLAN usage studies, 32
  channel access parameters, 46
  characterized, 2, 45
  802.11e provisions, 51–52
  HW queue, 50, 60, 63–64
  interference-aware, 84–90, 99
  legacy, 48–49, 51
  MANETs, 82–91, 99
  network sniffing, 12
  security issues, 288, 301–302
  Service Access Point (SAP), 55
  Simple Network Management Protocol
    (SNMP), 10–11
  storage management protocols, 267
  syslog messages, 8–10, 24
  transmission range and, 77
  VoIP, 55–56
Mesh networks, 406–407
Message authentication code (MAC),
  367–368
Metadata, storage management, 263
MICA, 284, 302–303
MICA nodes, 262
MICA-2, 262–263
MICA2DOT, 262
Microelectromechanical sensors (MEMs)
  functions of, 221, 405
  microsensors, 141
Microsoft MSN Messenger, 55
Microwave ovens, 95
Middleware
  architecture, 309
  categories of, 319
  computer-network-related, 309
  defined, 316

functions of, 318–319
IT infrastructure, 318, 388–389
MavHome, 310
on-demand business, 310–311, 387–388
RFID, 309, 313–325, 332
security issues, 310
support for sensor networks, 309
technologies, 316–319
web services, 319–320
WinRFID, 309, 313, 323–324, 328–331
Military applications
security issues and, 283, 294
sensor networks, 141
target tracking, 191–192
wireless sensor networks, 197, 221, 224, 260
Minimal cutset, 298
Minimax algorithm, 242–243, 248
Minimum-description-length principle, 350
Minimum energy consumed per packet routes, 95
Minimum-hop path, wireless sensor networks, 202
Minimum-hop routing algorithm, 118, 121
Minimum-hop routing protocol (MHRP), 78, 80, 89–90, 105–106, 121, 181
Minimum transmission power, MANETs, 84–85
MIPS R4000, 286
MIPv4, 361
MIPv6, security issues
authenticating binding update messages
cryptographically generated addresses (CGA) protocol, 361, 371–372, 378
home agent proxy (HAP) protocol, 361–362, 372–379
implications of, 365–366
return routability (RR), 361, 366–371, 378
cryptographic primitives, 364–365
future directions, 378–379
importance of, 310
location privacy, 379
operations, 361–362
public key infrastructure (PKI), 376–379
redirect attacks, 360, 362–364, 369, 378
Mirrored traffic, 32–33
Mixed-sensor networks, 243, 248, 251–253
MMBCR algorithm, 109

Mobile ad hoc networks (MANETs)
CDMA-based, 95–96, 100
characterized, 2, 73–74
power control design
complementary approaches, 91–96
energy-oriented approaches, 77–82
IEEE 802.11 deficiencies, 75–76
power saving modes (PSM), 96–98
standard characteristics, 73–75, 98–100
TPC, MAC perspective, 82–91
transmission range, 76–77
Mobile IP (MIP), 360–361
Mobile IP Working Group, 368
Mobile networks, security enforcement, importance of, 359–361. *See also* MIPv6, security issues
Mobile sensor networks, 238, 243–244, 247–248, 252
Mobile target tracking (MTT), using sensor networks
distributed tracking
network architecture design, 174, 187–194
protocol support for, 174, 182–187
implications of, 142, 173–174
information-driven dynamic sensor collaboration, 175–177
multiple target tracking, 182
power-efficient, 174–175, 179–180, 194–195
quality of surveillance, 189–191
robustness, 175
target localization methods
using binary sensors, 174, 177–180
sensor-specific methods, 180–182
track initiation and maintenance, 182–185
traditional, 174–177
warning messages, 181–182
Mobility
campus WLAN studies, 19–20
power control and, 90
traces/tracing, 16–18, 24
Modified dual queue (MDQ), quality of service (QoS)
characterized, 50, 58
comparative performance evaluation, 56, 61–66, 68–69
comparison with
EDCA, 63–66, 70
single queue, 58–63

Motorola MC68328, 286
Movement adjustment scheme, 243–244
Moving objects, tracking with binary sensor network, 177–178
MP3 music download service, 394
Multiaccess interference (MAI), 95
Multihop(s)
  ad hoc networks, routing algorithm, 107
  communication, security issues, 284
  field gathering wireless sensor network, 213
  path, transmission range and, 77
  RTS, 94
  security issues, 294
  sensor networks, 222–223
  wireless networks, *see* Multihop wireless networks
Multihop wireless networks, energy efficient reliable packet delivery
  characteristics of, 107–108
  on-demand routing protocols
    adaptations for, 125–135
    extensions for, 135–137
  related work, 108–109
  routing algorithms, minimum energy paths
    assigning link costs, 115–118
    energy cost analysis, 109–115
    overview, 105–107, 137–138
    performance evaluation, 118–125
    roadmap, 108
Multimodality, 389
Multipath key reinforcement scheme, 295–296
Multiple access interference, MANETs, 91
Multiple tracking, mobile target tracking, 184
Multiplicative increase, multiplicative/linear decrease (MIMLD) algorithm, 47
Multiresolution-based storage
  aging problem, 272–273
  constraints, 273
  drilldown queries, 272
  overview of, 271–272
  summarization, 272
Multisensor systems, 173
Multivendor networks, 304
μTESLA, 290–291
Mutual entity authentication, 360

Near-far problem, 95, 100
Neighbor/neighborhood
  discovery, 362
  field gathering wireless sensor network, 208
  list, 132
  MANETs, 84, 87–88, 96
  multihop wireless networks, 118
  one-hop, 276
  security issues, 292
  Voroni, 235
  wireless sensor networks, 240, 250, 285
.NET, 330–331
NetBSD kernel, 33
Net hop, 78
Network analysis, campus WLAN usage studies, 32
Network architecture, mobile target tracking
  broadband sensor networks, 191–194
  deployment optimization, 188
  hierarchical sensor networks, 191–194
  power conservation, 188–191
Network congestion, limitation strategies, 47–48
Network disruption, minimization of, 25
Network environments, significance of, 31–32
Network interface card (NIC), 10, 14–15, 22, 50
Network layers, MANETs, 82
Network partitions, 78
Network sniffing, 12–13, 17
Network sysadmin, 13
Network throughput, 209
Network topology, 33, 57
Network transport capacity, 209
Node density, 222
Node placement, sensor networks, 143
Node redundancy, 223
Node scheduling schemes, 252
Node-to-node communication, 223
Noise
  ambient, 120
  field gathering wireless sensor networks, 210
  multihop wireless networks, 135–136
  spectral density, 119
  wireless sensor networks, 146
Nonacademic WLANs, 20–21

Non-real-time (NRT) queues, 49–50, 61, 63
Normalized energy, 120–121
Ns-2 simulator, 118, 267

Occupancy maps, wireless sensor networks, 244–245
Omnidirectional antennas, 93–94, 210
On-demand business
  application aware, 397
  defined, 383
  end systems, 394
  evolution of, 386, 391
  flowchart, 394
  hardware abstraction layer (HAL), 395
  implementation of, 388–389
  mesh networks, 406–307
  network layering, standardization considerations, 402–404
  operating environment, 384–394
  OSS/BSS layers, 399–401
  overview of, 310–311, 401, 407–408
  pervasive computing ecosystem, 394–395
  sensor networks, 405–407
  service-oriented architecture (SOA)
    defined, 387
    principles of, 395–396
    service access domains, 396–399
    service domains, 399–401
  wireless security, 404–405
On-demand power-aware routing protocol, 79
On-demand routing protocols, multihop wireless networks
  adaptations for, 125–135
  extensions for, 135–137
One-hop
  clustering algorithm, 266
  neighbors, 250, 276
One-on-one communication, 211–212
One-way hash function, 364
Operation expenditures (OpEx), 399–400, 403
Optimal geographic density control (OGDC), 249
Optimal node density, 251
Optimal quantization algorithm, 153–154, 156–157, 159
Oracle RDBMS, 327
Orinoco (Lucent), 14, 22

Oscillation control scheme, 243
OSPF, 106
OSS/BSS networking, 399–401
Overhearing, 192, 194
Overprovisioning networks, 387

Packet(s)
  collision effect, 58
  drop rate, VoIP, 58
  energy-efficient delivery, 2
  error rate, multihop wireless networks, 120, 127
  header(s)
    analysis, 34
    campus WLAN usage studies, 33
    characteristics of, 13, 30–31
  loss rate, 1, 3
  sniffing, 12–13
  traces, 13
  transmissions, multihop wireless networks, 110–112
Packetization, VoIP, 55
Palm OS, 32
Pareto distribution, 20
Parsing, syslog messages, 8–9
Partitioning, 93, 259–260
Pebbles, defined, 293
Peer-to-peer caching systems, 31
Peer-to-peer (P2P) networks
  characteristics of, 406
  security issues, 286
  storage management, 273–274, 279
PEGASIS, 299
Perl modules, 11
Per node throughput, 209
Per node transport capacity, 209
Personal digital assistants (PDAs), 18–19, 105, 330, 372, 392
Phoneline networking alliance (PNA), 343
Physical layer (PHY)
  characterized, 45
  field gathering wireless sensor network, 210
  quality of service (QoS)
    implications of, 49
    VoIP admission control, 56
PHY/MAC layer, 21, 49
Playback analysis, 259–260
Plug-and-play technology, 392

Point coordination function (PCF), 48, 51
Point-to-point wireless links, long-range, 32
Poisson process, 20
Poll-and-response mechanism, 48
Port mirroring, 12
Position verification, 289
Postenergy, defined, 268
Potential field algorithm (PFA), 239–240, 247
Power amplification, 92, 210
Power-aware multiaccess protocol with signaling (PAMAS), 97, 106, 108
Power-aware routing (PAR), 74
Power-aware routing optimization (PARO), 79–80, 106, 108, 111
Power-aware routing protocols (PARPs), MANETs, 78–81, 99
Power consumption, 78, 223
Power control design, mobile ad hoc networks (MANETs)
  complementary approaches, 91–96
  energy-oriented, 77–82
  IEEE 802.11 approach, deficiencies of, 74–76, 95
  overview of, 73–75, 98–100
  power-saving modes, 96–98
  transmission power control (TPC), 82–91
  transmission range, selection factors, 76–77
Power-controlled dual channel (PCDC), MANETs, 85, 88–90
Power-controlled MAC protocols, 2
Power-controlled multiple access (PCMA), 87–88
Power failures, 31
Powerline control systems, 340, 343
Power-save polls, wireless-side measurement studies, 22
Power saving modes (PSM), 2, 10, 74, 96–99
Prediction by partial match (PPM), 349–350
Prediction models, campus WLAN studies, 20
Predictive caching systems, 31
Preenergy, defined, 268
Prepare mode, mobile target tracking, 190–191
Primitives, public-key-algorithm-based, 284
Priority access parameters, 55

Prism NIC, 22
Privacy, 13
Proactive routing protocols, 78–79, 82, 107
Proactive wakeup (PW) algorithm, mobile target tracking, 190–191
Probabilistic sensing model, 226–227
Probability distribution function (PDF), sensor network connectivity, 228
Probe(s)
  messages, MANETs, 97
  multihop wireless sensor networks, 126–128, 135–136
  response polls, wireless-side measurement studies, 22
Processors, 197
Propagation
  delay, 90
  field gathering wireless sensor network, 210, 212
  multihop wireless networks, 106
  wireless sensor network model, 215–216
Protocol model, field gathering wireless sensor networks, 211, 213
Pseudo-random-noise (PN) code, 95
Public key certificate, 365, 373, 376
Public key exchange, 374–378, 379
Public key infrastructure (PKI), 376–379
Pulse-coded modulation (PCM), 55
PVM, 309

QAP, 54–55, 66, 69
q-Composite random key predistribution scheme, 295–296
Q-learning, 351
QoS STA (QSTA), 52, 54–55, 66
Quad trees, storage management, 278
Quality of service (QoS)
  MANETs, 97
  on-demand business, 398
  significance of, 1–2
  positioning, *see* Quality of Service (QOS), IEEE 802.11 WLAN
  power control design, 77
  wireless sensor networks, 222–223
Quality of Service (QOS) positioning, IEEE 802.11 WLAN
  channel access parameters, 46–47
  comparative performance evaluation characterized, 56–57

comparison of MDQ and EDCA, 63–66
EDCA default access *vs.* PIFS access,
    66–68, 70
jitter performance comparison, 68–70
single queue and MDQ, 58–63
VoIP capacity for admission control,
    56, 58
dual-queue scheme for, 49–50
emerging IEEE 802.11e
    admission control, contention-based,
        54–55
    enhanced distributed channel access
        (EDCA), 51–54
    implications of, 50–51
legacy DCF, 48–49, 58
overview of, 45–46
related work, 46–48
voice over IP (VoIP), for admission
    control
    characterized, 55
    802.11b capacity for, 56
Query/queries
    drilldown, 272
    execution, data collection, 259–262
    overhead, 271
    wireless sensor networks, 285

Radio
    resource testing, 289
    transceiver, 197, 285
Radiofrequency (Rf)
    circuit design, 222
    wireless sniffing, 14
Radiofrequency identification (RFID)
    applications of, 309, 313–314, 331–332
    benefits of, 321–322
    challenges of, 322–323
    current technologies, 317
    data processing layer, 326
    ecosystem research at WINMEC, *see*
        WinRFID
    implementation of, 320, 325
    overview of, 314–315, 320
    physical layer, 324–325
    protocol layer, 325
    tags, types of, 316, 325
    web services, 320
Radius, wireless sensor networks
    of complete influence, 237

implications of generally, 226, 228
    of no influence, 237
Random pairwise key scheme, 295
Random path heuristic, 234–235
Range queries, 278
Rate control, 2
RC5, 290
Reachability grid, wireless sensor networks,
    245
Reactive routes, MANETs, 77
Real-time (RT)
    queues, 49–50, 58, 64, 66
    services, quality of service, 45–46
Real Time Streaming Protocol (RTSP), 40,
    42
Reassociation, wireless-side measurement
    studies, 21
Rebooting, automatic, 13
Receiver, multihop wireless networks, 106
Rectangular sensing field, 252
RedHat Up2Date, 13
Redirect attacks, 360, 362–364, 369, 378
Redundancy, 174, 223, 249, 266, 288
Reencryption, 294
Refreshing, 276–277, 293
Reinforcement learning, 351
Relays, wireless sensor networks, 198
Reliable communication, MANETs,
    84–85
Reliable delivery, multihop wireless
    networks, 111
Reliable packet delivery, 106
Remote access, 12
Remote activated switch (RAS), 97
Renegotiation, MANETs, 84–85
Replayed routing information, 299
Reprogramming, 285
Repulsive forces, 241
Request-to-send (RTS) packet
    MANETs, 75, 81, 83, 87–88, 90, 93–94,
        96–97
    multihop wireless networks, 127
Residual battery capacity, 109
ResiSim update, 352–353
Resource
    consumption, matchbox file system, 264
    management, wireless sensor networks,
        253
    testing, benefits of, 288–289

Retransmission(s)
  multihop wireless networks, 106–109,
    111, 113–114, 117, 121–125, 137
  potential, 2–3
Retransmission-energy-aware (RA)
    algorithm, 118, 122, 135, 136
Return routability (RR), 361, 366–371, 378
RFC 1812, 10
RF spectrum, wireless sniffing, 14
RIP, 106
Roam events, syslog messages, 9
Roaming patterns, 36–40
Robotics, 337, 340, 342
Robustness
  geographic hash table (GHT) system and,
    276
  sensor collaboration, 175
  sensor networks, 146, 166–170
Rockwell WINS sensor nodes, 303
Rogue packets, wireless sensor networks,
    223
Role assignment, field gathering wireless
    sensor networks, 203
Roundtrip delay, multihop wireless
    networks, 122–123
Route discovery process
  MANETs, 79
  multihop wireless networks, 132–135
Routed networks, 32
Route reply (RREP) packet
  MANETs, 79–80, 88
  multihop wireless networks, 129–135
Route request (RREQ) packets
  MANETs, 77, 79, 88–89, 91
  multihop wireless networks, 129–135
Routers
  network sniffing process, 12
  Simple Network Management Protocol
    (SNMP), 10
  wireless sniffing, 15
Routing
  algorithm, 2
  altered, 299
  dynamic source, 79, 91, 107, 109, 125
  energy-aware, 105, 118, 122, 135
  energy-efficient, 114
  greedy perimeter stateless, 276–277, 299
  "intelligent," 78
  IPv6, 361

layer, 2
load-aware, 213
minimum-hop, 78, 80, 89–90, 105–107,
    118, 121, 181
multihop networks, 105–125, 171
multipath, 299
power-aware, 74, 78–81, 99, 106, 108, 111
proactive protocols, 78–79, 82, 107
replayed information, 299
rumor, 299
shortest-path, 95, 176–177
wireless sensor networks, 146, 176–177,
    223, 298–299
RSA encryption, 286, 292
RS-232, 325
RS-485, 325
RTP transport, 55
RTS/CTS exchange
  MANETs, 81, 83, 87, 90, 93, 97
  multihop wireless networks, 127
RTSP, 42
Rumor routing, 299

Scalar quantizers, 214
Scheduling
  of nodes, 249
  phase, MANETs, 91
  schemes, 249–250, 252
Scientific data gathering, 197
Scientific monitoring, 259–260, 271, 279
Secondary routes, MANETs, 97–98
Secure socket layer (SSL), 376–377, 379
Security
  authentication logs, 11–12
  sensor networks, 143
  Simple Network Management Protocol
    (SNMP), 11
  surveillance, 221
  wireless sensor networks
    aggregation, 299–302
    applications, 284–285
    attacks, 287–289
    data encryption/authentication,
      289–291, 299
    implementation, 302–304
    intrusion detection, 284, 297–298
    key management, 287, 289–297
    overview of, 286–287
    resources, 285–286

routing, 298–299
significance of, 283–284
Self-detection algorithm, 244
Self-scheduling, 250
Sensing field, intensity of, 232, 234
Sensing gaps, 175
Sensing model, 225–226
Sensor deployment strategies
bidding protocol (BIDP), 243–244, 248
characterized, 238
comparison of, 246
distributed self-spreading algorithm
(DSSA), 241–242, 247
imprecise detections algorithms (IDA),
238–239, 247
incremental self-deployment algorithm
(ISDA), 244–245, 248
integer linear programming algorithm
(ILPA), 244, 246, 248
minimax algorithm, 242–243, 248
potential field algorithm (PFA), 239–240,
247
security issues, 284–285
uncertainty-aware sensor deployment
algorithm (UADA), 246, 248
vector-based algorithm (VEC), 242–243,
248
virtual force algorithm (VFA), 240–241,
247
Voronoi-based algorithm (VOR),
242–243, 248
Sensor networks, wireless
applications, 382–383
centralized option, analysis of
characterized, 147, 149–151, 158,
160–161
detection performance, 170
numerical results, 163–164
robustness, 166, 168, 171
characterized, 145–148, 170–171,
221–223
connectivity, 221–253
coverage
based on exposure paths, 230–237
based on sensor deployment strategies,
238–248
future research directions, 252
mathematical framework, 225–229
types of, 221–224

detection performance, 146–147,
155–158, 164, 170
development of, 197
distributed option, analysis of
characterized, 147, 149, 151–153, 158,
160–161
detection performance, 170
numerical results, 163–164
robustness, 166, 168, 170–171
energy-efficiency analysis
energy consumption model, 158–161,
171
numerical results, 161–166
energy optimization, 146, 197
field gathering
implications of, 197–199
lifetime limits, 199–208, 216
open problems, 215–216
throughput limits, 208–216
mobile target tracking (MTT)
distributed tracking, protocol support
for, 182–187
implications of, 173–174
network architecture design,
187–194
target localization methods, 174–182
multihop routing, 171
operating options, 149–150
protection of, 284
quantized option, analysis of
characterized, 147, 149, 163–164
optimal, 153–154, 156–157, 159
robustness, 166, 169
suboptimal, 154–157, 159–161
research, 141–143
robustness
implications of, 146, 171
node destruction, 166
observation data deletion, 167–170
routing algorithms, 146
security, 283–304
simplified model, 148–149
storage management, *see* Storage
management for wireless sensor
networks (WSNs)
Service differentiation, quality of service
(QoS), 46–47
Service-level agreements (SLAs),
398–399

Session
  diameter, 19
  hijacking, 363–364, 377–378
SHA1 algorithm, 33
SharePoint, 327
Shortest-cost path, 117
Shortest-delay (SD) AODV protocol, 135
Shortest-path
  algorithms
    MANETs, 78–79, 82–83
    routing, sensor network tracking,
      176–177
  heuristic, 235
  routing, 95, 176–177
  tree, 208
Short hops, 106, 112
Short IFS (SIFS), 49, 54
Short-range hops, 107
SIGCOMM, 21
Signal processing, 221
Signal strength, 10, 18, 78
Signal-to-interference-and-noise ratio
      (SINR), MANETs, 76, 84, 86–87, 91
Signal-to-noise ratio (SNR)
  multihop wireless networks, 126–129, 137
  wireless sensor networks, 158
Signature (misuse) detection, 297–298.
      See also Digital signature
Simple Network Management Protocol
      (SNMP)
  campus WLAN studies, 18–19, 30–31, 40
  wireless measurement, 10–11, 16–17,
      24–25
Simple Open Access Protocol (SOAP), 330
SIMPLE, MANETs, 77–78, 99
Simulation, wireless sensor networks,
      252–253
Simultaneous transmissions, field gathering
      wireless sensor networks, 212–214
Single-beam directional antennas, 95
Single-channel, signal transreceiver
      distributed systems, 84
Single-hop 802.11 deployment, 2
Single queue, comparison with MDQ, 58–63
Single-source shortest-path algorithm
      (SSSP), 232, 234
Sinkhole attacks, 299
Sink node, 198
Skype, 393

Sleep-awake-active pattern, 193
Sleep mode, 254, 284
Sleep period, mobile target tracking,
      189–190
Sleep-to-active transmission, MANETs,
      97–99
Slepian-Wolf model, field gathering wireless
      sensor network, 206–208
Smart dust sensors, 222, 286, 405–406
Smart environments
  defined, 337
  device communications, 338, 341
  enhanced services by intelligent devices,
      342
  networking standards, 342–343
  overview of, 337–340
  predictive decisionmaking capabilities,
      343–344
  remote control of devices, 340–341
  schematic view of, 338
  sensory information acquisition/
      dissemination, 341–342
  smart home illustration, *see* MavHome
  types of, 337–338
Smart sensor network, 180
Smart Sofa, 342
Snapshot, field gathering wireless sensor
      network, 198, 207–208, 214
SNEP, 290, 303
Sniffing
  in measurement studies
    network, 12–13
    wireless, 14–15
  nonacademic WLAN studies, 20
Software, up-to-date, 13
Software as a service (SaaS), 398–399
Source-destination pairs
  field gathering wireless sensor networks,
      212, 215
  MANETs, 79
  multihop wireless network, 107–108, 110
Sources, wireless sensor networks, 198
SPAN
  MANETs, 98
  multihop wireless networks, 138
Spatiotemporal data summarization,
      271–272
SPINS, 290–291, 293, 303
Sponsor nodes, 250

Spoofed routing information, 299
Spread-spectrum technology, 2, 95
SQL server, 327
ssh usage, campus WLAN studies, 17–18, 21
Stanford University, campus WLAN usage studies, 17, 31
Static field gathering networks, 209
Static sensor networks, 238–239, 243–244, 247–248, 252, 276, 285
Station (STA), VoIP
  dual-queue scheme, 51
  ED default vs. PIFs access, 66–68, 70
Stationary users, nonacademic WLAN studies, 20
Station up/down messages, 8
Statistical en route filtering (SEF), 301
Storage constraints, 273
Storage management for wireless sensor networks (WSNs)
  collaborative
    characterized, 257–258, 262, 264, 279
    cluster-based (CBCS), 265–267
    coordinated, 267–270
    design space, 265
    storage balancing effect, 270–271
    storage-energy tradeoffs, 261, 268–270, 279
    storage protocols, 265–266
  components of, 262
  data retrieval, 273–277, 279
  design considerations, 257, 260–261
  effective, 143, 260–261
  efficiency of, 258
  goals, 258, 261–262
  indexing, 273, 277–279
  load balancing, 278
  motivation for, 259–260
  significance of, 257–258, 279–280
  system support
    hardware, 262–263, 279
    Matchbox file system, 258, 263–264
Structured networks, storage management, 273
Subdividing, field gathering wireless sensor network, 211–212
Subnets
  campus WLAN usage studies, 31–32
  characterized, 25
  wireless, 11–12

Sub-Network Access Protocol (SNAP), 55
Suboptimal minimum connected sensor covers, 252
Suboptimal quantization algorithm, 154–157, 159
Subtrack mode, mobile target tracking, 190
Summarization, 271–272
Suppression, distributed mobile target tracking, 183–184
Surveillance
  military operations, 283, 294
  mobile target tracking, 189, 191
  using wireless sensor network, 197, 224
Sweep coverage, 224
Switched networks, 32
Sybil attacks, 288, 299
Synchronous orthogonal CDMA system, 95
Synthetic mobility models, 18
Sysadmins, 25
Syslog
  campus WLAN studies, 17, 19
  characterized, 24–25
  WLAN measurement studies, 6–10, 22

TAG, 271
Tampering, 287
Target tracketing, sensor networks, 142. *See also* Mobile target tracking (MTT)
Tcpdriv, 33
Tcpdump, 12–13, 18–19, 33
TEA, 290
TEEN, 299
Telnet, 17–18, 325
TESLA, 291, 302
Third-generation cellular systems, 93
Throughput capacity, field gathering wireless sensor networks, 211
Time-division multiple access (TDMA), 95, 215
Timescale, MANETs, 84
Time synchronization, 184, 302
TinyDB, 264
TinyOS, 263, 299
Topology control
  algorithms, MANETs, 82–83
  wireless sensor networks, 222
TORA, multihop wireless networks, 125
Total coverage, wireless sensor networks, 227

Total transmission energy, 2
Trace/tracing
    collection, 32–33
    data, 40–42
    gathering, campus WLAN usage studies,
        30, 33
    traffic, 33
Tracking stage, mobile target tracking, 189
Track maintenance, mobile target tracking
    (MTT), 184–185
Traffic
    authentication, 296
    campus WLAN usage studies, 32
    flow, wireless sensor networks, 252
    patterns, 17
    permutation attack, 369–370
    specification (TSPEC), 54
    streams (Tss)
        QoS provisioning, 54
        rate of, 48
Transistor–transistor logic (TTL), 325
Transmission control protocol (TCP)
    multihop wireless networks
        energy costs, 112–114
        reliable packet transmissions, 122–123
    QoS provisioning
        access, EDCA default vs. PIFs, 66, 70
        flow, single VoIP queue, 58–63
        packet headers, implications of, 30,
            56–57
    traffic, campus wireless networks, 40
    wireless sensor networks, 325, 328
Transmission energy, multihop wireless
    networks, 109–110
Transmission floor, 75
Transmission mode, MANETs, 75–76
Transmission opportunity (TXOP), 51
Transmission power
    control, see Transmission power control
        (TPC)
    field gathering wireless sensor network,
        210–211, 213, 215
    levels of, 2
Transmission power control (TPC)
    CDMA-based ad hoc networks, 95–96,
        100
    implications of, 2, 82
    MANETs
        data packets, 77–78

directional antennas, 93–95
        implications of, 73
        interference-aware MAC design/
            protocol, 84–91
        PARP/SIMPLE approach, 81–82, 99
        power-aware routing protocols
            (PARPs), 78–81
        SIMPLE approach, 78–80
        topology control algorithms, 82–83
    mobility issues, 90
    transmission rate control, 92–93
Transmission queue, 48
Transmission range
    critical, 229
    field gathering wireless sensor networks,
        210–211, 213, 215
    MANETs, 76–77
    wireless sensor network, 201
Transmission rate control, MANETs,
    92–93
Transmission schedule, field gathering
    wireless sensor networks, 215
Transport capacity, field gathering wireless
    sensor networks, 210, 212
Tree management, 185–187
Tribal Nations, 382
Trust routing for location aware sensor
    networks (TRANS), 299
Turnaround time, 90
Two-channel architecture, 85–86
Two-phase clustering, 253
Type loss, 14–15

UDP
    multihop wireless networks, 118, 121, 135
    packets, 30
    traffic, campus wireless networks, 40
    transport, 10, 55
Ultra-low-power RF radios, 270
Unauthorized traversal (UT) algorithm,
    233–234
Uncertainty-aware sensor deployment
    algorithm (UADA), 246, 248
Unicast packets, wireless networks, 127
Uninterrupted power supply (UPS), 13
U.S. Defense Advanced Research Projects
    Agency (DARPA), 406
U.S. National Science Foundation, 339
Universal data compression, 348

Universal description, discovery, and integration (UDDI), 329
University of California San Diego, campus WLAN studies, 18
University of Maryland, wireless-side measurement studies, 21–22
University of North Carolina at Chapel Hill, campus WLAN usage studies, 31
University of Saskatchewan, campus WLAN usage, 18, 29–30
Unix timestamp, 9
Unstructured networks, storage management, 274
URLs, 17
Usage peaks, campus WLAN studies, 17
User behavior, 17

Variable-power transmission, multihop wireless networks, 127–128, 136
Vector-based algorithm (VEC), 242–243, 248
Virtual force algorithm (VFA), 240–241, 247
Virtual MAC (VMAC), 48
Virtual polar coordinate space (VPCS), 277
Virtual source (VS) algorithm, 48
Viruses, 25
Vision Media Technologies, Inc., 382
Visual surveillance, 259
VLANs, 25
Voice codecs, 55
Voice over Internet Protocol (VoIP)
  for admission control
    characterized, 55, 70
    802.11b capacity for, 56, 58
  campus WLAN usage studies, 19, 32
  characterized, 1, 393
  quality of service, 45–46, 50
Voltage controlled oscillators (VCOs), 92
VOR algorithm, 242–243, 248
Voronoi-based algorithm (VOR) 242–243, 248
Voronoi diagrams, 233, 235–236, 242, 252
Voronoi neighbors, 235
Voronoi polygons, 242–243

Waiting mode, mobile target tracking, 190–191
Wakeup
  rate, 251

  signals, 97
  state, 251
WAKEUP message, wireless sensor network, 193
Warehouses, inventory management, 197
Weather
  conditions, impact of, 259, 341
  monitoring, 383
Web Service Description Language (WSDL), 330
Web services, 319–320
Weighted moving average, 128–129
Weighting, cooperative mobile target tracking, 179–180
WiFi (wireless fidelity), 31, 382, 404–405
Wildlife tracking sensor network, 259
Windows Update, 13
WINMEC, 309, 332
WinRFID
  architecture of, 323, 330
  functions of, 309, 313, 324, 328–332
  rule engine, 330
  runtime plugins, 330–331
  services
    reader web service, 329–330
    reader windows services, 328
    remote object based service, 328
Wireless ad hoc networks, 2
Wireless clients, misconfigured, 25
Wireless-enabled devices, 105
Wireless Internet service providers, 382
Wireless intrusion protection system, 14
Wireless inventory tracking devices, 32
Wireless LANs (WLANS)
  802.11 infrastructure network, 6
  measurement studies of traffic, 1, 5–24
  popularity of, 1
Wireless measurement checklist, 24–25
Wireless sensor networks, multipath routing, 299
Wireless sniffing, WLAN measurement studies, 14–15, 22, 24
Wireless transceiver, 286
Wireless usage studies
  benefits of, 5
  campus WLANs, 16–20
  data collection, 5–6
  long-term, 16
  manually obtained data, 16

Wireless usage studies (*Continued*)
  measurement tools
    authentication logs, 11–12
    client-side tools, 15–16
    network sniffing, 12–13
    SNMP,10–11, 17
    syslog, 6–10
    wireless sniffing, 14–15
  methodologies, 22–24
  nonacademic WLANs, 20–21
  wireless measurement checklist, 24–25
  wireless-side measurement studies,
    21–22

Wireline network, campus WLAN usage
    studies, 31–32
Working node, wireless sensor network, 251
World Wide Web (WWW)
  browsing, 17, 21
  traffic, 19
Wormholes, 299
Worms, 25

X10, smart environment features, 340, 343,
  352

ZoomAir, 22